CW01262082

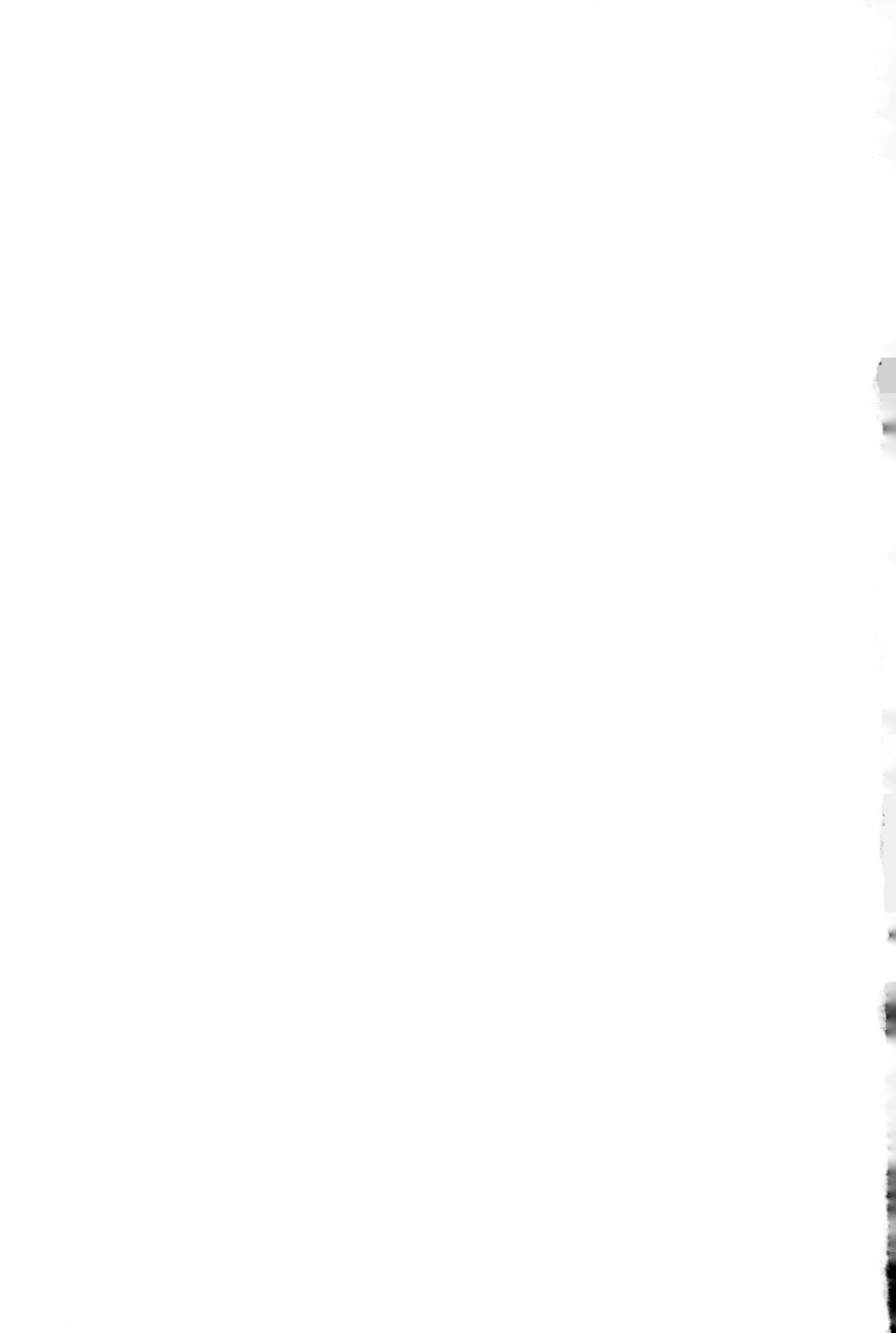

HUMAN FORMS

Human Forms

THE NOVEL IN THE AGE
OF EVOLUTION

Ian Duncan

PRINCETON UNIVERSITY PRESS
PRINCETON & OXFORD

Copyright © 2019 by Princeton University Press

Published by Princeton University Press
41 William Street, Princeton, New Jersey 08540
6 Oxford Street, Woodstock, Oxfordshire OX20 1TR

press.princeton.edu

All Rights Reserved

Library of Congress Control Number 2019931719
ISBN 978-0-691-17507-2

British Library Cataloging-in-Publication Data is available

Editorial: Anne Savarese and Jenny Tan
Production Editorial: Jill Harris
Jacket Design: Layla Mac Rory
Production: Merli Guerra
Publicity: Alyssa Sanford and Keira Andrews
Copyeditor: Hank Southgate

Jacket image: George Baxter, *The Crystal Palace from the Great Exhibition, installed at Sydenham: sculptures of prehistoric creatures in the foreground.* Process print, ca. 1864. Courtesy of the Wellcome Collection

This book has been composed in Miller

Printed on acid-free paper. ∞

Printed in the United States of America

10 9 8 7 6 5 4 3 2 1

For Ayşe

CONTENTS

Acknowledgments · ix

INTRODUCTION	The Human Age	1
CHAPTER 1	The Form of Man	31
	Conjecture, History, Science, Fiction	31
	The Faculty of Perfection	41
	The Formation of Humanity	44
	The Paragon of Animals	49
CHAPTER 2	The Form of the Novel	55
	Novelistic Revolution	55
	Bildungsroman	61
	Infinity or Totality	67
	The Classical Form of the Historical Novel	71
	The Dignity of the Human Race, the Glory of the World	75
	Dark Unhappy Ones	82
CHAPTER 3	Lamarckian Historical Romance	86
	Of Paris	87
	Retrograde Evolution	100
	Reading in the Dark	109
	Le grotesque au revers du sublime	113
	The Great Book of Mankind	118

[vii]

CHAPTER 4	Dickens: Transformist	123
	No Humanity Here	123
	The Poetry of Science	129
	Dickens's Teratology	133
	The Prose of the World	141
	Visionary Dreariness	145
	The Noise of the World	152
CHAPTER 5	George Eliot's Science Fiction	158
	We Belated Historians	158
	Knowledge and Its Languages	163
	Species Consciousness	169
	An Intellectual Passion	177
	Involuntary, Palpitating Life	184
	An Inherited Yearning	189
	Shadows of the Coming Race	199

Notes · 201
Bibliography · 249
Index · 279

ACKNOWLEDGMENTS

I BEGAN RESEARCH for *Human Forms* during my term as a visiting scholar at Boğaziçi University, Istanbul, in the winter of 2009. Subsequent visiting appointments, at the Department of English (2012) and the Center for Advanced Studies (2015) at Ludwig-Maximilians-Universität, Munich, and at the English Department and Council of the Humanities, Princeton (2017), afforded precious time for thinking and writing, as well as opportunities to try out emerging ideas on new constituencies of students and colleagues (bless them). I thank my hosts for their unstinting generosity and kindness: Cevza and Alpar Sevgen at Boğaziçi; Julia Schreiner, Anna Jakubowska, and (especially) Christoph Bode at LMU; Deborah Nord and Sarah Chihaya at Princeton. Thanks, too, to Steffi Fricke, Sabrina Kessler, Katharina Pink, Felicitas Meinhard, and other participants in the Romanticism Colloquium at LMU; Isabel Schneider, Anna Kunde, and Doris Haseidl at the LMU English Department; Moritz Baumstark, Annette Meyer, and Susanne Schaffrath at CAS; my LMU students; the members of my Princeton graduate seminar; and Catie Crandell and Camey VanSant, who organized the Victorian Colloquium at Princeton. The core of my research and writing took place here at Berkeley, an intellectual environment that continues to demand the best from all of us: I don't think I would have attempted this project anywhere else. So thanks, first, to my fabulous graduate students: those who helped out as research assistants, Monica Soare, Slavica Naumovska, Ella Mershon, Jesse Cordes Selbin, Katherine Ding, Wendy Xin, and Veronica Mittnacht, as well as others whose work it has been my privilege to supervise (say rather, to follow) over the past decade: Ruth Baldwin, Catherine Cronquist Browning, Marisa Palacios Knox, Ben Cannon, Charity Ketz, Margaret Kolb, Luke Terlaak Poot, Lauren Naturale, Jessica Crewe, Oya Erez, and Tim Heimlich, as well as Batya Ungar-Sargon, Jessica Ling, Claire Marie Stancek and Lise Gaston; Tanya Llewellyn and Mark Taylor at Stanford; and visiting postdoctoral fellows Juan Sanchez and John Savarese. Thanks also to my incomparable English Department colleagues in Romanticism (however we define it!): Steve Goldsmith, Kevis Goodman, Celeste Langan, Janet Sorensen, Anne-Lise François, Elisa Tamarkin, and Amanda Goldstein; and in Victorian studies: Cathy Gallagher, David Miller, Kent Puckett, and Grace Lavery; also to Ann Banfield, Mitch Breitwieser, Dori

Hale, Kristin Hanson, Lyn Hejinian, Abdul JanMohamed, Donna Jones, Sam Otter, and James Turner, who all pointed me in helpful directions or sowed seeds they may not have realized have borne fruit here; and not least, in memoriam, Janet Adelman and Alex Zwerdling, without whom I would not be at Berkeley. I continue to rely on the brilliance, energy, and goodwill of my students in classes and seminars, undergraduates as well as graduates, for keeping my wits honed and my propositions honest. Thanks, also, to Tony Cascardi, dean of arts and humanities, and my department chairs, Sam Otter, Katherine O'Brien O'Keeffe, and Genaro Padilla, for accommodating my leaves of absence. The resources of the Florence Green Bixby Chair (testimony to the generosity of my colleagues) have provided essential material support.

At Princeton University Press, it has been my great good fortune to work with Anne Savarese, who ushered me through the publication process with clarity and tact; Lauren Bucca and Jill Harris; and Hank Southgate, my incomparably attentive and meticulous copyeditor. I have benefited from the careful and generous scrutiny of the press reviewers, Nancy Armstrong and Simon During: the book is vastly better for their advice, even where they might not always be satisfied with my response to it. Vital interlocutors elsewhere have included Penny Fielding (who first encouraged this to be a book); Jim Chandler (who brainstormed the title); Jon Klancher, Noah Heringman, Sarah Winter, Elaine Freedgood, Mike Goode, Cannon Schmitt, Sebastian Lecourt, Maureen McLane, Wendy Lee, and Paula McDowell (who read bits of the work in progress); and those who have provided guidance and encouragement over the years, from all sorts of angles: John Barrell, Christoph Bode, Marshall Brown, Jim Buzard, Jill Campbell, Joe Carroll, Gianni Cianci, Claire Connolly, Adriana Craciun, David Duff, Angela Esterhammer, Peter Garside, Denise Gigante, Lauren Goodlad, Rae Greiner, Devin Griffiths, Nancy Isenberg, Tony Jarrells, Claire Jarvis, Priti Joshi, Priya Joshi, Catherine Jones, Maggie Kilgour, Scott Klein, Nigel Leask, Yoon Lee, Clare Lees, Deidre Lynch, the late Susan Manning, Caroline McCracken-Flesher, Jerry McGann, Randy McGowen, Silvia Mergenthal, Steven Meyer, Franco Moretti, Matt Ocheltree, Murray Pittock, John Plotz, Padma Rangarajan, Jonathan Sachs, David Simpson, Katie Trumpener, Julian Weiss, Matt Wickman, the late Sandy Welsh, Mark Wollaeger, and Susan Zieger.

The original germ of *Human Forms* (now part of chapter 3) was a keynote lecture for the 2007 International Scott Conference, for which I found myself trying to make sense of *Count Robert of Paris*. Over the next decade, various invited lectures, colloquium presentations, and conference papers provided a public or semipublic forum for testing and developing my arguments. I thank my hosts at those events, as well as their colleagues and students, and audience members for their generosity and (on occasions) forbearance: Caroline Jackson-Houlston, the International Scott Conference; Cairns Craig, David Duff, and Jan Todd, Centre for the Novel, University of Aberdeen; Tony Jarrells, Rebecca Stern, and Patrick Scott, the Victorians Institute, University of South Carolina; Nick Mason and Matt Wickman, 2011 NASSR Conference; Nigel Leask, 2011 British Association for Romantic Studies; Jens Gurr and Frank Pointner, Gesellschaft für Englische Romantik, 2011; Meg Russett, University of Southern California; Priya Joshi and Rachel Buurma, Greater Philadelphia Nineteenth-Century Forum; Enrica Villari, University of Venice, Ca' Foscari; Adriana Craciun and Mario Biagioli, University of California Multi-Campus Research Group "Material Cultures of Knowledge," and my respondents, Sarah Kareem (the Huntington), Stan Robinson (UC Davis), and Dana Simmons and Sherryl Vint (UC Riverside); fellow panelists Ina Ferris, Matt Ocheltree, and Margaret Kolb, the World Congress of Scottish Literatures and 2014 Scott Conference; Elaine Freedgood and Cannon Schmitt, "Denotatively, Literally, Technically," at NYU and the UC Berkeley Townsend Center; Terry Robinson and Alan Bewell, the 2015 Vincent A. De Luca Lecture in Nineteenth-Century Studies, University of Toronto; John McCourt, the 2015 Giorgio Melchiori Lecture, Università degli Studi Roma Tre; Nathalie Vanfasse, organizer, and my fellow speaker John Gardiner, Colloquium on Victorian Literature and the Human Sciences, LERMA, Aix-Marseille University; Norbert Lennartz, University of Vechta; Ali Behdad, Saree Makdisi, Jonathan Grossman, and their colleagues, the 2016 Barbara L. Packer Lectures, UCLA; Matthew Ocheltree, the Long 18th-Century Colloquium at Harvard; Chenxi Tang, LMU–Berkeley joint conference on "Fictions of the Human"; Mark Bevir, "Historicism and the Victorian Human Sciences," UC Berkeley Center for British Studies; Penny Fielding, Susan Manning Memorial Symposium, Institute for Advanced Studies in the Humanities, Edinburgh; Mike Goode and Claudia Klaver, 19th-Century Studies Working Group of the Central New York Mellon Humanities Corridor, Syracuse University; Elsie Michie, Helena Michie, and John Jordan, 2012 Dickens Universe; Peter Garside, the Edinburgh Walter Scott Club; Rob Mitchell and Nancy Armstrong, "The Biological Turn" symposium, Duke University; Jonathan Sachs, "Romantic Ephemerality," and Bill Warner, "Literary Science," 2015 MLA Convention, Vancouver; Rae Greiner and Jesse Molesworth, "New Approaches to the Rise of the Novel," Indiana University; Dorothy Hale,

Vicki Kahn, and Alan Tansman, Free Indirect Discourse Workshop, UC Berkeley Townsend Center for the Humanities; Chuck Rzepka and Joe Rezek, Boston Area Romanticist Colloquium; Catie Crandell, Princeton University Victorian Studies Colloquium; Elizabeth Bejarano and colleagues, NYU Eighteenth-Century Colloquium; Nathalie Vanfasse, Fanny Robles, and participants in the Aix-Berkeley collaborative workshop on Victorian literature and science; and David Duff and his colleagues at the London-Paris Romanticism Seminar. I also read papers touching on this project at meetings of the MLA Convention, the GER conference, and the James Hogg Society, and I thank colleagues and audience members there for their questions and comments.

Some of these talks and papers, early drafts of segments of this book's argument, have appeared in print. A portion of chapter 1 was published as "Kant, Herder and the Anthropological Turn," in *Romanticism and Knowledge*, edited by Stefanie Fricke, Felicitas Menhard, and Katharina Pink (Wissenschaftlicher Verlag Trier, 2015). Parts of chapter 2 appear as "*Bildung* versus *Roman*: Germaine de Staël's *Corinne*," in *Narratives of Romanticism*, edited by Sandra Heinen and Katharina Rennhak (Wissenschaftlicher Verlag Trier, 2017), "The Bildungsroman, the Romantic Nation, and the Marriage Plot," in *Replotting Marriage in Nineteenth-Century Literature*, edited by Jill Galvan and Elsie Michie (Ohio State, 2018), and "Against the Bildungsroman," in *MLQ* 80: 1 (2019). "The Trouble with Man: Scott, Romance and World History," in *Romantic Frictions*, edited by Theresa Kelley (*Romantic Circles: Praxis Series*, 2011), is an early draft of part of chapter 3, as is a section of "The Form of the City and the Form of Man," in *Romantic Cityscapes*, edited by Jens M. Gurr and Berit Michel (Wissenschaftlicher Verlag Trier, 2013). A discussion of the double narrative in *Bleak House* (chapter 4) appears as "The Novel and the Romantic 'Moment,'" in *Romantic Ambiguities: Abodes of the Modern*, edited by Sebastian Domsch, Christoph Reinfandt, and Katharina Rennhak (Wissenschaftlicher Verlag Trier, 2017). Chapter 5 evolved via "George Eliot and the Science of the Human," in *A Companion to George Eliot*, edited by Amanda Anderson and Harry E. Shaw (Blackwell-Wiley, 2013), and "George Eliot's Science Fiction," in a special issue of *Representations*, "Denotatively, Technically, Literally" (2014), edited by Elaine Freedgood and Cannon Schmitt. Early, fragmentary sketches of the general argument appear as "We Were Never Human: Monstrous Forms of Nineteenth-Century Fiction," in *Victorian Transformations: Genre, Nationalism and Desire in Nineteenth-Century Literature*, edited by Bianca Tredennick (Ashgate, 2011); and "Literature," in *Historicism and the Human Sciences in Victorian Britain*, edited by Mark Bevir (Cambridge, 2017). I thank the editors of these volumes for accommodating my work in its early (and sometimes quite inchoate) stages.

My parents, Hamish and Maureen Duncan, died as I was bringing this book toward completion; I thank them, too late, and also Sheana Miller, Rory, Victoria, Fern, and Aeneas Duncan. My best thanks—best for coming last—go as ever to Ayşe Agiş, without whom none of this would be possible, or would matter much.

HUMAN FORMS

INTRODUCTION

The Human Age

1

Human Forms explores a commonplace that has largely escaped critical attention. In the classical era of European realism, circa 1750–1880, novels invoke human nature as their topic and ground, the theme they are uniquely equipped to realize and the scientific basis for their art—at the same time that the human species becomes the subject of a new discourse, natural history, and its logic of an organic transmutation of forms and kinds.[1] Densely entangled with the rise of evolutionary science, from Buffon through Lamarck to Charles Darwin, the "natural history of man" was that science's critical occasion, its original and ultimate scandal. The major intellectual revolution of the age did not so much lay a foundation for literary innovation, upon which a genre might settle into a repertoire, as provoke a rolling earthquake of speculation and controversy—conditions that favored the loose, capacious, fluctuating forms of nineteenth-century fiction. The novel's supposed aesthetic disability, its lack of form, now marked its fitness to model the changing form of man. (This book will retain period usage of the universal particular, "man," for reasons to become explicit later.)

"Before the end of the eighteenth century, man did not exist," Michel Foucault declares in *The Order of Things*. The waning of metaphysics, and with it "the pre-critical analysis of what man is in his essence," made way for man's appearance "as an object of knowledge and as a subject that knows"—the vanishing point of a secular, scientific and historical, natural order.[2] Calls for a reorganization of knowledge upon the basis of anthropology, issued early in their careers by Johann Gottfried Herder and Immanuel Kant, displaced the Enlightenment science of man—synthetic,

[1]

synchronic, universal—for a new discourse, the natural history of man, and beyond it the modern disciplines of the human sciences.[3] A double or divided nature, physical life plus an immortal soul, had set humans apart within the premodern (theological) order of nature. Rather than closing that ancient schism, the natural history of man compounded it, with a new set of dyads: nature and history, individual and species, human life and biological life, or life as such. These dyads fractured outward from the new crux intended to separate humans from other creatures. Born into the world "with few or no special instincts,"[4] in other words lacking a predetermined nature, the human has instead the freedom to find or make its nature in time, through experience. Being as fixed form, "preformation," gives way to an open, plastic, developmental conception of form as becoming. In man alone does the species as well as the individual have a history, according to late Enlightenment philosophical anthropologists, who translated the old providential plot of human destiny into a civilizational progress toward perfection, charged with the energies of European industrial and geopolitical expansion. A new "epigenetic" biology provided that progress with its internal logic in the hypothesis of a formative principle, the *Bildungstrieb*, driving organic development from simple to more complex (advanced) states.

But once that developmental principle invested all of nature, it subverted the human exception it was meant to save. After 1800, an emergent evolutionist or transformist natural history sank human species being within the general, irregular flux of terrestrial life. All natural forms, including the human, were subject to mutation. Modern revolutions in astronomy and geology vastly extended the scale of natural history, from abysmal pasts before humanity (and the histories of life and of the earth) to inconceivable posthuman futures. Far from being settled and finished, a solution to the questions raised by the Enlightenment's secularization of knowledge, human nature became—more urgently than ever—a question, a problem, a fault line of philosophical disturbance in the terrain of a supposed "anthropologism" or "anthropologization,"[5] well before that terrain's postmodern subsidence.

The novel, the ascendant imaginative form in nineteenth-century Europe, did more than broadcast the anthropological turn of secular knowledge: it helped steer it and—under the license of fiction—it pressed it to its limits. As the history of man broke up among competing disciplinary claims on scientific authority after 1800, the novel took over as its universal discourse, modeling the new developmental conception of human nature as a relation between the history of individual persons (the traditional subject of the novel) and the history of the species (the contentious subject of the new anthropology). Novels could offer a comprehensive representation of human life—a Human Comedy—in a general writing accessible to all readers, mediated not by specialist knowledge or technical language but by the shared sensibilities that constitute "our common nature."[6] Novels became active instruments in the ongoing scientific revolution, advancing its experimental postulates—that human nature may not be one but many, that humans share their nature with other creatures, that humans have no nature, that the human form is variable, fluid, fleeting—as well as developing a technical practice, realism, to defend humanity's place at the center of nature and at the end of history. Realism mounts its defense even as natural history evicts man from that privileged station: delivering us to a world that is ours, not as an inheritance, but as a colony, a ground we have come late to as invaders and settlers, exterminating or enslaving the prior inhabitants, transforming the physical terrain—reconstituting nature as a byproduct of human wants.

Enlarging the plan of individual *Bildung*, or moral and spiritual formation, to a species-scale project, a formation of humanity or *Bildung der Humanität*, late Enlightenment anthropology opened the philosophical matrix for a new kind of novel. The Romantic Bildungsroman provided not so much an exemplary form for nineteenth-century realism as a pervasive principle, a "conceptual horizon," to the extent that all modern novels are in some sense novels of development, in which individual progress and the achievement of humanity dialectically produce each other.[7] The new anthropology supplied a new, and newly gendered, novelistic subject—unformed, malleable, finding himself in time—as well as a new conception of the novel's alleged formlessness, now legible as a technical equipment—temporal extensiveness, openness to contingency, internal heterogeneity and variability—for the representation of evolutionary becoming. At the same time, the conjunction of individual and species histories opened a discrepancy between the bounded, purposive, progressive order supposed to regulate a particular human life and the potentially interminable, differentiating, entropic drift of life as such—with the latter exerting warping pressure on the former, as instantiated in the aleatory, unfinished career of the prototypical Bildungsroman protagonist, Goethe's Wilhelm Meister.

That strong variant of the Bildungsroman, Walter Scott's "classical form of the historical novel,"[8] proposed an influential solution. Invoking national history as the medium between individual and species histories, it recast the discrepancy as one of temporal scale, across which human nature might be maintained as singular and stable. Insofar as national history reproduced a universal developmental plan (as in the Scottish Enlightenment historiography of a progress of economic and cultural stages), it could stand in for species history. That mediation, or rather substitution, relied however on the novel's tactical limitation of its mise en scène to a proximate or intermediate historical setting, between the familiar present and an exotic, alien deep past, and to a confined regional range: canonically, in Victorian realism's reorientation of Romantic historicism toward the present, English provincial life a generation since. Non-European peoples, concomitantly, dwelt beyond the pale of realism and its amenities, such as history and *Bildung*: stranded in a static or regressive "time of the other."[9]

Later Romantic fiction, responsive to the disruptive surge of transformist natural history, reopened the gap between individual and species history: by a drastic expansion of the historical distance, hence morphological difference, between past and present, and—more consequentially—by a turn from regional or provincial to metropolitan and cosmopolitan settings. The world-city (Constantinople, Paris, London), a total human environment, offered an experimental laboratory for evolutionist speculation, one in which universal history decomposes into a tangle of incommensurable temporal states, variously accelerating, stagnant, or retrograde developmental stages, and divergent evolutionary trajectories. As the cosmopolis ingests its imperial frontiers, the distorting pressures of urban life generate new mutations, new forms and kinds, contingent adaptations, monstrous births. In Dickens's allegorical realism, the mid-century apogee of this urban-transformist aesthetic, man has reconstituted nature into an artificial, evolving, self-organizing system that—transcending its human genesis—recursively reconstitutes "man." Dickens's achievement throws into relief the identification of a mainstream, mimetic rather than allegorical mode of Victorian realism with provincial life, a tranquil enclosure within which "the history of man" (that "mysterious mixture")[10] can appear to remain stable. Its major artist within the English tradition, George Eliot, pitches the realism of provincial life against its limit. "Species consciousness," the moral achievement of the *Bildung der Humanität*, falters before the inhuman expanse of "involuntary, palpitating life,"[11] beyond which strange mutations of race and kind and (eventually) organic life itself occur, no longer accessible to human recognition.

Human Forms seeks to recapture the experimental rather than the finished or perfected aesthetic of George Eliot's mature novels—and the experimental energy of the novel across the period generally. Realism reverts to its primal strangeness, a tense grapple of weird science and wild fiction, not much like the set of pedagogical programs into which some criticism has calmed it. Reading Eliot's practice alongside Dickens's, *Human Forms* makes the case for alternative, rival realisms within the British tradition, as instantiated by its major nineteenth-century novelists—as opposed to reducing both to a unified aesthetic, or flattening one into the other's false or failed shadow. Before turning, in its last two chapters, to Dickens and George Eliot, my argument ranges across developments in the history of man, the life sciences, and the history of the novel in France, Germany, and Scotland: to suggest that the British case was anomalous rather than typical for the modern history of the novel, not least in its assumption of national history as a progressive frame for its plots of development. Germaine de Staël's early challenge to the consolidating norms of the Bildungsroman in *Corinne* (1807), aimed at the gendered alignment of individual (masculine) progress with universal human equivalence, models a history of the novel that breaches not just national boundaries but the ideological adequacy of modern nation-state formation, within which not all humans qualify as human. Staël's philosophical romance opens onto a history of the genre that moves not smoothly along the rails of national history but by lateral, irregular jolts and swerves of attachment, assimilation, and antagonism—rather like processes of biological variation, although according more to the accounts of Jean Baptiste Lamarck and Geoffroy Saint-Hilaire (saltationist, catastrophic, fitfully and irregularly charged with purpose) than Darwin's.[12]

Human Forms does not attempt comprehensive coverage of the interaction between the novel and the sciences of man in the century that stretches from Buffon, Herder, and Goethe to Dickens, Eliot, and Darwin. Its argument comes to bear, instead, on a selective succession of crisispoints. The first two chapters consider the contentious formation of a human natural history in the late Enlightenment and the emergence of new Romantic genres, the Bildungsroman and historical novel, from that philosophical cauldron. Chapter 3 analyzes late Romantic mutations of historical romance that imagine the dissolution of the Enlightenment idea of a universal human nature via the transformist hypotheses of Lamarck and Geoffroy. Dickens's mid-century recourse to a pre-Darwinian evolutionist natural history and his contestation of the anthropomorphic techniques of Victorian realism occupy chapter 4. Chapter 5 addresses the full-on reckoning with the nineteenth-century revolutions in the human and natural sciences and with realism's formal legacies by George Eliot.

Although my argument attends to the history of the evolutionist science of man and the history of the novel, in continental Europe as well as

Great Britain, the aim of *Human Forms* is more critical than historical. Its concern is with human nature as a formal question for the novel rather than as a primarily thematic or ideological one (although of course these cannot be disentangled). Versions of a developmental dialectic—between individual and species life, between inner aspiration and outward reality, between self and society or world—articulate the form of the novel as well as of human nature: a dialectic under persistent threat of the dislocation or collapse of its constituent terms. My scrutiny of the novel's claim on human nature during the expansive movement of thought I am calling the age of evolution aims to upset the normative status still granted to a mimetic version of realism and its techniques in critical discourse on the novel by insisting on their contingent rather than necessary formation, played out in the recurrent debates and doubts about the order of nature and the form of man from Herder and Kant to Darwin. In the works of its most ambitious English practitioner, George Eliot, realism is a sophisticated holding action, the protective reassertion of a formal anthropomorphism in the light of new kinds of knowledge of an inhuman and indeed posthuman world (as Eliot imagines it in a late fable, "Shadows of the Coming Race"). Man bears within himself the germs of other histories, other grammars of being—past, future, speculative, subjunctive—in which (as among these novelists Dickens, perhaps, most darkly intuits) he and his works, along with nature itself, grow alien and inimical.

The book's chapters unfold a loosely dialectical scheme. The broad topic of chapter 1, "The Form of Man," is resumed in chapter 3, and that of chapter 2, "The Form of the Novel," in chapter 4; chapter 5 considers the attempt to synthesize emergent forms of scientific knowledge with novelistic form across a literary career. Each chapter (while maintaining a combination of wide-angle historical description with close reading of case studies) offers a different register of analysis. The first sketches the foundation and development of the new scientific history of man through the early debates that animated it; the second studies the contested formation of a genre, the Bildungsroman, through the central crux of gender; chapter 3 is a comparative study of two contemporaneous novels, British and French, in light of the controversial resurgence (and politicization) of transformist science circa 1830; chapter 4 examines a single work, *Bleak House*, to connect its evolutionism with its experiments in narrative and lyric form; and chapter 5 considers a single author, George Eliot, who engaged more decisively than any of her contemporaries with the ongoing scientific revolutions of the age. While the first half of *Human Forms* looks at French and German developments in the novel and the history of man, its argument comes to bear in the last two chapters on the major British novelists of the mid-nineteenth century. A global account of the topic is well beyond the scope of this book (and beyond my expertise); a complementary or competing work might orient its argument, for instance, around the great nineteenth-century French experiments in a systematic scientific project of literary anthropology, by Balzac (briefly touched on in these pages) and then Zola; or track the subject into the new, emerging domains of the novel, establishing different aesthetics and protocols from the British or French traditions, in North America, in Russia, and—critically—across imperial and postcolonial sites of writing into our own time, when at last the adequation of human nature to a European standard loses its grip.

This introduction goes on to review the argument of *Human Forms* in greater detail, sketching some of its critical and historical contexts and implications.

2

Writing in the late 1930s, Georg Lukács aligned the key tenets of European realism with the Romantic idea of history as an "unbroken upward evolution of mankind" founded on "the organic, indissoluble connection between man as a private individual and man as a social being."[13] Lukács's affirmation rings poignantly against what Mark Greif calls "the crisis of man" in the literary institutions of mid-twentieth-century liberal democracy, driven to reassert an indomitable human nature in the face of ruinous totalitarian projects of human reinvention.[14] Cold war polemic on both sides wielded Man as an ideological slogan, "the unperceived and uncontested common ground of Marxism and of Social-Democratic or Christian-Democratic discourse."[15] Its utopian iterations recalled the discourse of man in its emergent, generative phase: on the one hand, Lukács's Soviet-humanist echo of the *Bildung der Humanität*, on the other, the recapitulation of the program of the Romantic Bildungsroman in postwar North Atlantic human rights discourse, in tandem with Kant's vision of the end of history as a worldwide peace-keeping confederation of nation-states:[16] claims on a universal humanity challenged with the decolonizing agenda of the Bandung Conference of Asian and African nations in 1955.

A diagnosis of the modern intellectual regime of "anthropologism" drove the poststructuralist critique of the late 1960s, anticipated by the programmatically antirealist *nouveau roman*. Foucault branded late-born man "a kind of rift in the order of things," destined to fade from view "as soon as [our] knowledge has discovered a new form": already, in 1966, "'anthropologization' . . . is disintegrating before our eyes, since we are beginning to recognize and denounce in it, in a critical mode, both a forgetfulness of the opening that made it possible and a stubborn obstacle standing obstinately in the way of an imminent new form of thought."[17] Two years later, Jacques Derrida called for the arduous rethinking of "an end of man which would not be organized by a dialectics of truth and negativity, an end of man which would not be a teleology in the first person plural."[18] Arguably, the decisive disassembly of "man" has taken place through recent work in critical race, gender, and sexuality studies, while the *nouveau roman*'s antirealism has proven a dead end, as alternative realisms proliferate in novels written outside as well as within the old European and Atlantic core.[19] Current modes of posthumanist thought are responding to the more obdurately material crisis of the Human Age, renamed the Anthropocene, a new geological epoch marked by the indelible signature of human civilization on world history. Impersonal, unwitting, that signature encodes not our species' artistic and philosophical achievements but the onset of the fossil-fueled industrial revolution.[20] The subjection of the world as geophysical system to human activity, in short, propels current efforts to think beyond and outside the human, in dour admission of the literal, denuding force of that "teleology in the first-person plural." The "end of man" has given way to an "end of nature," a crisis too immense (in Amitav Ghosh's recent polemic) for the representational capacity of "literary fiction" (realism) and its fine calibrations of agency and event.[21]

"One of the striking features of the discourse of man to modern eyes is how unreadable it is, how tedious, how unhelpful," writes Greif: a discourse that was already "somewhat empty in its own time, even where it was at its best; empty for a reason, or, one could say, meaningful because it was empty."[22] That tedious emptiness—the banality and bombast fogging "man" by the late twentieth century—accounts, no doubt, for the aversion of critical eyes from the novel's attachment to the category. (Such once certified-fresh titles as *Of Human Bondage* [Somerset Maugham, 1915] and *Man's Fate* [*La condition humaine*, André Malraux, 1933] now give off a mildewed whiff.) Consequently we lack a sustained consideration of the novel and its avowed subject in the era of that subject's revolutionary transformation, when—it is this book's key claim—the novel reorganizes itself as the literary form of the modern scientific conception of a developmental, that is, mutable rather than fixed human nature. We have thriving recent studies of the novel's interactions with Darwinian (but far less with pre-Darwinian) evolutionism,[23] with themes articulated in Victorian anthropology and ethnography, the physiology of mental life, and an emergent biopolitics of race and population;[24] studies of literature and the science of man in its earlier Enlightenment phase;[25] and a fleet of critical approaches—object-oriented ontology, new and speculative materialisms, actor-network theory, thing theory, animal studies—engaged in destabilizing, downgrading, and displacing human agency within and from our assumptions and hermeneutic procedures.

Human Forms descends into the man-shaped hole (a grave?) around which these critical projects arrange themselves, not to revive a critically endangered anthropocentrism but to investigate the vicissitudes of its formation. My argument attends to the discourse of man when it was "at its best," tracing its upward arc in natural history and the novel, from Buffon and Herder to Darwin, from *Wilhelm Meister's Apprenticeship* (1795–96) to *Daniel Deronda* (1876)—to read its all too meaningful emptiness, the "white shield" of man's open, multivalent, metamorphic potential.[26] In the major key of early Victorian triumphalism: "He thrives in all climates, and with regard to style of living, can adapt himself to an infinitely greater diversity of circumstances than any other animated creature."[27] "[Man] opposes himself to Nature as one of her own forces, in order to appropriate Nature's productions in a form adapted to his own wants," Karl Marx writes in volume 1 of *Capital*: "By thus acting on the external world and changing it, he at the same time changes his own nature."[28] Coming into the world without form, unbound to a special task or function, man takes on all forms and functions, fills the world and consumes it, making his own nature and making nature his own: making it, as we see too clearly now, his own waste product. Emptiness, then, as a (positive) formal principle, rather than a lack of content: since man, unformed, can absorb all contents as he changes and remakes his form, in the heady era (at least) of the West's seemingly unlimited imperial and industrial expansion. *Human Forms* argues that the cultural instrument fitted to this modern anthropological conception is the novel, the form without form that likewise assimilates all forms: the *literary form* of the human.

3

Early in Mary Shelley's novel *The Last Man* (1826), Lionel Verney recalls how friendship reclaimed him from a feral state:

> I now began to be human. I was admitted within that sacred boundary which divides the intellectual and moral nature of man from that

which characterizes animals. My best feelings were called into play to give fitting responses to the generosity, wisdom, and amenity of my new friend.[29]

Sympathetic socialization admits him to full membership in his species. "Friendship is the offspring of reason," the great natural historian, the Comte de Buffon, had written: "thus friendship belongs only to man."[30] Verney's progress makes for a striking contrast with the repeated failure of friendship that seals the monster's exile from the sacred boundary of humankind in Shelley's earlier novel *Frankenstein*. Nor will the monster's creator, refusing to give him a mate, open an alternative way to species being—since its condition (again according to Buffon) is biological reproduction. In retaliation, the monster inflicts his plight on his creator: exterminating Victor Frankenstein's friends and family, reducing him to an inhuman solitude. And this will be Lionel Verney's fate too. By the end of *The Last Man*, he finds his matriculation into humanity mocked by the extinction of his species, and hence, the disappearance of the condition of possibility for being human.

"I now began to be human": in her novels, Shelley pits the Romantic ethical and pedagogical ideal of *Bildung*, or culture as the means to becoming human, against the ascendant biological conception of species life.[31] Pressing antinomies of the late Enlightenment natural history of man—human nature as organically given versus human nature as historically formed—to their breaking point, Shelley unsettles the philosophical basis claimed for the novel since its "rise," its culturally acknowledged consolidation as a genre, in mid-eighteenth-century England. "The provision, then, which we have here made," Henry Fielding promised his readers at the opening of *Tom Jones* (1749), "is no other than HUMAN NATURE":

> Nor do I fear that my sensible reader, though most luxurious in his taste, will start, cavil, or be offended, because I have named but one article. The tortoise, as the alderman of Bristol, well learned in eating, knows by much experience, besides the delicious calipash and calipee, contains many different kinds of food; nor can the learned reader be ignorant, that in *Human Nature*, though here collected under one general name, is such prodigious variety that a cook will have sooner gone through all the several species of animal and vegetable food in the world than an author will be able to exhaust so extensive a subject.[32]

When Fielding had claimed to describe "not Men, but Manners; not an Individual, but a Species" in his earlier novel *Joseph Andrews*, he meant by "species" a moral and social type rather than a natural-philosophical entity.[33] Now, with the appeal to human nature, Fielding annexes his "new province of writing"[34] to the new science designated by David Hume, ten years earlier, as the foundation of secular knowledge:

> There is no question of importance, whose decision is not compriz'd in the science of man; and there is none, which can be decided with any certainty, before we become acquainted with that science. In pretending therefore to explain the principles of human nature, we in effect propose a compleat system of the sciences, built on a foundation almost entirely new, and the only one upon which they can stand with any security.[35]

The novel too, an upstart genre in want of legitimacy, makes its claim on this new foundation. Fielding articulates it as a *formal* claim: to curate a "prodigious variety" of phenomena "under one general name," to subdue a manifold to a unity.

From the rise of the novel to its Victorian zenith . . . In the opening sentence of *Middlemarch* (1872), George Eliot addresses a reader who "cares much to know the history of man, and how the mysterious mixture behaves under the varying experiments of Time."[36] The science of man has become a history of man, and human nature, so capacious as to contain without loss of definition a global diversity of living forms, has become a contingent and heterogeneous phenomenon that shapeshifts over time. Where Fielding's contemporary was Hume, George Eliot's is Charles Darwin, whose work completes the delivery of nature to the determinations of geography and history initiated in the "Buffonian revolution" of the generation following Hume and Fielding.[37] Recasting species as fluid effects of biological succession, Buffon's *Natural History* planted the seeds (Buffon called them "internal moulds" [*moules intérieures*], genetic principles of organic development) of an evolutionary or transformist account of life on earth. Darwin's *The Descent of Man* (1871)—absorbing into its argument the various branches of the human sciences, from anthropology to aesthetics—accomplishes the full subjection of mankind to the new natural history, which Buffon himself had balked at.

Conceiving of her novels as interventions in scientific discourse, not mere applications of it, George Eliot kept abreast of what she and her circle (George Henry Lewes, Herbert Spencer) called "the Development Hypothesis" across a broad range of European disciplines, from comparative mythology to cell biology, in the age of the so-called second scientific revolution. Eliot's attunement of *Middlemarch* to "the varying experiments of Time" announces a strong resumption of the project of the English novel during its emergence in the half-century before *Tom Jones*. The early eighteenth-century novel shared the empirical premises and exploratory mission—to represent the sensible world we inhabit—of the contemporaneous (first) scientific revolution, even as (Tita Chico contends) "early science formulated itself through literary knowledge."[38] But by mid-century, intent on stabilizing its truth claims, scientific discourse pulled away from literary discourse, even as the novel was refining its own techniques and protocols. The signpost of their divergence was a hardening antagonism between verifiable hypothesis and avowed fiction, which masked (according to John Bender) a shared *techne*, the positing of "a provisional reality, an 'as if,' that possesses an explanatory power lacking in ordinary experience."[39] Hume and Fielding configure a complementary antithesis: Hume founds the science of man upon a strong affirmation of fictionality, regulated by customary conjunction, as the fabric of empirical reality, while Fielding's claim on that scientific foundation allows in turn the blithe acknowledgment of his work's fictional status.[40] By the time George Eliot is writing, sophisticated accounts of scientific method (by William Whewell, Thomas Henry Huxley, and John Tyndall) insist upon the vital role of the imagination—"a constant invention and activity, a perpetual creating and selecting power"—in the making of new knowledge, while nevertheless worrying over the relation—porous, transitive, reversible—between technical and figurative language, fictive invention and verifiable fact.[41] In that relation, George Eliot finds the opening for her experimental practice. Darwin, relying (as recent commentary has insisted) on fiction-generating devices such as conjecture, analogy, plotting, and personification to mount his "one long argument," characterizes "the considerable revolution in natural history" it will accomplish as a revolution in language—a simultaneous conversion of metaphors (of kinship and descent) into "plain signification" and of technical terms (taxonomic categories) into "merely artificial combinations, made for convenience."[42]

Hypothesis and fiction, scientifically disciplined and imaginatively licentious iterations of "as if," converged to form the new discourse, the history of man, that took over the science of man in the late eighteenth century. Chapter 1 of *Human Forms* considers this vexed conjunction. Buffon imported a "literary" stylistics of analogy, metaphor, conjecture, and probabilistic reasoning into his scientific argument, while Jean-Jacques Rousseau, following Buffon's discussion of the human species in the *Natural History*, opened his *Discourse on the Origins of Inequality among Men* by "[setting] aside all the facts"[43] for a provocative exercise in free-ranging speculation on the relation between human nature and history. Scottish philosophers such as Adam Ferguson sought to refute Rousseau by reasserting scientific protocols for analogical reasoning and hedging the role of conjecture in their own work, which would nevertheless be given the generic title "conjectural history."[44] Conjectural history became a preeminent literary genre of the late Enlightenment, the medium of the new natural history of man and its subdivisions, "the history of languages, of the arts, of the sciences, of laws, of government, of manners, and of religion."[45]

The devolution (or disintegration) of hypothesis into invention, science into fiction, would remain a persistent scandal of the history of man and the evolutionist natural history that shadowed it. In a defining controversy, Kant accused Herder of abandoning science and philosophy to produce a work of "mere fiction" in the guise of a "conjectural history" in the latter's most ambitious of all essays in philosophical anthropology, *Ideas for a Philosophy of the History of Mankind*—the work that, more than any other, implicated human history with an evolutionist natural history at the close of the eighteenth century. Kant's complaint would generate the methodological distinction, crucial for the completion of his own critical philosophy, between the constitutive "as if" of verifiable hypothesis and the regulative "as if" of teleological judgment; Herder's fault was to have muddled the two, irresponsibly positing an occult organic force as a causal principle in the evolution of natural forms, from minerals to plants to people. Conjectural history was only a novel with a collective protagonist, Man, instead of an individual one, Tom Jones or Clarissa Harlowe. The reiteration of the charge, and the controversy, throughout the nineteenth century would confirm the conceptually monstrous status of the natural history of man: a grotesque hybrid of science and fiction, mutating between both, settling into neither.

4

As controversy and scandal shake the history of man, so human nature fails to become a settled, singular entity, a scientific fact, and its subject, man, remains (as he was in the old regime of theological knowledge) a "mixture of two realms, the animal and the human."[46] Giorgio Agamben argues that an internal "mobile border," an "intimate caesura," has always defined the Western philosophical category of man: human nature bears within itself the cleft between "what is human and what is not," in the suturing together of physical and metaphysical natures, animal body and immortal spirit.[47] The "anthropological machine" of Enlightenment knowledge reproduces this chimera as Buffon's *Homo duplex*, "Man the Double," a taxonomic reinvention meant to preserve human nature within the wilds of natural history. Man's physiological constitution might place him among the "brutes," Buffon allows, but the soul, a divine implant manifest in the uniquely human work of reason, sets him absolutely apart.[48]

Buffon saw that the consignment of man to nature required a redrawing of the boundary of human exceptionalism, all the more urgently now that metaphysical walls were down. Other thinkers, dissatisfied with the resort to a supernatural prosthesis of reason or language, looked elsewhere for a uniquely human principle.[49] Rousseau proposed the most consequential solution: the "faculty to perfect oneself," the capacity for progressive development as a species, sets humans apart from other creatures. This developmental capacity is predicated on humans' freedom from instinctual predetermination or (to use the term from Enlightenment genetic theory) preformation. Lacking strong sensory bonding to a fixed behavioral repertoire, human nature is plastic, mobile, open—free to mold itself, in time and through experience, by observation and imitation. Rousseau's denial even of a basic social instinct supplied a moral crux for the ensuing debates over the history of man, with the scientific empiricists (the Scottish philosophers, Herder, eventually Darwin himself) insisting on a social foundation for human development—whence the appeal of Shelley's characters to the humanizing force of sympathy and friendship—and Kant, notably, following Rousseau, although turning in a quite different direction. Man, in Rousseau's and Kant's (and, later, Marx's) radical declension of the new anthropology, has *no given nature* and hence must *make his nature*. Unformed in nature, man achieves form developmentally, in history—a history doubled between the individual life and the life of the species. Here philosophical anthropology generates founding principles for the new kinds of fiction (considered in chapter 2 of *Human Forms*) that will define nineteenth-century practice: the Bildungsroman, which tracks "the harmonious formation of the purely human"[50] through a character's sentimental education; and the historical novel, which makes national history the progressive medium that regulates discrepant scales of individual and species life. In both, a developmental narrative shapes the fortunes of a sensitive, susceptible protagonist, a Wilhelm Meister or Waverley—the new human subject given by the new anthropology.

The rise of transformist natural history after 1800 turns this solution back into a problem. Once the principle of development becomes universal, covering all organic beings, the human exception disappears. The formal qualities supposed by late Enlightenment philosophers to set man apart in nature—plasticity, perfectibility—embed him more deeply within it. Johann Friedrich Blumenbach affirmed the unity of the human species and (following Buffon) the absolute distinction of man from other animals as interdependent principles, each guaranteeing the other.[51] Transformism threatened both, raising twin specters of a polygenetic dispersal of man among biologically distinct kinds (excessive diversity) and an abominable kinship between humans and brutes (excessive unity). While polygenetic racial theory simmered around debates on the abolition of slavery (coming to a toxic boil at mid-century[52]), the "Orang-Outang Hypothesis" of Rousseau and Lord Monboddo gained scientific strength through the resurgence of interest in Lamarck's zoological philosophy in the late 1820s. Lamarck's attribution of species transformation to the inheritance of physical modifications acquired through an organism's habitual response to environmental conditions would blend with Geoffroy Saint-Hilaire's argument for the genesis of new species through deformations of embryological development—"monstrosities"—in popular syntheses such as, in Great Britain, Robert Chambers's *Vestiges of the Natural History of Creation* (1844), which was not only eagerly read by poets and novelists but outsold most contemporary fiction. The "teratology" of Geoffroy and his son Isidore naturalized the monster as the origin of species—a hypothesis accepted by Chambers as well as by his formidable opponent, Richard Owen. We find it organizing the character system as well as narrative procedures of the popular urban-gothic fiction of Victor Hugo and Charles Dickens, in a lingering affront to mimetic-realist canons of novelistic form and character.

Darwin, resolutely monist and materialist in his stricter accounting of evolution by natural selection, banishes monsters (like a modern Saint Patrick) from the order of nature.[53] Metamorphosis happens all but imperceptibly on the timescale of human experience, as singular mutations graduate into varieties, and varieties into species, by the slow accumulation of individual differences within a fluid continuum of natural forms. In contrast, the archetypal order of Geoffroy's natural system necessitated a saltationist mechanism—the birth of monsters, an abrupt leap from given form—for species change.[54] *The Descent of Man* seeks to close the "ceaseless divisions and caesurae" of which man, according to Agamben, is both the site and the effect.[55] It resolves the chimerical figure of *Homo duplex* through a biological accounting of all aspects of the human, in which reason, sentiment, morality, taste, and religious belief are subsumed under natural drives and processes anterior to humanity's emergence as a species. Reversing Blumenbach's postulate, Darwin makes unity with nature at large the condition of a monogenetic human nature in *The Descent of Man* and its sequel, *The Expression of Emotions in Animals and Men*. We are one species—therefore, because, we share our life with other creatures.

5

That key question, of the unity or multiplicity of the human species, was made urgent by two major historical developments of the late Enlightenment. "Now the Great Map of Mankind is unrolld at once," Edmund Burke wrote to the philosophical historian William Robertson in 1777: "there is no state or Gradation of barbarism, and no mode of refinement which we have not at the same moment under our View."[56] Burke's tribute to Robertson's *History of America* also alludes to the imminent completion of the map of the world's coastlines by the state-sponsored circumnavigations of Louis Antoine de Bougainville, James Cook, and other explorers, culminating in Cook's third voyage, then in progress (1776–1780). The reduction of the earth to a closed cartographical system (trussed for European projects of resource extraction and colonization) accompanied a series of first encounters with a bewildering diversity of hitherto unknown peoples across the South Pacific. Crucially, these late eighteenth-century expeditions were licensed as scientific projects, and (beginning with Bougainville's) they carried professional astronomers, botanists, and geographers with them. First contact with new peoples took place, in other words, within the widening—fracturing—horizon of scientific knowledge. The voyages had an immediate, forceful impact on Enlightenment philosophical discourse, in speculations on universal history and human diversity by Rousseau and Diderot (whose "Supplement to Bougainville's Voyage" is an early experiment in biopolitical conjecture), Robertson, Lord Kames and other Scottish philosophers, Kant (debating the unity of the human species with Georg Forster, who had sailed on Cook's second voyage with his father, the expedition's naturalist), and Herder (defending multiple paths of cultural development within a universal *Bildung der Humanität*). These discussions yielded trenchant critiques of the spread of European empire, as Sankar Muthu has shown, as well as justifications of it.[57]

These discussions were also charged by, as they charged in turn, intensifying debates around the Atlantic slave trade: the second of the major historical developments that shaped the new natural history of man. Both proslavery and antislavery writers "[endeavored] to traverse the porous boundaries between human and commodity," in Lynn Festa's analysis, with the latter party "emphasizing those aspects of the human that are inalienable," generating an early version of human rights discourse.[58] Former slaves, writing their autobiographies, authored their own emergence from the condition of property, a thing to be bought and sold, into full humanity; meanwhile, abolitionist writers mobilized sentimental rhetoric to activate sympathetic bonds of universal brotherhood and sisterhood.[59] Against this, in an ominous escalation, proslavery apologists began to scramble the biological boundary between species, as in Edward Long's argument (drawing on early evolutionist conjecture) that "[negroes] are a different species of the same genus" as Europeans, and that "the oran-outang and some races of black men are very nearly allied," in his *History of Jamaica* (1774).[60] For now, however, monogenesis—the doctrine of a unified human nature—remained orthodox, the premise of projects of *Bildung* in the novel as elsewhere.

Even when (or, perhaps, especially when) not making racial difference its theme—it largely went without saying that *Bildung*, like property, was a white male privilege—the novel confronts the formal problematic of human nature as the relation between a uniform framework of kind or species and the degrees of variability it can tolerate. How much difference can man contain before he himself becomes different? So sure is Fielding of a universal human nature that he finds in its "prodigious variety" a source of festive enjoyment: "a cook will have sooner gone through all the several species of animal and vegetable food in the world, than an author will be able to exhaust so extensive a subject." The analogy looks forward to the anthropological characterizations of Herder, Lamarck, and Chambers: man the imperial animal, world-occupying, world-devouring. That a Caribbean product is being consumed—the turtle, with its "delicious calipash and calipee," standing in for sugar and the human bodies that produce it—sharpens the analogy's edge.[61]

Fielding's mock-heroic domestication of the novelist as chef admits him to the experimental vocation of the philosopher and scientist described a year earlier in Hume's *Enquiry Concerning Human Understanding*, a popular reworking of the *Treatise of Human Nature*:

> It is universally acknowledged that there is a great uniformity among the actions of men, in all nations and ages, and that human nature remains still the same, in its principles and operations. The same motives always produce the same actions: The same events follow from the same causes. Ambition, avarice, self-love, vanity, friendship,

generosity, public spirit; these passions, mixed in various degrees, and distributed through society, have been, from the beginning of the world, and still are, the source of all the actions and enterprises, which have ever been observed among mankind. . . . Mankind are so much the same, in all times and places, that history informs us of nothing new or strange in this particular. Its chief use is only to discover the constant and universal principles of human nature, by showing men in all varieties of circumstances and situations, and furnishing us with materials, from which we may form our observations and become acquainted with the regular springs of human action and behaviour. These records of wars, intrigues, factions, and revolutions, are so many collections of experiments, by which the politician or moral philosopher fixes the principles of his science, in the same manner as the physician or natural philosopher becomes acquainted with the nature of plants, minerals, and other external objects, by the experiments which he forms concerning them.[62]

With these principles (twisted to misanthropic satire by Fielding's Man of the Hill[63]), Hume outlines a practical program for the natural history of man. It would be adopted by his countryman Walter Scott, sixty years hence, as a scientific basis for his own epoch-making experiment in historical fiction: Scott all but cites the *Enquiry* in the introductory chapter to *Waverley*. Opening human life to the prospect of wholesale historical change, Scott seeks to regulate its potentially infinite variability, first, by invoking "the passions common to men in all stages of society,"[64] and second, by limiting the range of temporal difference to the intermediate past, "sixty years since," the span of individual human life and personal memory. Within this range, "the constant and universal principles of human nature" may hold steady.

Also addressing Hume's program, Jane Austen's narrator in *Northanger Abbey* alludes to a key principle of the conjectural history assumed in *Waverley*, the global contemporaneity of different developmental states of human society:

Charming as were all Mrs. Radcliffe's works, and charming even as were the works of all her imitators, it was not in them perhaps that human nature, at least in the Midland counties of England, was to be looked for. Of the Alps and Pyrenees, with their pine forests and their vices, they might give a faithful delineation; and Italy, Switzerland, and the south of France might be as fruitful in horrors as they were there represented. Catherine dared not doubt beyond her own country, and even of that, if hard pressed, would have yielded the northern and western extremities. But in the central part of England there was surely some security for the existence even of a wife not beloved, in the laws of the land, and the manners of the age. . . . Among the Alps and Pyrenees, perhaps, there were no mixed characters. There, such as were not as spotless as an angel might have the dispositions of a fiend. But in England it was not so; among the English, she believed, in their hearts and habits, there was a general though unequal mixture of good and bad.[65]

Austen's irony cuts both ways. Of course human nature must be a universal constant—in populating foreign countries with fiends and angels, Gothic romancers expose their own as well as their readers' ignorance of the world. The human nature found at home in England, however, is an entity shaped by "the laws of the land, and the manners of the age," so that "mixed character," its essential ingredient, not only resembles that historically contingent formation, the British constitution, but has coevolved with it—is shaped by it. Hume had described the English as having "the least of a national character, unless this very singularity may pass for such," by which he meant that the English national character expressed England's historical achievement of civil and religious liberty and a mixed constitution.[66] If human nature amounts to national character, then it might well be different among the Alps and Pyrenees, or even in Wales or Scotland, let alone in generations or centuries past. Meanwhile, with "the Midland counties of England," Austen claims for nineteenth-century British fiction the homeland of realism that *Middlemarch*, supreme achievement in the mode, will call in its subtitle *Provincial Life*: a geographical realization of that cosmic "middle nature" Herder had identified as the habitable zone of the "noble middle creature," man, where human nature can be maintained at the center of its world by a refined literary technology—mixed character, omniscient narration, free indirect speech, and so on. Notoriously, as though fulfilling Austen's prescription, realism gives way to other aesthetic modes, such as melodrama, romance, and Gothic, when the English novel ventures out to the imperial periphery, or even below the middle classes.[67]

Human nature has become a historical discourse and man a historical problem by the time Austen and Scott are writing, in the wake of the French Revolution: a shocking acceleration of historical change made more shocking by claims upon it as a historical event that would *change human nature*. "The crisis in the conception of 'Man,' propelled most notoriously by the French Revolution, coincided with a reconception, perhaps even an invention, of 'literature,'" which took the form, writes Maureen McLane, of "a literary anthropology—a conscious conjunction of the literary and the human."[68] Major English poets (Wordsworth, Coleridge, Percy Shelley) exalt poetry as a reparative "discourse of the species" that, originating in human nature, may thus restore it: "Poetry models itself as a totality for man, a synthesis of his faculties and powers, a return of human language to the human body."[69] McLane turns to *Frankenstein* for its exemplary demonstration of the breakdown of the new synthesis under the stresses it is required to bear. The monster is an impossible subject who makes his eloquent claim upon human reason, human sentiment, and human rights from outside biological generation. Accordingly, and following the slave's cue, he resorts to the Enlightenment endeavor to make himself human—to make his own nature—through the new technology of *Bildung*, that is, through letters: learning to read, receiving and relating stories. And he almost makes it. . . . When the creature narrates his own history, at the center of the novel, Frankenstein is moved to recognize his humanity—until the spectacular evidence of his obscene genesis, his botched-together body, blocks the turn to sympathetic inclusion:

> His words had a strange effect upon me. I compassionated him, and sometimes felt a wish to console him; but when I looked upon him, when I saw the filthy mass that moved and talked, my heart sickened, and my feelings altered to those of horror and hatred.[70]

Aesthetic sensibility, fixed in the visual register, throws up "an insuperable barrier" to "humane assimilation."[71] Simultaneously burdened with an excess of feeling (diverted into suffering) and an excess of physical life (his grotesque body), the monster can neither be human, the product of culture, nor a species, the product of nature.

Anomalous if not monstrous, like its subject, Shelley's novel is a singular experiment in British Romantic fiction: singular too in being the only novel featured in McLane's study, which addresses poetry as the period's "literary absolute," the privileged bearer of "the promise of a totality for man."[72] Complementing *Romanticism and the Human Sciences* (an early inspiration for this book), *Human Forms* turns to the novel, the new genre that coevolved with the scientific revolution, shared its empiricist protocols, and, as itself a mode of experimental history, set out to compose and shape, not merely reproduce, the natural history of man. Lacking the preformed components (metrical, stylistic, stanzaic) of poetic genres, the novel finds its form in time—in development. The genre's supposed aesthetic fault, its lack of form, becomes its asset. Fictional prose narrative—emergent, extensive, open, speculative, combinatory—provides an analogue for the new, epigenetic model of biological development and the human figure it gestates. Where conjectural histories track the development of the race or species through the succession of modes of production and social institutions, novels articulate particular life stories against that collective scale, bringing its determinations home to personal experience, to individual thought and feeling.

In short: the Bildungsroman. The new kind of novel, born in the "novelistic revolution" of European Romanticism, becomes regulative—the "symbolic form of modernity"—to the extent that realist novels are all in some sense novels of development.[73] *Frankenstein* staged the calamitous failure of the radical (Rousseauvian, Kantian) version of *Bildung* as man's artificial remaking of man, in which humanity and species life are torn asunder to leave the wretched subject without either. The normative model of *Bildung*—considered in chapters 2 and 5 of *Human Forms*—appeals to organic growth for a narrative of gradual development through trial and error, divagation and discovery (good experimental practice), rather than by sudden revolutionary reinvention (bad). Its narrative tendency, played out in Goethe's prototype *Wilhelm Meister's Apprenticeship*, is toward a potentially endless deferral of the horizon of realization, a substitution of human being with everlasting becoming. The tension is thematic in the two major competing critical accounts of the new form, defining the theory of the novel into our own time, by G. W. F. Hegel and by Friedrich Schlegel: the former deploring the modern novel's betrayal of an "epic" aesthetic of unity and totality, the latter celebrating a new, "Romantic" aesthetic of perpetual emergence and infinite differentiation.

In a sharpening dialectic, the human is to become universal by becoming individual, shedding the local and contingent shackles of social type. But social integration remains the impassable horizon of the Bildungsroman, entailing (for Franco Moretti, after Hegel and Lukács) the hero's renunciation of individual autonomy for the "symbolic gratification" that attends "the happy belonging to a harmonious totality."[74] The tension between these imperatives is radicalized, early on, in the case of women, whose artificial exclusion from the social paths of *Bildung* enforces an a priori exclusion from full humanity (as advertised in the name of the universal particular, man). Germaine de Staël's feminist deconstruction of the Bildungsroman in *Corinne* (a project resumed later by George Eliot) presses its critique through a turn to the nation as *Bildung*'s secular horizon, the site of human totality, in a preemptive anticipation of the solution offered in the contemporaneous Irish and Scottish genres of national tale and historical novel. Binding together individual and national destinies, the national tale's "allegory of union"[75] charges a marriage plot with the sentimental and libidinal contents of the political union between former nations that constitutes the modern state: legislating, in Staël's scathing gloss, the confinement of women to the infantine half-life of domesticity. The dissenting heroine, exiled from domesticity as well as from *Bildung*, suffers death without biological issue—extinction: to claim, instead, the melancholy sovereignty of memory and imagination, enshrined in the novel that bears her.

The historical novel makes national history the middle term that correlates the history of the individual with the history of the species—a crux opened in philosophical anthropology, and exacerbated in the nineteenth century not only by the confounding event of revolution, splitting history apart, but by sublime extensions of the chronology of earth history beyond human species life. That vastly dilated temporal range affords the new transformism one of its conditions of possibility. If natural history now affords "no vestige of a beginning,—no prospect of an end," in the phrase of geologist James Hutton,[76] then development—mutation—may likewise have no end, no final station or perfect type. This intimation troubles the later historical novels of Scott and his successors, as they break through the boundary of an intermediate past (sixty years since) to imagine remoter, stranger scenes of human and not-quite-human life. The nation, an only ever temporary bridge between the divergent paths of personal history and natural history, no longer holds.

6

Mary Shelley stages the crisis of the monster's humanity as a crisis of form. He repels sympathetic acceptance because his body appears in others' eyes as a "miserable deformity," a "filthy mass that moved and talked,"[77] form on the brink of deliquescence into matter—foul, formless, Bataille's *informe*.[78] The crisis of the human form is an insurmountable aesthetic offense.

Fielding's conceit of the author as cook invokes taste—aesthetic judgment—as the faculty that subdues the "prodigious variety" of human nature to a form. *Tom Jones* is an enormous work, comparable in length to the serial novels of Dickens; but where critics would deplore the sprawling shapelessness of those, they praised the beautiful order of Fielding's. "Uniformity amidst variety is justly allowed in all works of invention to be the prime source of beauty, and it is the peculiar excellence of *Tom*

Jones," wrote Arthur Murphy in 1762.[79] Modern critics debate whether the form of *Tom Jones* might be regulated by classical ratios of geometric proportion, derived from Vitruvian architectural theory. In an influential article, Frederick W. Hilles proposed that *Tom Jones* was "shaped like a Palladian mansion," and cited as its "ground plan" John Wood's design for Prior Park, the country house of Fielding's patron Ralph Allen.[80] An English edition of Andrea Palladio's *Four Books of Architecture*, a byproduct of the vogue for Palladian villas, appeared in successive volumes from 1715, along with its ancient model, Vitruvius's *De Architectura* (*Vitruvius Britannicus*, also 1715). The design of a temple, wrote Vitruvius, should reproduce the proportions of the human body, since those epitomize the harmonious symmetry of the cosmos. Leonardo da Vinci's great cartoon "Vitruvian Man" (c. 1490) exhibits the classical placement of the human form at the center of the universe, measuring its rational order.

Tempting though it is to detect a human geometry encrypted in *Tom Jones*, it seems unlikely Fielding had Prior Park in mind as a blueprint.[81] Fielding draws the reader's attention, in any case, to his work's temporal dimension, tracking the uneven relation between the pace of the narrative (*discours*) and the time of its narration (*récit*) as measured in the length of its successive parts: book 3, "From the time when Tommy Jones arrived at the age of fourteen, till he attained the age of nineteen" (29 pp.); book 4, "Containing the time of a year" (49 pp.); book 5, "Containing a portion of time somewhat longer than half a year" (51 pp.); book 6, "Containing about three weeks" (49 pp.); book 7, "Containing three days" (62 pp.); book 8, "Containing about two days" (75 pp.); books 9 and 10, "Containing twelve hours" (30 pp.); and so on. (The pattern of narrative acceleration is not sustained.) Far from reproducing a geometrical order, perceptible (like a sonnet's) at the glance of an eye, novelistic form unfolds in time.[82] Famously, *Tom Jones*'s apparent temporal drift is reined in, in the closing chapters, to an efficient Aristotelian equipage of plot: classical form after all.

The neoclassical aesthetic standard—according to which artistic forms represent ideal ratios of cosmic order—was being displaced, at mid-century, by a sensationalist aesthetics that rooted taste in the common faculties of the human physiology.[83] Explicitly rejecting the canons of proportion, Edmund Burke identifies aesthetic effects with "some quality in bodies, acting mechanically upon the human mind by the intervention of the senses."[84] Burke's signature aesthetic category, the sublime, entails the dissolution of form as a spatial property, measured by sight (since its qualities include immensity and obscurity), and the accession, instead, of durational effects of interminability, succession, repetition, interruption, and intermittence. Burke's treatise looks forward to the escalating scandal of the novel's formlessness in the following century. Critics bewail novels, especially popular novels such as Dickens's, for the indefiniteness of their dimensions and proportions and their excessive internal heterogeneity: they are too long, they are crammed with too many characters, incidents, and settings, they promiscuously mix discourses, registers, and styles—or they lack style altogether. Dispersed across serial installments, generating "mechanical" (automatic) rather than "organic" (purposive) rhythms of interruption, suspense, and repetition, they play to readers' susceptibilities to sensation rather than their capacity for reflection. In a more receptive spirit, the Victorian critical movement that Nicholas Dames calls "physiological novel theory" made "the problem of elongated artistic forms— forms whose length makes continuous, heightened attention impossible and acts of recollection difficult" central to its inquiry. Interested in the "pitch, intensity and duration of readerly attention," the oscillation between nervous states of tension and relaxation, and the "engrossment" of individual consciousness into a mass reading experience, critics such as E. S. Dallas and G. H. Lewes recognized the long Victorian novel as a sublime genre in the Burkean sense.[85] This did not mean that Lewes (for instance) allowed Dickens's novels to be works of art. Rather, they were "phenomena of hallucination" that short-circuited human rationality:

> The writer presents almost a unique example of a mind of singular force in which, so to speak, sensations never passed into ideas. Dickens sees and feels, but the logic of feeling seems the only logic he can manage. Thought is strangely absent from his works. . . . Compared with that of Fielding or Thackeray, his was merely an *animal* intelligence, i.e., restricted to perceptions.[86]

As though the novelist is himself one of his inhuman creatures. Dickens's twenty-part serials of the mid-century dislocate or abandon altogether the axis of fictional biography with which the Bildungsroman sought to control its "bad infinity."[87] Henry James's well-known putdown of serial novels as "large loose baggy monsters [with] queer elements of the accidental and the arbitrary" branded them not with a sublime but with a grotesque aesthetic—the label above all others that would stick to Dickens's works.[88]

Dickens was the foremost English practitioner of a distinctively modern kind of fiction (Richard Maxwell calls it the novel of urban mysteries[89]) condemned by critics for being formally excessive, monstrous itself, and trafficking in monstrous deformations of humanity. Incubated in new industrial-era audience demographics and modes of production, the early

Victorian monster-novel feeds on a mass reading public; its natural habitat is the serial, from penny-dreadful to weekly miscellany and monthly shilling number. This new fiction emerges around 1830, in the turmoil of a second French Revolution and Reform agitation in Britain—the cheering or dismal prospect, depending on one's politics, of revolution no longer as a historical singularity but as a chronic condition of modern life, rooted in metropolitan experience and the urban crowd. *Frankenstein* returns, in a popular edition revised by the author, in 1831; no longer solitary, it is flanked by novels by Scott (*Count Robert of Paris*) and Victor Hugo (*Notre-Dame de Paris*) that feature weird quasi-human monsters at the center of their labyrinths. Hugo had theorized the grotesque as the aesthetic mode of modern life in the preface to his 1827 drama *Cromwell*: the signature of fractured, hybrid, mutant form, of nature as perpetual metamorphosis, its figureheads are the giant orangutan, Scott's Sylvan, and the deformed bell-ringer, Hugo's Quasimodo. Unlike Frankenstein's monster, man-made unman, they are natural born. The grotesque, mingling horror with pathos, is the emanation of their intimate proximity to—and ambiguous encroachment upon—the human form.

Neither *Count Robert of Paris* nor *Notre-Dame de Paris* (discussed in chapter 3) appeared as serials, although their progeny would, the *romans-feuilletons* of Eugène Sue and G. W. M. Reynolds, and Dickens's monthly numbers. Both Scott's and Hugo's novels attune their grotesque aesthetic to the controversial resurgence of transformism in the late 1820s, in the renewed attention to Lamarck's "Orang-Outang Hypothesis" and the teratology of Geoffroy and Isidore Saint-Hilaire, broadcast via the high-profile public debate over morphological principles between Geoffroy and Cuvier and inflamed by the association of "radical science" with reformist and revolutionary politics. Not for nothing does transformist speculation find fertile soil in the new forms of urban popular fiction. Dickens's art, I argue in chapter 4, channels the popular diffusion of evolutionist thought in Great Britain in the decade and a half before the appearance of *On the Origin of Species* via Chambers's *Vestiges of the Natural History of Creation*. *Bleak House*, the supreme transformist thought experiment in English before Darwin, poses a massive affront to an ascendant aesthetic of novelistic realism predicated on the constitutive centrality of the human form, scale, and perspective in the world. The affront is more powerful for Dickens's identification of the world with the city: in the total man-made environment, human nature comes undone, speciating into morbid and pathetic fragments rather than sustaining coherent forms of life.

The Dickens World, the world as city, consumes and metabolizes nature. Its dark totality excludes the domain of realism claimed by the "great tradition of the English novel," a modern critical canon assembled as a bulwark against the demographic ascendancy of a mass reading public serviced by popular fiction. The great English novelists, wrote F. R. Leavis, "are significant in terms of the human awareness they promote; awareness of the possibilities of life."[90] We are to look to Jane Austen and George Eliot, not to the allegorical realism of Dickens and Hugo, for a faithful mimesis of human nature, faithful not only in content but in the perfection of techniques (modern innovations such as free indirect discourse as well as classical properties such as unity of plot) that conform "the possibilities of life" to the scale of "human awareness." Although size matters, it need not be decisive: in *Middlemarch*, Eliot subdues the enormity of Dickensian serial form to the proportions of *Emma* or *Mansfield Park*. Eliot sought to emulate the popular format of Dickens's fiction but at the same time to dignify it, issuing *Middlemarch* and *Daniel Deronda* in eight books appearing at two-month intervals, statelier than the spate of nineteen monthly pamphlets.[91] Eliot reaffirms provincial life as realism's world, but critically rather than reactively, in full cognizance of its possibilities and limits, investing it with what Lauren Goodlad calls (after Fredric Jameson) the "Victorian geopolitical aesthetic." George Eliot, most European of the great Victorian novelists, reckons not only with current scientific and philosophical movements in her work but with the broad modern tradition of the novel—specifically, with those key forms of nineteenth-century realism, the Bildungsroman and the historical novel, which she recombines, as Goodlad argues, to realize "the historical novel of our time."[92]

If *Middlemarch* brings the project of English realism to formal completion, *Daniel Deronda* ruptures it. George Eliot's last novel abandons the temporal and geographical bounds of provincial life for the oceanic flux of contemporary world history and the deep time of racial history. The pressures of that chronotopic abyss warp humanity into strange new forms, along with its representational apparatus—the relations between figure and letter, metaphor and event, allegory and mimesis, that have persistently vexed the discourse of man and that realism sought to stabilize. *Middlemarch*, far from resting content within its domain, vouchsafes an intuition of a reality beyond the realist novel and its conditions, the human form and scale, in the heroine's late epiphany of "involuntary, palpitating life"—an intuition of the immanence of life as such, a dynamic material process exceeding human consciousness. Exceeding, that is, not only Dorothea's consciousness of her own being but the "consciousness of species" Ludwig Feuerbach had designated (in a work translated by George Eliot) as the uniquely human faculty: that "consciousness which man has of his own—not finite and limited, but infinite nature."[93] Disclosing the far horizon of the topic named in the novel's opening sentence, "the history of man," and hence the far horizon of her fictional project, Eliot invokes the Darwinian conception of life in which, writes Elizabeth Grosz, "the human is one species among many, one destined itself to be overcome, as are all the forms of life on earth."[94] The great work of English realism folds open to admit—at this culminating moment—a glimpse of its outside.

7

Sinking human in animal life, Darwin brings to theoretical completion the modern displacement of the classical figure of man—the individual human form as universal cynosure and standard—for a distributed, dynamic conception of "life as such." (At the same time, faithful to inductive and empiricist principles, Darwin participates in a contemporary disengagement of life from Romantic vitalism, which recasts it as a *combinatoire* of chemical and physical processes.[95]) Writing in 1860, Herbert Spencer comments on the obsolescence of the traditional figure of the body politic: the human form no longer offers a microcosm of the natural or social order. Instead, "the indefiniteness of form, the discontinuity of the parts, and the universal sensitiveness" are "peculiarities of the social organism ... to which the inferior classes of animals present approximations."[96] At the turn of the nineteenth century, Thomas Malthus's *Essay on the Principle of Population* had undone the traditional equivalence "between individual and social organisms by tracing social problems to human vitality itself," Catherine Gallagher argues, as Romantic and early-Victorian intellectuals "relocated the idea of ultimate value from a realm of transcendental spiritual meanings to organic 'Life.'"[97] Before Malthus, Buffon made population a key determinant in "a biological and demographic model for differentiating and evaluating human cultures."[98] Natural historians as well as political economists recalibrated life to the macroscopic scale of populations, statistical reckonings of historical probability, and a biopolitical regime of value that swamped personal experience and meaning. Conversely, the discovery of a new basis of life in the nucleated cell (by German philosophical anatomists Theodor Schwann and Matthias Schleiden) reconceived individual bodies, including human bodies, as aggregates of microscopic biological entities. By the close of the century, "life" would swallow the figure of man—along with the dialectic between individual and social life that (according to Lukács) sustained the realist project—to become the scientific principle of the novelist's art. Henry James described the novel as not just "a personal impression of life" but itself "a living thing, all one and continuous, like every other organism," and its raw material, "experience," as "an immense sensibility, a kind of huge spider-web, of the finest silken threads, suspended in the chamber of consciousness and catching every air-borne particle in its tissue."[99] Monstrosity returns, dandiacally bedecked, in James's declension of Eliot's "involuntary, palpitating life."

Human Forms pauses with the consummation of realism (followed by its exemplary breach) in Eliot's late fiction and with Darwin's resolution (unfinished, still contentious) of the natural history of man. After Darwin, the game changes. Scientific racism and other biological determinisms, tributary until the last third of the century, flood the cultural field. New discourses, psychoanalytic and sexological, reopen human nature even as those surge in to close it down.[100] Emile Zola's naturalist manifesto "The Experimental Novel," published in the year of Eliot's death, is symptomatic in its literal-minded appeal to scientific methodology (citing Claude Bernard's *Introduction to the Study of Experimental Medicine*) and to scientific themes, such as the interaction of heredity and environment, for determining the novelist's art. A decisive philosophical step beyond Darwin, according to Grosz, is taken by Henri Bergson: "If Darwin demonstrates man's immersion in and emergence from animal (and ultimately plant) life (or even life before plants and animals separated)," Bergson "demonstrates man's immersion in and emergence from the inhuman, the inorganic, or the nonliving."[101] And also, by implication, his reimmersion and dissolution: a step George Eliot foresaw in her eerie late essay in speculative fiction, "Shadows of the Coming Race," which imagines the supersession of human and indeed organic life on earth by the evolution of intelligent machines.

"A new humanities becomes possible," Grosz suggests, "once the human is placed in its properly inhuman context . . . within the animal, within nature, and within a space and time that man does not regulate, understand, or control."[102] Her proposal echoes other philosophers of science:

> The essential function of science is to devalorize the qualities of objects that comprise the milieu proper to man; science presents itself as the general theory of a real, that is to say, inhuman milieu. . . . In all rigor, the qualification *real* can be applied only to the absolute universe, the

universal milieu of elements and movements disclosed by science. Its recognition as real is necessarily accompanied by the disqualification, as illusions or vital errors, of all subjectively centered proper milieus, including that of man.[103]

Georges Canguilhem's appeal to science as discourse of the real by virtue of its access to the "inhuman milieu" of an "absolute universe" is amplified by Quentin Meillassoux and Ray Brassier in their critique of the anthropocentric doctrine (decisively installed by Kant) of "correlationism," which maintains "we only have access to the correlation between thinking and being, and never to either term considered apart from the other."[104] Reality, according to correlationism, can only ever be given to human cognition as the effect of a reciprocal relation between thought and world: a tenet that finds its literary form in nineteenth-century realism's correlations of human spatial and temporal scales and, via free indirect discourse, of world and subject. But the mathematical techniques of natural science—radiometric dating, spectrographic analysis, and so on—yield knowledge of a reality outside the "horizon of correlation": a history of the earth, and of the universe, "anterior to the emergence of thought and even of life," and thus *"anterior to every form of human relation to the world."*[105] To this Brassier adds the necessary condition of "posteriority," constituted by the eventual extinction of all life on earth, solar death, and the collapse of space-time: a cascade of catastrophes that more decisively *"disarticulates* the correlation," since it establishes extinction as a transcendental condition: "Terrestrial history occurs between the simultaneous strophes of a death which is at once earlier than the birth of the first unicellular organism, and later than the extinction of the last multicellular animal."[106]

Whether or not we go along with these critiques (with their rumble of *après nous le déluge*), we may read the historical opening of a deep time of inhuman anteriority (Meillassoux's "ancestrality") in our period in a series of natural historical writings, from Buffon's *Epochs of Nature* (1778) through Charles Lyell's *Principles of Geology* (1830–33) to *On the Origin of Species*.[107] It is Darwin who poses the decisive challenge to correlationism, Julián Jiménez Heffernan argues in a recent essay, by virtue of his insistence on extinction as a key determinant in the history of life. Darwin's theory infers not only a biological deep past before human emergence but an all-but-infinite futurity beyond it, shadowed by the prospect not only of transmutation (whether progressive or degenerative) but of extinction: a shadow that falls, in Heffernan's provocative reading, across the racial theme of *Daniel Deronda*.[108] Deep time, extending before and after human life, makes up the conceptual outside—the impassable limit—of the history of man. Its eruption into scientific knowledge coincides historically with the onset of humanity's takeover of the history of the earth, with what we now call the Anthropocene. Human beings, wrote Karl Marx in 1845, "begin to distinguish themselves from animals as soon as they begin to *produce* their means of subsistence," by that "indirectly producing their actual material life"—remaking nature, in other words, as a human product.[109] By the mid-1840s, according to recent commentary, the industrial—imminently global—scale of that remaking was becoming evident.[110] Both prospects, of an inhuman earth history and of a human conquest of earth history, constitute too the outside of the nineteenth-century novel, the horizon against which novels strive to think their reality. If George Eliot's involuntary, palpitating life signals a biological continuum beyond realism's human precinct, the opening of *Bleak House*—with its visionary conceits of a megalosaurus on Holborn Hill and urban soot as snowflakes mourning the death of the sun—overcasts Victorian London, that total human environment, with a more drastic intimation of the before and after of the conditions for life, all life. "Far from lying in wait in for us in the far distant future, on the other side of the terrestrial horizon, the solar catastrophe needs to be grasped as something that has already happened," writes Brassier, trading Dickensian whimsy for existential portentousness and citing Jean-François Lyotard: "*Everything is dead already*."[111] This is the aspect of the Dickens World that John Ruskin, also writing from the imperial capital, saw clearly, and detested:

> The thoroughly trained Londoner can enjoy no other excitement than that to which he has been accustomed, but asks for *that* in continually more ardent or more virulent concentration; and the ultimate power of fiction to entertain him is by varying to his fancy the modes, and defining for his dulness the horrors, of Death.[112]

Involuntary, palpitating life, transcendental death: the ends of man . . .

CHAPTER ONE

The Form of Man

Time has changed everything so much that one often needs a magic mirror in order to recognize the same creature beneath such diverse forms. The form of the earth, its surface, its condition, has changed. Changed are the race, the manner of life, the manner of thought, the form of government, the taste of nations—just as families and individual human beings change.

—J. G. HERDER, "ON THE CHANGE OF TASTE" (1766)[1]

Under domestication, it may be truly said that the whole organisation becomes in some degree plastic.

—CHARLES DARWIN, ON THE ORIGIN OF SPECIES (1859)

Conjecture, History, Science, Fiction

"There is no question of importance, whose decision is not compriz'd in the science of man," David Hume announced in his introduction to *A Treatise on Human Nature* (1739): "In pretending therefore to explain the principles of human nature, we in effect propose a compleat system of the sciences, built on a foundation almost entirely new, and the only one upon which they can stand with any security."[2] After Hume, across the Atlantic Enlightenment, a series of major thinkers sought to realize this totalizing project—the science of all sciences—as a philosophical historical anthropology. The science of man became the natural history of man, a history not of individuals or nations but of the human species.[3] A new biological conception of species "as an entity distributed in time and space," released from the synchronic grid of Linnaean taxonomy as well as (eventually) from a providential cosmology, comprised what Philip Sloan has called the "Buffonian revolution." That revolution would be as consequential for literary genres, especially the novel, as it was for the natural and human sciences, in part due to Buffon's recourse to a literary style and techniques of "speculative thought experiment," probabilistic reasoning, "analogical reasoning, and divination" in his scientific method.[4] Buffon's *Natural History, General and Particular*, published in successive volumes from 1749 to 1788, opened philosophical space for a "zoological history of the human species, abandoned to the vicissitudes of time and space," and a "radical historicizing and naturalizing" of the Enlightenment figure of man (even as the latter step was one Buffon himself was hesitant to take).[5]

Looking back from the following generation, Edinburgh philosopher Dugald Stewart identified "the natural or *theoretical history* of society in all its various aspects," comprising "the history of languages, of the arts, of the sciences, of laws, of government, of manners, and of religion," as "the peculiar glory of the latter half of the eighteenth century." The unity and stability of human nature, its expressions diversified by geographical and historical circumstance, constituted the "fundamental and leading idea" of the new science.[6] Despite such affirmations, it was immediately riven by controversy. Leading exponents of the natural history of man— among them Adam Smith, Adam Ferguson, John Millar, and Lord Kames in Scotland, Johann Gottfried Herder and Immanuel Kant in Germany— responded to the challenge issued to its premise, the conjunction of nature and history, by Jean-Jacques Rousseau, whose *Discourse on the Origin and Foundations of Inequality Among Men* (1755) argued that the onset of history entailed the divorce of human beings from nature. Rousseau's intervention helped ensure that the natural history of man would remain a field of philosophical contention rather than consensus over the next hundred years. The ensuing debates heralded the post-Enlightenment breakup of the putatively unified science of man, the dispersal of human nature across rival, professionalizing disciplines, and the partition between the natural and human sciences (and between "science" and "literature") that would divide the terrain of knowledge by the latter half of the nineteenth century.[7] From its outset, the new anthropology was vexed by a critical question: Could a conjectural history of the human species be truly philosophical? Was it science or was it fiction? Or worse, was it some monstrous hybrid, a kind of science fiction?

The fault line opened, with an explicit articulation of rival principles, in the debate over the history of man that broke out in the mid-1780s between Kant and Herder.[8] One of the great intellectual quarrels of the late Enlightenment, it signposted the forking paths of Kant's critical philosophy, on the one hand, and the scientific project of natural history, on the other, which would culminate in Charles Darwin's programmatic digestion of the human sciences (from anthropology to aesthetics) nearly a century later in *The Descent of Man*. A generation after the *Treatise on Human Nature*, both Prussian thinkers echoed Hume's ambition to reconstitute all knowledge upon the ground (or through the figure) of "man." (My discussion will adopt period usage of the universal particular, "man," for reasons that become explicit in the following chapter.) "What fruitful new developments must not occur here," Herder wrote in 1765, "if our whole philosophy becomes anthropology."[9] Four years later, he sketched a plan for a "history of the human soul in general, by ages and peoples," which would eventually issue in his magnum opus, *Ideas for a Philosophy of the History of Mankind* (*Ideen zur Philosophie der Geschichte der Menschheit*), published in five parts from 1784 to 1791.[10] This was the first of Herder's works to be translated into English, as *Outlines of a Philosophy of the History of Man*, in 1800, under the imprint of Joseph Johnson, arguably the most influential publisher in Romantic-period England. Widely acclaimed, *Outlines* went into a second edition (1803), and was reprinted and excerpted in scientific journals: "Herder's *Outlines* brought to England the Continental tradition of historical-philosophical-anthropological thought of which other key instances are eighteenth-century language-origin treatises by Rousseau and Condillac, Volney's *Ruins*, Kant's *Anthropology*, and the writings of Mme de Staël."[11] Meanwhile, in 1773, Kant—Herder's former professor at the University of Königsberg—proposed making anthropology "a proper academic discipline," and by doing so to "disclose through it the sources of all the sciences, ... and thus of everything that pertains to the practical" (that is, to ethics and politics).[12] After his retirement, in 1798, Kant published the annual anthropology course he had continued to teach throughout his career, as *Anthropology from a Pragmatic Point of View* (*Anthropologie in Pragmatischer Hinsicht*).

Notoriously, Kant and Herder fell out over the protodiscipline. Kant's review of the first volume of Herder's *Ideas for a Philosophy*, published in the *Allgemeine Literatur-Zeitung* in January 1785, offended his former student and sparked a public controversy between them. In the second volume of the *Ideas*, Herder rejected key propositions of Kant's own short essay in conjectural anthropology, "Idea for a Universal History from a Cosmopolitan Perspective" ("Idee zu einer allgemeinen Geschichte in weltbürgerlicher Absicht," 1784), notably, that human nature will find its vocation "only in the species, but not in the individual."[13] Herder argued, instead, that the life of the species is immanent in individual bodies, in the genetic force that informs matter and joins all beings in a progressive evolutionary scale. Kant replied to Herder's criticisms in a review of that second volume (November 1785) and then in another, provocatively ironical essay, "Conjectural Beginning of Human History" ("Mutmaßlicher Anfang der Menschengeschichte," January 1786). The refutation of Herder would inform the great work Kant was preparing in these years, the *Critique of the Power of Judgment*, and he would reaffirm his scientific principles in *Anthropology from a Pragmatic Point of View* (107–75).

At stake in the debate over human nature were not only philosophical principles (discussed in the next section of this chapter) but also methods and protocols of representation. In his first review of *Ideas for a Philosophy of the History of Mankind*, Kant complains of the work's literary and figurative rather than properly scientific character. Poetical fancy and not philosophical reason drives Herder's argument. Kant exhorts "our spirited author" to "put his lively genius under some constraint," and be guided "not [by] conjectured but observed laws, not by means of a force of imagination given wings whether through metaphysics or through feelings, but through a reason which is expansive in its design but cautious in the execution" (133). In his second review, Kant speculates, in the same vein, whether "the poetical spirit that animates his expression has not sometimes also invaded the author's philosophy; whether here and there synonyms have not been allowed to count as explanations and allegories for truths . . . and whether in many places the fabric of bold metaphors, poetic images, mythological allusions, has not served rather to conceal the body of thoughts as under a *farthingale* than to let it shine forth" (138). Kant opens his own "Conjectural Beginning of Human History" by drawing a distinction between legitimate and illegitimate uses of conjecture in historical method:

> In the *progression* of a history it is indeed allowed to *insert* conjectures to fill up gaps in the records, because what precedes as a remote cause and what follows as an effect can provide a quite secure guidance for the discovery of the intermediate causes, so as to make the transition comprehensible. Yet to let a history *arise* simply and solely from conjectures does not seem much better than to make a draft from a novel. Indeed, it would not be able to support the name of a "conjectural history" [*einer muthmaßlichen Geschichte*] but that of a mere fiction [*einer bloßen Erdichtung*]. (163)

Kant then embarks on what he calls a "pleasure trip," his own exercise in the genre, which turns out to be a parody of Herder's gloss of the Book of

Genesis as an allegory of human history in book 10 of *Ideas for a Philosophy of the History of Mankind*.[14]

The dispute between Kant and Herder, to which I return later in this chapter, marks a persistent crux of the natural history of man: bedeviled from the start by the charge that it amounted to little more than conjecture, and that its exponents were authors, in effect, of works of fiction.[15] Conjectural history belonged to the repertoire of "analogical reasoning," combining "a close investigation of empirical phenomena" with "the cultivation of creative imagination," imported into natural scientific discourse by Buffon—most flamboyantly, in his *Epochs of Nature* (1778), "the first attempted secular history of the Earth" on natural historical principles.[16] In Peter Hanns Reill's summary:

> Analogical reasoning, ever active, not clearly defined, yet always productive, would encourage interaction between observation and imagination and generate a form of understanding superior to discursive reasoning. [Buffon] called this type of understanding divination or intuition, and regarded it as a form of mediation that resulted in the heightened ability to perceive simultaneously form and force, structure and process, to discern the resemblances between sign and significant, and to mediate between the particular and the general.[17]

In 1795, looking back at the recent generation of Scottish philosophers, Dugald Stewart coined the English term "conjectural history," synonymous with "natural or theoretical history," to designate the "particular sort of inquiry ... entirely of modern origin" that unified the different branches of Adam Smith's research, "whether moral, political or literary":

> To this species of philosophical investigation, which has no appropriated name in our language, I shall take the liberty of giving the title of *Theoretical* or *Conjectural History*: an expression which coincides pretty nearly in its meaning with that of *Natural History*, as employed by Mr. Hume.[18]

The historicization of human nature, Stewart explained, was necessarily conjectural. Since the origins of human history lie outside our empirical purview, "we are under a necessity of supplying the place of fact by conjecture," reckoning—via analogical reasoning—"in what manner [human beings] are likely to have proceeded, from the principles of their nature, and the circumstances of their external situation."[19]

The late Enlightenment debate on human nature thus came to bear on conjecture—Buffon's "divination"—as the hinge (soft, but not necessarily weak) between knowledge-production and mere invention. Conjectural history became one of the flourishing if controversial discourses of the age, marked by recurrent topoi and even, as Frank Palmeri notes, by a distinctive grammar, the "'conjectural necessary' form of the past" tense.[20] "In framing our account of what man was in some imaginary state of nature, we overlook what he has always appeared within the reach of our own observation, and in the records of history," Adam Ferguson asserts at the opening of his *Essay on the History of Civil Society*: "In every other instance, however, the natural historian thinks himself obliged to collect facts, not to offer conjectures. . . . It is only in what relates to himself, and in matters the most important, and the most easily known, that he substitutes hypothesis instead of reality, and confounds the provinces of imagination and reason, of poetry and science."[21] Ferguson aims this uneasy disavowal, minimizing the margin of conjecture in his own work, at Rousseau's preemptive challenge to the natural history of man in the *Discourse on the Origins of Inequality*, published at the very inception of the genre (1755), one year after Buffon's treatment of the human species in the fourth volume of his *Natural History*. "The Philosophers who have examined the foundations of society have all felt the necessity of going back as far as the state of Nature, but none of them has reached it," Rousseau avers in his "Exordium": "Let us therefore begin by setting aside all the facts, for they do not affect the question. The Inquiries that may be pursued regarding this Subject ought not to be taken for historical truths, but only for hypothetical and conditional reasonings."[22] Whereas the Scottish philosophers will seek an accommodation between evidence and conjecture to reconstruct the original state of human nature, Rousseau insists on their separation: raising conjecture, in Kant's phrase, to the dubious power of "mere fiction." "The impossibility of reaching a rationally enlightened anthropology accounts for the necessary leap into fiction," in Paul de Man's summary of Rousseau's strategy, "since no past or present human action can coincide with or be under way towards the nature of man."[23]

This scandalous outbreak of fictionality within the scientific project of a human natural history expresses the instability Giorgio Agamben finds intrinsic to the "anthropological machine" of Enlightenment knowledge, which, in attempting to separate man from the nature in which he is embedded, separates man from himself. Anthropology is "an ironic apparatus that verifies the absence of nature proper to *Homo*, holding him suspended between a celestial and a terrestrial nature, between animal and human—and thus . . . always less and more than himself."[24] Its specific mark or symptom is "a mobile border within living man," an "intimate caesura" that divides human from animal nature within human nature.[25] Agamben traces the division back to Aristotle's distinction between nutritive and organic (i.e., developmental) principles of life; formulations of the Chain of Being, pervasive across eighteenth-century thought, defined man as the "middle link" between celestial and terrestrial orders, "torn by conflicting desires and propensities, . . . [wavering] between both, and . . . not quite at home in either"—a "strange hybrid monster," at once inside and outside nature.[26]

The Buffonian revolution of the mid-eighteenth century (a stage of what Lovejoy calls "the temporalization of the Chain of Being"[27]) brought a newly critical scrutiny to this anomalous condition. Buffon's *Natural History* relocated the idea of species from an archetypal form or abstract essence, fixed on an immutable scale or taxonomic grid (as in Linnaeus's *Systema naturae*), to living creatures' "faculty of reproducing beings similar to themselves, [in a] successive chain of individuals which constitutes the real existence of the species," by a "generative power which is perpetually in action"—dissolving the distinction between species from ontological to nominal status.[28] Far from abolishing it, this delivery of man to secular history and terrestrial geography made it more necessary than ever to redraw the ancient boundary between man and nature. Buffon himself balked at the full naturalization of human species being his work entailed. He sought to protect human exceptionalism by appealing to a divinely implanted rather than organic faculty of reason, reinstalling the traditional (metaphysical) division between man and nature within human nature, and giving *Homo sapiens* a new name, *Homo duplex*:

> The internal man is *double*. He is composed of two principles, different in their nature, and opposite in their action. The mind, or principle of all knowledge, wages perpetual war with the other principle, which is purely material.[29]

Buffon recasts taxonomic division as an internal, ontological antagonism, and baptizes a key figure of literary Romanticism, the double.[30] In doing so, he reconstitutes man as another allied Romantic figure, the monster, characterized—in Foucault's summary—by "the mixture of two realms, the animal and the human."[31]

Not only, then, was human nature fractured at its foundation (*Homo duplex*): so too was the project of a human natural history, which bifurcated into rival tendencies or traditions, already established in eighteenth-century debates over the constitution of society.[32] One, an empiricist tendency, with ancient, Aristotelian roots (comprising the

Scottish philosophers, Herder, and nineteenth-century scientific naturalists, including Charles Darwin), constitutes the secular figure of man upon a principle of innate or organic sociability, a cohesive instinct that binds together not only individual beings but the human species and its environments, and human nature and history. The historical progress of mankind, that is, remains continuous with, and is driven by, an organic human nature. The rival, critical tendency (Rousseau, Kant, Nietzsche, eventually Agamben and Foucault, who wrote his secondary doctoral thesis on Kant's *Anthropology*[33]), in contrast, divorces nature from history via a rejection of innate sociability. Instead, human progress entails the forging of an artificial or second nature, in which man's original nature is erased and overwritten. This critical account, inaugurated by Rousseau, came first, while the empiricist account, reacting to it, was infected by it, casting the naturalization of man as a remedial project: invoking, in Herder's phrase, a *Bildung der Humanität*, a "formation of humanity," based on the epigenetic theory of development (from undifferentiated to ever more complex organic structure) proposed in the contemporary German life sciences.

Rousseau's critical intervention, so early in the history of the genre, reset its terms: positing the state of nature as (in de Man's phrase) a "radical fiction," opening (as David Bates argues) "the radical possibility that we were never natural."[34] Dividing human nature from human history so as to forestall ideological appeals to a natural foundation of inequality, Rousseau deploys natural man as a heuristic figure, affording critical leverage for the political project he goes on to outline in the *Social Contract*: "The man who dares to undertake the establishment of a people has to feel himself capable of changing, so to speak, the nature of man, . . . of altering man's constitution in order to strengthen it; of substituting a morally dependent existence for the physically independent existence that we have all received from nature."[35] Man must be remade, in an ontological break with prior regimes, for a reciprocal remaking of political society. The imperative cuts to the heart of the scientific and philosophical project of the history of man. "Rousseau did not really want the human being to *go* back to the state of nature, but rather to *look* back at it from the stage where he now stands," Kant observes in the closing section (on "the character of the species") of *Anthropology from a Pragmatic Point of View* (422). Kant reiterates the ontological split within human nature as a disciplinary split within anthropological knowledge: "Physiological knowledge of the human being concerns the investigation of what *nature* makes of the human being; pragmatic, the investigation of what *he* as a free-acting being makes of himself, or can and should make of himself" (231). Man is an "animal endowed with the *capacity of reason (animal rationabile)*" that must, hence, "make out of himself a *rational animal (animal rationale)*" (417). Human nature becomes man's peculiar praxis, his work of self-invention, "until perfect art again becomes perfect nature, which is the ultimate goal of the moral vocation of the human species" (171).[36]

In Kant's account, man rebuilds himself on the rubble of his instincts. Kant seeks to make productive the Rousseauvian negative principle (discussed in the next section of this chapter) he calls "unsociable sociability": humans are innately, spontaneously averse to one another, but reason subdues that aversion and brings them together for mutual preservation and self-governance.[37] Thus Kant converts the conceptual antagonism within human nature, which he poses between reason and instinct, into the dialectical engine of human history, where it reiterates itself in a moral opposition between individual being, condemned to a restless, anxious striving, and species being, which is less the medium of human progress than its ever-receding bright horizon. Kant posits human perfection as a universal submission to civil law within a future "cosmopolitan society (*cosmopolitismus*)," a worldwide confederation of states. But this condition, he admits, is "an unattainable idea"—in philosophical terms, not "a constitutive principle," subject to empirical verification, but "a regulative principle," informing the moral imperative "to pursue [perfection] diligently as the destiny of the human race, not without grounded supposition of a natural tendency toward it" (427). The distinction between constitutive and regulative principles, worked out in his critical philosophy, clarifies Kant's management of human perfectibility as a heuristic fiction, and his desire to distinguish the rigorously regulative "as if" (*als ob*) that articulates teleological judgment from the loose, indiscriminate "as if" of "mere fiction."[38] Kant's commitment to the regulative principle of teleological judgment conditions his scorn for a conjectural history that invokes the rhetorical boundary of "as if" only to rush recklessly across it. Indeed, the severity of that scorn may imply that the philosophical stability of the regulative "as if," a crucial technique of teleological reason, is under threat— vulnerable to deflation as, itself, mere make-believe.

Kant's disavowal of fiction in the name of a strict standard of regulative usage stands against Herder's equally determined—strategic rather than inadvertent—commitment to the scientific efficacy of a literary stylistics. "Rather than unintentionally confusing philosophy with literature, as Kant charges, Herder is intentionally pushing philosophy in the direction of literature," notes Frederick C. Beiser.[39] Buffon's *Natural History*, as we saw, set the precedent, advocating an intuitive mode of analogical reasoning that overrode the Cartesian duality between observer and object, mind and body; Buffon's critics, accordingly, dismissed his more radical speculations, in *The Epochs of Nature*, as mere effusions of style.[40] In his essay "On the Cognition and Sensation of the Human Soul" ("Vom Erkennen und Empfinden der menschlichen Seele," 1774/1778), which constitutes, together with "This Too a Philosophy for the Formation of Mankind" ("Auch eine Philosophie der Geschichte zur Bildung der Menschheit," 1774), a manifesto for *Ideas for a Philosophy of the History of Mankind*, Herder defends analogy—a primary target of Kant's critique—as an interpretive technique that meshes the forms of human cognition, inextricable from their physiological medium, with the forms of the world. Amanda Jo Goldstein argues that Herder understands "poetic tropes as intrinsically linked to the physical activity of sensory experience, and proceeds to advocate figurative language as sensory knowledge's fittest mode of expression."[41] Herder reaffirms these principles in a later essay, "On Image, Poetry, and Fable" ("Über Bild, Dichtung, und Fabel"), published in the wake of the quarrel with Kant in 1787:

> In the forest of sensible objects that surrounds me, I find my way to becoming master of the chaos of the sensations assailing me only by separating objects from others, by giving them outline, dimensions, and form; in short, by creating unity in diversity and vividly and confidently designating these objects with the stamp of my inner sense, as if this were a seal of truth. Our whole life, then, is to a certain extent poetics: we do not see images but rather create them. . . . Hence it follows that our soul, like our language, allegorizes constantly.[42]

In Goldstein's summary, "the form of representation scientifically adequate to sensory experience is figuration—image, metaphor, allegory, personification."[43] Figuration accommodates the vitalist and epigenetic turns in late eighteenth-century natural philosophy,[44] the key terms of which—the life-force or genetic force (*Lebenskraft, genetische Kraft*), formative drive (*Bildungstrieb*), and so on—have not yet concretized, via scientific consensus, into a literal, technical vocabulary, and so retain their metaphoric multivalence and evocative power. Herder's figurative mobilization of concepts emerging from the new life sciences for a naturalization of the history of man would prove to be the key development, after Rousseau's critical intervention, in the genre's early formation: opening the way not only to the nineteenth-century study of cultural anthropology, as scholars of Herder have long recognized, but to the full immersion of the human in a natural history of infinite formal variability.

The Faculty of Perfection

Homo duplex signals an exceptional if persistent revulsion within the general logic of natural history. To save human nature, Buffon splits it, reconfiguring a traditional duality of body and spirit—a foundational principle of Christian religious anthropology, established by Paul and Augustine and influentially reaffirmed by Blaise Pascal. This rearguard action would not exempt man from the new knowledge unleashed by the "Buffonian revolution." The apprehension of nature as historical process turns on a temporal distinction between individual and species life:

> An individual is a solitary, a detached being, and has nothing in common with other beings, excepting that it resembles, or rather differs from them. . . . It is neither the number, nor the collection of similar individuals, but the constant succession and renovation of these individuals, which constitutes the species. A being, whose duration was perpetual, would not make a species. Species is an abstract and general term, the meaning of which can only be apprehended by considering Nature in the succession of time, and in the constant destruction and renovation of beings.[45]

The transience of particular, embodied lives is a necessary condition of the "abstract and general" life of the species. Buffon's opening of this distinction evacuates the medium of collective or social life: an individual is solitary and detached, having "nothing in common with other beings" except its difference from them, while the species is manifest "in the succession of time, and in the constant destruction and renovation of beings"—that is, in the existentially solitary events of birth and death (or rather—the order is telling—death and birth). Man differs from other creatures in his ability to reflect on and reason about this condition, which may indeed be the very condition of reflection, of human consciousness as such, as the more radical iterations of late Enlightenment philosophical anthropology suggest.

Decisively, Rousseau's. The *Discourse on the Origins of Inequality* sharpens Buffon's distinction between individual and species being so as to sever the constituent terms of the natural history of man. Seeking also to define the unique trait that sets humans apart from the rest of the natural order, but without resorting to a divine prosthesis of language or reason, Rousseau develops an analogous distinction of Buffon's, between "two kinds of education": "the education of the individual, which is common to man and the other animals; and the education of the species, which appertains to man alone."[46] The "very specific property that distinguishes" man from other creatures, according to Rousseau, is "the faculty of perfecting oneself" (*la faculté de se perfectionner*):

> [This faculty], with the aid of circumstances, successively develops all the others, and resides in us, in the species as well as in the individual, whereas an animal is at the end of several months what it will be for the rest of its life, and its species is after a thousand years what it was in the first year of those thousand.[47]

Humans are the only animals that dwell in history—or, rather, they dwell in or between two histories, an individual life history and the history of the race or species.

Rousseau's redefinition (adapting an Enlightenment commonplace) would be momentous, not just for perfectibilitarian enthusiasts of the revolutionary era such as Priestley, Godwin, and Condorcet, but for his ideological opponents and for the empiricist tradition of natural history.[48] Ferguson echoes it at the opening of his *Essay on the History of Civil Society*, a work that otherwise rejects the Rousseauvian account of human nature:

> In other classes of animals, the individual advances from infancy to age or maturity; and he attains, in the compass of a single life, to all the perfection his nature can reach: but, in the human kind, the species has a progress as well as the individual.[49]

In optimistic declensions, such as the early stages of Ferguson's stadial scheme and Herder's "chain of culture" (see below), the two histories occupy a recursive relation, in which each developmental track boosts the other, and the immanent, universal character of species-history precipitates the singularity of the individual life.

Rousseau, however, converts the doubling of individual and species histories into a fundamental antinomy between nature and history. Perfectibility unmakes the human nature it is supposed to constitute. As soon as man forms social relations—the medium of perfectibility—and enters history, he forfeits his nature and becomes an artificial being:

> In thus discovering and retracing the forgotten and lost paths that must have led man from the Natural state to the Civil state ... [the Reader] will sense that the Mankind of one age is not the Mankind of another age, the reason why Diogenes did not find a man is that he was looking among his contemporaries for the man of a time that was no more.... [He will understand] how the human soul and passions, by imperceptible adulterations, so to speak change in Nature; why in the

long run the objects of our needs and of our pleasures change; why, as original man gradually vanishes, Society no longer offers to the eyes of the wise man anything but an assemblage of artificial men and factitious passions which are the product of all these relationships, and have no true foundation in Nature.[50]

The history of man is not—and cannot be—a natural history.

Amplifying Buffon's account of the individual as "a solitary, a detached being," Rousseau bases his argument on a denial that sociability, the motor of human progress, is an original principle of human nature, which is, accordingly, instinctually impoverished. Ferguson aims his critique of Rousseau's general thesis at that denial.[51] In common with the other Scottish philosophers, Ferguson asserts that human nature *is* innately social, and hence it is realized, not betrayed, in the historical development of forms of social life:

> If we are asked therefore, Where the state of nature is to be found? we may answer, It is here; and it matters not whether we are understood to speak in the island of Great Britain, at the Cape of Good Hope, or the Straits of Magellan. . . . If we admit that man is susceptible of improvement, and has in himself a principle of progression, and a desire of perfection, it appears improper to say, that he has quitted the state of his nature, when he has begun to proceed. (14)

All the same, Ferguson goes on to follow Rousseau in diagnosing a fall from nature, although he locates it at a later stage in human history. The imperial expansion of the state and the division of labor ("separation of arts and professions") dissolve the organic (synecdochic) bond between a society and its citizens and, in Ferguson's resonant phrase, "dismember the human character" (218). It is not the state of nature but the "commercial state" that renders man "a detached and solitary being": "The mighty engine which we suppose to have formed society, only tends to set its members at variance, or to continue their intercourse after the bands of affection are broken" (24). For both Rousseau and Ferguson, then, "man" becomes a fugitive entity, unable to be fully present in his own history, save in strange, oneiric fits of retrospection, when that history doubles back on itself. Inverting the standard scenario of the contact zone (the empirical topos, in this great age of circumnavigation, of the "map of mankind"), Ferguson imagines the savage as a visitor to modern commercial society, beholding in astonishment the division of labor:

> The savage, who knows no distinction but that of his merit, of his sex, or of his species, and to whom his community is the sovereign object of

affection, is astonished to find, that in a scene of this nature, his being a man does not qualify him for any station whatever: he flies to the woods with amazement, distaste, and aversion. (173)

Man, like Minerva's owl, flies at twilight.

The faculty of perfectibility or improvement, launching both individual and species into history, divides man from himself as well as from other creatures. In Rousseau's more stringent reckoning, man falls out of the equilibrium of nature into a state of unevenness (*inégalité*), which entails a temporal difference from himself as well as a hierarchical difference from others. He becomes an anachronism, lost in a time that is not his. Denying an instinctive principle of sociability, Rousseau refuses to identify a necessary cause that expels us from nature into history. Free agency and perfectibility are empty potentials, absent from the solitary, mute, unreflective life of natural man, unless activated by a "fortuitous concatenation of several foreign causes which might never have arisen and without which he would eternally have remained in his primitive condition" (159).[52] Rousseau speculates on the origins of reason and language only to insist on the "immense distance" that separates them from the state of nature (146). They are artificial inventions that "might never have arisen."

The Formation of Humanity

In his early *Treatise on the Origin of Language* (1772), Herder seeks to answer Rousseau's challenge: to account for the uniquely human developments of language and culture without severing man from nature (and without resorting, thus, to "some Rousseauian *élan* or leaping").[53] Herder develops an early observation made by Rousseau in the *Discourse on Inequality*, "that each species has but its own instinct, while man perhaps having none that belongs to him, appropriates them all" (135).[54] Man activates a mimetic capability of observation and imitation that generates social relations, in Rousseau's view, as a fatal condition of reflection, or internal division and bondage to one's imaginary figure in the eyes of others. For Herder, the lack of an inherited instinct that specifies human nature—that gives human nature a *form*—supplies a solution: one that will turn out to be crucial for the new kinds of Romantic-period fiction, as the next chapter will show.

Herder defines instinct as the strong sensory bond of a creature to a special task or function: the spider spinning its web, the bee building its cell. In proportion as a creature's instinct is strong, the horizon (*Kreis*, circle: 78) of its world-apprehension is narrow: the spider weaves its web without experience or instruction, but that is all it will ever be able to do. The human being, at the other extreme, comes into the world without organic predetermination to a special task, indeed, without being able to do anything by itself. Compared with other creatures, its senses are weak and diffuse. Instead, the human is cognitively open to the world—to the full tide of sensory impressions that floods it. "The whole of nature storms at the human being in order to develop his senses until he is a human being" (129): for this, the human being must sort the influx of impressions through acts of attention and reflection (*Besinnung*, becoming-aware or taking-aware) that are, Herder insists, linguistic acts, acts of "marking" or signification.[55] Language already occurs within the human, with the first reflexive stirring of consciousness, as this internal act of attention and signification—a prototype of writing more than of speech (which emerges at a later developmental stage). At the same time, the physical weakness of the human being as it enters the world makes socialization a necessary condition for bare survival and the subsequent activation of its faculties.[56]

Man, in Herder's account, is truly the world animal. "His forces of soul are distributed over the world"; unfettered by instinct to a specialized task, "he has free space to practice in many things, and hence to improve himself constantly" (79). The human being's freedom from cognitive preformation (to adapt the contemporary embryological term), combined with physiological porousness to his environment, affords the plasticity through which he is able to develop—to take on all forms, to occupy all spaces and functions, to engage the world as a totality. Herder recapitulates an enlightened claim on human exceptionalism that reaches back (at least) to Pico della Mirandola's late quattrocento Neoplatonic *Oration on the Dignity of Man*.[57] The difference lies in the convergence of a modern, secular conception of open-ended historical time with an emergent medico-physiological conception of the body as sensuously open to its world.[58] Together these accommodate the "epigenetic" model of immanent organic development, currently displacing preformation theory in the German life sciences:

> [He is] never the *whole human being*; always in development, in progression, in process of perfection.... We are always growing out of a childhood, however old we may be, are ever in motion, restless, unsatisfied. The essential feature of our life is never enjoyment but always progression, and we have never been human beings until we—have lived out our lives. By contrast, the bee was a bee when it built its first cell. (130–31)[59]

Herder reinserts the disjunction between the life of the individual and the life of the species, but redirects its horizon of unity from a hypothetical (inaccessible) past to an optative future, converting it from an ontological predicament to a historical one. Nature does not reside in an original, mythic condition from which we have fallen but in a potential that realizes itself in time, in a continuous, never-ending process of emergence: "we have never been human beings until we have lived out our lives."

Herder's essays and treatises generated a hospitable philosophical climate for the new ecology of the novel that would begin to flourish in the Romantic period, as I shall argue in the following chapter. Herder would presently retract his naturalistic account of linguistic origins in response to criticism from another of his former teachers, Johann Georg Hamann.[60] Nevertheless, he continued to develop an epigenetic and providential thesis of human development, dialectically open to environmental and historical contingencies. His 1774 essay "This Too a Philosophy for the Formation of Mankind" applies the word *Bildung*, hitherto "used almost exclusively of individuals, connoting an individual development and education," for the first time to "humankind as a whole in its historical development."[61] The contemporaneous "On the Cognition and Sensation of the Human Soul" draws on the physiological principle of a vital force (*Kraft*) interfusing the sensitive and cognitive faculties. The body itself is the medium of "the great secret of the progressive formation, renewal, refinement of all beings": "The sensing human being feels his way into everything, feels everything from out of himself, and imprints it with his image, his impress."[62] Herder amplifies this aesthetic and developmental conception, enfolding the species as well as the individual, in *Ideas for a Philosophy of the History of Mankind*, where he makes liberal use of the new hypothesis of a vital force (*Lebenskraft*) articulated by contemporary natural philosophers, notably Johann Friedrich Blumenbach, who drew on the sensationist physiology of Albrecht von Haller and the epigenetic embryology of Caspar Friedrich Wolff to posit a self-organizing force or formative drive (*Bildungstrieb*) inherent in organic matter. "Every creature seeks *to acquire form, and forms itself*"[63]: Herder characterizes nature as a wellspring of organic or genetic forces (*organische Kräfte, genetische Kräfte*), deriving ultimately from solar heat, that joins all living beings in a continuous, progressive scale of historical development, culminating in humankind—which will go on evolving, in Kant's hostile summary, into "yet higher stages of life in the future, and so on to infinity."[64]

It was the invocation of the life force, specifically, that spurred Kant into accusing Herder of abandoning philosophy for fiction: charging him with reliance on a mere analogy between "the natural formations of matter" and a putative "invisible universal nature that works within it and animates it," and an endeavor "to explain *what one does not comprehend* from *what one comprehends even less*."[65] As several commentators have shown, Kant also made use of Blumenbach's concept of the *Bildungstrieb*, the organic formative drive, in what became a mutual relation of influence.[66] However Blumenbach posited the *Bildungstrieb* as (in the Kantian terminology discussed above) a constitutive principle, "a teleological cause fully resident in nature," which "endowed the homogeneous, formless mixture of male and female semen with its most essential character—form, organization—and set the various parts so articulated into mutually harmonious operation"; whereas Kant viewed it as a regulative principle, bearing heuristic rather than mechanical efficacy, "[allowing] the biologist to pursue the study of organisms *as if* they had developed under the aegis of a directive, vital force, while yet restricting the researcher to explaining organic activity by appeal only to mechanistic laws."[67] Kant, consequently, held "that biology could not really be a science, but at best only a loose system of uncertain empirical regularities, not a *Naturwissenschaft* but a *Naturlehre*."[68] The cause "lies not in nature but in the limitations of the human faculty of understanding," hence, "the organic realm ... must therefore necessarily transcend the explanatory or theoretical constructive capacity of reason."[69] The principle would become crucial for Kant's demonstration of the organic formation not of any entity in nature but of reason itself.[70]

Herder followed Blumenbach in granting the vital force an immanent, causal role in natural history. Human *Bildung* is the climax of a self-organizing developmental drive that surges through the entire series of natural history, from the formation of the earth ("a grand manufactory for the organization of very different beings," 26) to the emergence of mankind:

> But what every kind of earth and stone possesses, is certainly a general law of all the creatures of our Earth: *conformation*, determinate *figure*, distinct *existence*.... From simple laws, as from ruder forms, [Nature] proceeds to the more complex, artful, and delicate: and had we a sense, enabling us to perceive the primitive forms and first germes of things, perhaps we should discover in the smallest point the progress of all creation. (26–27)

The early books of *Ideas for a Philosophy* imagine a dynamic interaction of geophysical forces in which primordial matter is imbued with vitality and evolves into ever more complex organic forms, eventually giving rise to man: "the end of our present existence is the formation of *humanity* [*Bildung der Humanität*], to which all the meaner wants of this Earth are subservient, and which they are all contrived to promote" (123).[71] Human emergence may have been a unique event in terrestrial history, after which "the door of creation was shut" (114): but it was nevertheless, as John H. Zammito insists, an event that took place *"within* nature."[72]

"Humanity is the End of Human Nature" ("Humanität ist der Zweck der Menschennatur"), proclaims the title of book 15 (438), as Herder marshals the full force of his argument against Kant's insistence that perfection belongs to the destiny of the species but not to historical human lives. Acknowledging a temporal division between "mankind," empirical species being, and "humanity," the fullness of moral and spiritual development, but at the same time anticipating their convergence, Herder insists that humanity resides as a potential within the existing faculties of our nature. Herder also insists on the plurality of developmental paths, shaped by geophysical and sociohistorical "climate" (*Klima*), so that each tribe, race, or nation unfolds its humanity in its particular way, at its own rate. The "chain of improvement" or "cultivation" (*Kette der Bildung, Kette der Kultur*), binding individuals and generations over time, makes history the organic medium for the realization of species being: "Every where man is what he was capable of rendering himself, what he had the will and the power to become" (441).[73]

Kant, however, remained adamant, wrenching open the gap Herder's thesis sought to close:

> But if "the human species" signifies the *whole* of a series of generations going (indeterminably) into the infinite, ... and it is assumed that this series ceaselessly approximates the line of its destiny running alongside it, then it is not to utter a contradiction to say that in all its parts it is asymptotic to this line and yet on the whole that it will coincide with it, in other words, that no member of all the generations of humankind, but only the species will fully reach its destiny.[74]

Kant spiked his rebuttal with a suggestion that Herder's rhapsodic conjuration of a unitary organic force misconstrued metaphoric or analogical relations between species—mere resemblances—into genealogical or homological affinities; and in so doing opened the door to the scandalous hypothesis of transformation. If life itself, organic matter, is imbued with a developmental drive, if natural forms are historical, then only an arbitrary fiat can exempt the human species from historical change—from mutation into future forms, just as it has evolved from past ones. "Life in its plenitude—in the sheer gusto of its living power—threatens to overwhelm formal containment" and generate "a new mode of monstrosity," in Denise Gigante's summary.[75] To admit the possibility that "either one species [could] have arisen from the other, or perhaps from a single procreative womb, would lead to *ideas* which are so monstrous that reason recoils before them," Kant complains.[76] He adds, "one may not ascribe such things to our author without doing him an injustice"—transformism is an inadvertent byproduct of Herder's poetical raptures, rather than a considered hypothesis.

The Paragon of Animals

It is a truth to which Herder has already pointed in his Ideen, *viz. that the distinctness of man from the brute is not to be looked for in any single point of difference. On the contrary, man is most closely related to the brute. Every creature is what it is through the unity of its whole being, and man is man as much by the form and nature of his upper jaw, as by the form and nature of his little toe. And so, too, every creature is a note, a shade, in a great harmony, and the study which apprehends this harmony as a whole and in its vastness is alone meaningful.*

—J. W. GOETHE (1784)[77]

Herder does not explicitly advocate the evolution of man from other species, and throughout the *Ideas* he swerves away from the more heterodox implications of his argument. Like Buffon before him, he defends the barrier between humans and brutes, memorably evoking the pathos of the orang-outang, close enough to the celestial horizon of reason to intuit his exile from it.[78] In Herder's memorable formulation, man occupies a middle station in the solar system and (hence) among the animals:

> Thus Nature has placed us on one of the three middle planets; in which, as a mean degree and more moderate proportion with respect to time and space apparently prevail, a middle order of beings may be supposed to dwell. (5) ... *Man is a middle creature among animals, that is, the most perfect form, in which the features of all are collected in the most exquisite summary.* (40)

Man's "middle order" is central, synthetic, and teleological, rather than median, horizontal, that of an animal among animals.[79] It turns out the perfection of humanity will take place, after all, outside the realm of natural history altogether:

> Man alone is in contradiction with himself, and with the Earth: for, being the most perfect of all creatures, his capacities are the farthest from being perfected, even when he attains the longest term of life before he quits the World. But the reason is evident: his state, being the last upon this Earth, is the first in another sphere of existence, with respect to which he appears here as a child making his first essays. Thus he is the representative of two worlds at once; and hence the apparent duplicity of his essence. (128)

Homo duplex redivivus. With this metaphysical turn, Herder wards off the specter of a transformation of the human species within natural history.

Kant was not wrong to glimpse the dissolution of species boundaries in Herder's vision of an organic developmental force flowing through nature and culminating in the ascent of man. The prospect opens in Herder's rhetoric rather than in the overt logic of his argument. As Goldstein shows, Herder claimed figuration—analogy and metaphor—as a technology of scientific cognition, rooted in the sensory apparatus of the human body; John K. Noyes, arguing along similar lines, finds Herder's early essays discovering "the poetic . . . at the heart of factuality," founded in the organic bond between sensuous experience and reason; and Stephen Gaukroger reconstructs the genealogy of Herder's linkage of language, thought, and sensation through the aesthetic writings of Baumgarten, Lessing, and Winckelmann.[80] Throughout *Ideas for a Philosophy of the History of Mankind*, Herder's writing activates figuratively what remains latent or potential in his philosophical commitment to the *Lebenskraft*. Kant took particular issue with Herder's contention, in books 3 and 4 of *Ideas*, that the fundamentally distinctive character of the human being, setting him apart from other animals, is a physiological one: his erect bipedal figure (127–28).[81] To posit reason as a functional consequence of man's upright posture was to degrade it from teleological to merely etiological and contingent status.

Meanwhile, Herder's prose enacts—as though virtually demonstrating— the metamorphosis of a lowly, groveling quadruped into a noble, upright, anthropomorphic figure:

> Look up to Heaven, O man! and tremblingly rejoice at thy vast superiority, which the creator of the world has connected with such a simple principle, thy upright form. Didst thou walk prone like a brute; were thy head gluttonously formed for the mouth and nose, and the structure of thy limbs answerable; where would be thy higher powers of mind? to what would not the image of the divinity in thee be degraded? The

wretch who ranks with the brutes has lost it: as his head is misshapen, his internal faculties are debased, and the grosser senses drag the creature down to the earth. But the fashioning thy limbs to an erect posture has given thy head its beautiful outline and position, whence the brain, that delicate ethereal germe [sic] of Heaven, has full room to extend itself and send out its branches. The forehead swells rich in thought; the animal organs recede; it is the form of a man. As the skull rises higher, the ear is seated lower; it becomes more closely connected with the eye, and the two senses have more intimate access to the sacred apartment in which ideas are formed. (81)

It would not be long before natural philosophers mobilized organic development to undermine the grounds of human exceptionality. Jean-Baptiste Lamarck's *Zoological Philosophy* (1809), developing the hypothesis of a "scale or graduated chain among living bodies,"[82] fully submerges man in the tides of terrestrial life—comprising the interaction of an organic progressive drive with geographical constraints and opportunities—such that the human form, like all natural forms, is open, plastic, shaped by environmental pressure, and reflexively self-shaping through heritable habit. Extending the "orangutan hypothesis" of Rousseau and Monboddo (see chapter 3), and defying Buffon's and Herder's bar against the creature's accession to humanity, Lamarck rehearses the scenario of a "quadrumanous animal" that descends from the trees and aspires, following an innate cognitive drive (a desire "to command a large and distant view"), to walk upright:

> If the individuals of which I speak were impelled by the desire to command a large and distant view, and hence endeavoured to stand upright, and continually adapted that habit from generation to generation, there is again no doubt that their feet would gradually acquire a shape suitable for supporting them in an erect attitude; ... there is again no doubt that their facial angle would become larger, that their snout would shorten more and more, and that finally it would be entirely effaced so that their incisor teeth became vertical.[83]

It takes little more than a grammatical shift, to the hypothetical "as if," to convert Herder's hortatory evocation of the ascent of man into a conjectural evolutionary history.

Once it is understood to pervade the whole system of nature, the very condition that sets man apart from other creatures—developmental capacity, plasticity, perfectibility—now dissolves the vaunted permanence and uniqueness of human nature. This time the countercharge of conjecture would be launched by authorities on both sides of the English Channel. The great comparative anatomist Georges Cuvier, who maintained a determined opposition to transformist ideas throughout his career, used his eulogy of Lamarck (1832) to expel the latter from the rank of true scientists, classing him among those "less favoured philosophers" who, "believing themselves able to outstrip both experience and calculation, ... have laboriously constructed vast edifices on imaginary foundations, resembling the enchanted palaces of our old romances, which vanished into air on the destruction of the talisman to which they owed their birth."[84] Cuvier also accused Lamarck's protégé, the philosophical anatomist Etienne Geoffroy Saint-Hilaire, with whom he was locked in controversy, of promoting "a doctrine substantiated only in the imagination of some naturalist, more poet than observer."[85] Meanwhile Charles Lyell inserted a detailed refutation of Lamarck in the second volume of *Principles of Geology* (1832), thereby securing Lamarck's arguments an English readership. (*Zoological Philosophy* would not be translated until the twentieth century.) In rehearsing the transformation of species, Lyell complained, Lamarck "gives us names for things, and with a disregard to the strict rules of induction, resorts to fictions, as ideal as the 'plastic virtue,' and other phantoms of the middle ages."[86] Among Lyell's more attentive readers were the young Herbert Spencer and Charles Darwin. Already acquainted with Lamarck's work from his studies with Robert Edmond Grant in Edinburgh, Darwin read *Principles of Geology* on the voyage of the *Beagle*, while Spencer later averred that his "reading of Lyell, one of whose chapters was devoted to a refutation of Lamarck's views concerning the origin of species, had the effect of giving me a decided leaning to them."[87] Kant's critical reason may have won the local engagement, by virtue of superior philosophical rigor; but Herder's flight into conjecture, charging the natural history of man with a biological drive, would seem to have won the war.[88]

More astounding visionary horizons unfold, meanwhile, in book 15 of *Ideas for a Philosophy of the History of Mankind*, "Humanity is the End of Human Nature." We begin to see that the history of man—the history of the species—has *a human form*. Man, the creature born, according to Herder's *Treatise on the Origins of Language*, without an innate form, that is, without instinct, has, instead, the macrocosmic form of his history as a species. It is not only that human history, rising ever upward to perfection, models the human being's erect figure, which thus embodies the immanence of the teleological principle of "humanity" within human history.

Herder also sketches an audacious temporal equivalent. Perfection, he argues, is the dynamic geometrical form of human history through time. It is measured by the oscillation of particular national histories between alternating states of stability and chaos. This actual empirical history of civilizations vibrates asymptotically around the ideal line of perfection[89] — never exactly coinciding with it, but following its path, lurching to either side of it, bound by the "chain of culture":

> Thus Rome disturbed the peace of the Globe for more than a thousand years; and half a world of savage nations was requisite for the slow restoration of its quiet. . . . The channel of cultivation [*der ganze Gang der Kultur*] on our Earth, with its abrupt corners, its salient and reentering angles, scarcely ever exhibits a gentle stream, but rather the rushing of a torrent from the mountains, such are the effects of human passions. It is evident, too, that the general composition of our species is calculated and established on such alternating vibrations. As our walk is a continual falling to the right and to the left, and yet we advance at every step; so is the progress of cultivation in races of men, and in whole nations. . . . Thus, as in the machine of our body, the work of time proceeds to the good of the human race by necessary opposition, and acquires from it permanent health. (456)

Herder takes up Kant's principles of human nature (notably the antagonistic principle of "unsociable sociability") and sublimates them into a grand cosmic figure of the human form striding across universal history. The history of man is not only upward-rising, looking toward a higher self-realization; it is driven by the passions (rather than by reason); and it is bipedal. The new science of life affords Herder this exorbitant updating of the ancient figure of Vitruvian Man, for which he converts the geometric proportions of the universe from a spatial to a temporal axis. Science fiction, perhaps, but nevertheless affording—as science fiction can—a vision of future knowledge, clothed in splendid metaphor.

In investing "the *Bildung* of a modern European individual" with the "*Bildung* of humankind," Herder created the philosophical plan of historical *Bildung* that Hegel would later elaborate in the *Phenomenology of Spirit*.[90] Herder also created a philosophical plan for the new, experimental forms of the novel that would arise in European Romanticism and become regulative for nineteenth-century realism. The Bildungsroman, pioneered in

Germany, plots a "harmonious formation of the purely human" through a protagonist's moral and sentimental development; the historical novel, established in Scotland, makes national history the synthetic medium of that formation. Like the natural history of man, this novelistic project is disputed at its foundation—this time, around the constitution of the universal particular, "man." Philosophical anthropology was shaped by European scientific encounters with other peoples in the Pacific discoveries and other voyages of the early-modern era and Enlightenment; Diderot, William Robertson, Herder, Kant, and others assimilated the growing archive of global ethnographic diversity into an expansive vision of "the great map of mankind" that accommodated racial diversity within a global human kinship. Abolitionist writings against the slave trade generated the liberal discourse of a universal humanity bound to a destiny of moral and spiritual progress. (Polygenesis, the doctrine of biological difference among the races of man, remained a heterodox position until later in the nineteenth century.) *Bildung*, like realism, would in effect become a European, "provincial" domain in the nineteenth-century novel; other populations fell beyond its pale, as even the case of Ireland appeared to demonstrate.[91] Instead, the key difference within "man" registered closer to home—within the home: within the very constitution of home as a domestic enclave sequestered from the public, homosocial and masculine arena of *Bildung*. This is the topic of the following chapter.

CHAPTER TWO

The Form of the Novel

The human container is capable of no full perfection all at once; *it must always* leave behind *in* moving further on.

—J. G. HERDER, "THIS TOO A PHILOSOPHY
FOR THE FORMATION OF HUMANITY" (1774)[1]

Novelistic Revolution

In the chapter "On Novels" in her critical treatise *On Germany* (*De l'Allemagne*, 1813), Germaine de Staël considers Johann Wolfgang von Goethe's *Wilhelm Meister's Apprenticeship* (*Wilhelm Meisters Lehrjahre*, 1795–96), "a work famous in Germany but little known elsewhere."[2] *Wilhelm Meister* exemplifies a new, anthropological mode of "philosophical novel" that is distinctively German: "a picture of human life altogether impartial, [in which] different situations succeed each other in all ranks, in all conditions, in all circumstances, and the writer is present to relate them" (2: 55). But Goethe's experiment fails to cohere. What should have been "a philosophical work of the first order," conveyed in fine sociological descriptions and "ingenious and lively" dialogue, is lumbered with a diffuse plot and an obtrusive protagonist (*un tiers importun*). Consequently "there is no other interest in the *tout-ensemble* but what we may feel in knowing the opinion of Goethe on every subject" (2: 56). Staël's complaint ignores the generic frame invoked by later criticism for reading *Wilhelm Meister*: that of the Bildungsroman, the "novel of formation" (*Bildung*) or of moral and sentimental development. Evidently the poles of plot and hero are too weak to generate the dialectic between them that is the Bildungsroman's narrative motor.[3]

[55]

Instead, Staël praises Goethe's realization of a secondary character, Mignon, the orphaned Italian girl Wilhelm rescues from a seedy traveling circus. Occupying the void left by the attenuation of plot and hero, Mignon moves the reader with exquisite songs of romantic longing—echoes, fragments, of a shattered idyllic world:

> We cannot represent to ourselves without emotion the least of the feelings that agitate this young girl; there is in her I know not what of magic simplicity, that supposes abysses of thought and feeling; we think we hear the tempest moaning at the bottom of her soul, even while we are unable to fix upon a word or circumstance to account for the inexpressible uneasiness she makes us feel. (2: 56–57)[4]

Lyric purity sequesters Mignon outside the narrative's developmental track. Dying on the threshold of adult sexuality, she embodies the "poetry of the heart" that must be sacrificed for the novel to claim its proper domain, everyday common life, or what Hegel, memorably, will call "the prose of the world."[5] In his *Aesthetics*, Hegel notes the condition of that claim. Marriage, the essence of compromised accommodation to the forms of social life, displaces the higher plot-principle of *Bildung*: "at last [the hero] gets his girl and some sort of position, marries her, and becomes as good a Philistine as others" (2: 593). *Wilhelm Meister's Apprenticeship* ends with the hero's betrothal—but at an exorbitant cost: Mignon's death consummates a pathos that overwhelms everything else in the novel.

Staël did more than pay critical homage to Mignon. A few years before writing *On Germany*, as though to realize a potential *Wilhelm Meister* could not find room for, she magnified Goethe's character into the protagonist of her own sensationally popular philosophical romance *Corinne, or Italy* (1807).[6] Here, the enigmatic Italian girl has survived adolescence and grown up to inhabit her art as a fully realized, triumphantly public vocation.[7] Reciting her rhapsodies to adoring crowds on the Roman Capitol, Corinne establishes, for the century to come, "the myth of the famous woman talking, writing, performing, to the applause of the world."[8] Her public career as bardic performance-artist flouts one of the ideological tasks assumed by nineteenth-century fiction, the creation of "a private domain of culture," in Nancy Armstrong's summary, "independent of the political world and overseen by a woman," whose soft command of taste proscribes "self-display."[9] Transvaluing the eighteenth-century stock figure of the Female Quixote,[10] Staël inaugurates the European novel's claim on *Bildung* for a female protagonist through the medium of art—the realization of a Schillerian ideal of aesthetic education in creative play, dedicated to "the dignity of humankind and the glory of the world."[11] Even so, Corinne turns out to be as vulnerable as her prototype to the engine of novelistic plot. She too is immolated on the altar of a sanctioned marriage, her genius wasted.

Recent feminist criticism, salvaging *Corinne* and its author from twentieth-century condescension, has addressed its challenge to the emergent genre typified by *Wilhelm Meister* as well as to the more settled conventions of domestic fiction. Corinne samples Mignon's most famous lyric ("Kennst du das Land, wo die Zitronen blühn?") in her Capitoline rhapsody ("Connaissez-vous cette terre où les orangers fleurissent?"),[12] inviting us to "reread Goethe's novel from the perspective of its sacrificial victim," writes Kari Lokke, who sets *Corinne* at the head of a dissenting tradition of female *Bildungsromane*.[13] The present chapter reconstructs Staël's reckoning with the "new genres or sub-genres characteristic of realism," the Bildungsroman and its British analogues, the Anglo-Irish national tale and Scottish historical novel, formed in the "novelistic revolution" of European Romanticism.[14] Modeling the scientific conception of human nature as a developmental entity or emergent phenomenon, discussed in the previous chapter, these new genres or subgenres rehearse a universal formation of species being—a *Bildung der Humanität*—through the ontogenetic narrative of subject formation. *Corinne* draws utopian energy from German Romantic prescriptions for *Bildung*: as well as Friedrich Schiller's aesthetic education, the novel's heroine realizes Friedrich Schlegel's conception of a "progressive universal poetry," a dynamic interplay of heterogeneous forms and discourses expressive of the infinite potential of human becoming. *Corinne* narrates a collision between the ideals of *Bildung* and a hardening repertoire of novelistic topoi—masculine vocation, national destiny, the marriage plot—as these are coalescing into a set of norms for nineteenth-century practice. Where the historical novel dialectically configures the components of the Bildungsroman and national tale into a synthesis, *Corinne* sets them in destructive mutual antagonism. National-historical formation and the marriage plot combine not to mediate the heroine's *Bildung* but to crush it.

Staël's broad target is the structural exclusion of women from the category that underwrites the new forms of the novel: the Enlightenment's grand universal particular, "man." Walter Scott, in the introduction to *Waverley*, claimed a universal human nature, manifest in the "passions common to men in all stages of society," as the historical novel's condition of enunciation, while early commentary on the Bildungsroman identified "the harmonious formation of the purely human" as its purposive principle.[15] The absorption and naturalization of the principle, making all nineteenth-century novels in some sense "novels of development," established the realist novel as the anthropomorphic genre *par excellence*—the genre of *The Human Comedy*, of human life viewed as a historical totality. Philosophical anthropology supplied a scientific matrix for these ascendant novelistic forms. Its decisive interventions, I argued in the preceding chapter, were Rousseau's opening of a radical division between human nature and human history, via the doubling of individual and species being around the axis of perfectibility, and Herder's naturalization of history by a recourse to emergent—hence figurative, poetic—concepts from the new life sciences: bonding the destinies of individual and species in the epigenetic force field of *Bildung*, accommodating a plurality of developmental paths within a universal progressive drive. Herder's writings, in particular, modeled the new anthropological figure of man that would be generative for the modern forms of the novel, rendering a new type of protagonist as well as a new philosophical conception of the novel's form—or rather, the aesthetic scandal of its formlessness—as a developmental modality.[16] "With the (modern) novel, literary form becomes a matter no longer of poetical forms but of the form of life," writes Rüdiger Campe. "The novel is always in demand of its own form": like Goethe's hero Wilhelm Meister, "novels don't have form, they are in quest of it."[17] Lack of form as the condition of the human, and of a freedom, an infinite potential, realized in development; a doubled history of the individual life and the life of the species. These are the uniquely human attributes around which the novel reorganizes itself in the Romantic period.

Crucially, they are claimed for a male protagonist. *Wilhelm Meister* and *Waverley*, generic prototypes, share a new mutation of the hero: susceptible, vacillating, drifting with the story rather than driving it, sensitive to aesthetic impressions of environment and atmosphere.[18] These conventionally feminine traits signal a takeover of the role of novelistic protagonist that women had largely come to occupy by the late eighteenth century—before the young man's journey of self-cultivation, there was the young lady's entrance into the world. Hegel, programmatically refining the *Bildung der Humanität*, made the gender of its universal subject explicit. In Toril Moi's summary: "Men, but not women, can achieve self-consciousness, that is to say, can become fully individualized human beings," thanks to their competition and collaboration with other men in the public sphere; whereas women, relegated to family life, "have no understanding of the universal, that which serves the common good," and hence "remain generic creatures."[19] In a formulation that comes to dominate nineteenth-century development theory (see chapter 5), individuation is supposed to attain universal equivalence—species being—by shedding the contingent trappings of the merely generic, the socially typical.

And yet, excluded from the new conception of humanity, women were most fully expressive of it. In the "Avant-propos" to *La Comédie humaine*, Balzac found the difference between the sexes injecting dialectical complication into his novelistic zoology of society:

> The limits set by nature to the variations of animals have no existence in society. When Buffon describes the lion, he dismisses the lioness with a few phrases; but in society a wife is not always the female of the male. There may be two perfectly dissimilar beings in one household. The wife of a shopkeeper is sometimes worthy of a prince, and the wife of a prince is often worthless compared with the wife of an artisan. The social state has freaks which Nature does not allow herself; it is nature plus society. The description of social species would thus be at least double that of animal species, merely in view of the two sexes.[20]

Woman, that is, is at once exceptional and exemplary. She embodies the human supplement, "nature plus society," that doubles, as it divides, the order of natural history. The bearer of both sex and gender ("in society a wife is not always the female of the male"), she inhabits an existential difference that is closed and naturalized as an identity for men. Where men are fixed in a social taxonomy, like animals in the system of nature, women possess the plasticity and fluidity, the capacity to move up and down the scale of being, that are specific markers of the human in late Enlightenment anthropology. But as Balzac's argument proceeds, this feminine character (*varium et mutabile*), now an influx of "animal nature," infects the species:

> Though some savants do not yet admit that the animal nature flows into human nature through an immense tide of life, the grocer certainly becomes a peer, and the noble sometimes sinks to the lowest social grade.... [T]he habits of animals, those of each kind, are, at least to our eyes, always and in every age alike; whereas the dress, the manners, the speech, the dwelling of a prince, a banker, an artist, a citizen, a priest, and a pauper are absolutely unlike, and change with every phase of civilization.[21]

Civilization feminizes: making men, too, flexible, mobile forms, products of their changing social environments.

Franco Moretti argues that the novelistic revolution of European Romanticism was functionally counterrevolutionary, a project to imaginatively contain the destabilizing energies released in the age's political upheavals. "In the traumatic, fast-moving years between 1789 and 1815, human actions seem to have become indecipherable and threatening; to have—quite literally—lost their meaning," so that it becomes the novelist's task to recover "the anthropomorphism that modern history seems to have lost."[22] Developments in the natural history of man provided for both loss and recovery. In Rousseau's wake, perfectibility became a watchword for projects of political transformation that sought at once to liberate and to harness the dynamism of human nature. Buffon himself, at the close of *Époques de la nature* (1778), pitched it as a utopian project:

> What could [man] not do upon himself, I wish to say upon his own species, if the will was always guided by intelligence? Who knows to what point man could perfect his nature, either moral or physical? Is there a single nation that can boast to have arrived at the best government possible, which would make all men not equally happy, but less unequally unhappy? . . . Here is the moral goal of all society that seeks to better itself.[23]

The Marquis de Condorcet maintained "that the perfectibility of man is absolutely indefinite; that the progress of this perfectibility, henceforth above the control of every power that would impede it, has no other limit than the duration of the globe upon which nature has placed us."[24] Revolution would tear away the artificial bonds and hierarchies that constrained the infinite plasticity of human nature, coterminous with the history of the earth.

That infinite plasticity would then be available for remolding—but how, and by whom? Rousseau had argued that the reinvention of the state must entail the reinvention of man:

> The man who dares to undertake the establishment of a people has to feel himself capable of changing, so to speak, the nature of man; of transforming each individual, who in himself is a perfect, isolated whole, into a part of a larger whole from which the individual, as it were, receives his life and his being; of altering man's constitution in order to strengthen it; of substituting a morally dependent existence for the physically independent existence that we have all received from nature. In a word, he must deprive man of his own strength so as to give him strength from outside, which he cannot use without the help of

others. The more completely these natural strengths are destroyed and reduced to nothing, the more powerful and durable are those which replace them, and the firmer and more perfect, too, the society that is constituted.[25]

Human nature, far from spontaneously, organically remaking itself from below, abides the enlightened rule of the legislator who will realize the General Will: an all but impossible, messianic task, Rousseau concedes. Revolution would encompass "not merely the reform of political, moral, and social life, but the reshaping of physical nature": in their 1790 address to the Assemblée Nationale, the Jacobin naturalists "declared themselves regenerated: they were 'new men.'"[26] "It is time to dare to practice on ourselves what we have practiced with such success upon many of our fellow creatures," proclaimed physiologist Pierre Jean Georges Cabanis, a dozen years later: "we must *dare to revise and correct the work of nature.*"[27] Cabanis and his fellow Ideologues (Destutt de Tracy, Constantin Volney), liberal rather than radical revolutionaries, followed Condorcet in promoting an "applied science of man," an "anthropological medicine" founded on "the modifiability of temperament," which "would produce organic improvements through the inheritance of acquired characteristics," inculcated in regimens of hygiene and education.[28] Against these dizzying prospects, of a "bad" or "empty" infinity of human nature available for artificial cultivation, the novel seeks to capture a dialectic whereby the unformed protagonist, acquiring a form organically through history, might—freely and spontaneously—give history back a human form.

Bildungsroman

The unformed protagonist gestates in the formless form of the work all but universally designated as the foundational novel of development. Marshall Brown calls *Wilhelm Meister's Apprenticeship* a "breakthrough text, for both fiction and the theory of fiction"; for Fredric Jameson, it marks "the true beginning of the nineteenth-century novel."[29] Its critics remark a mismatch between the novel itself and its archetypal status: "a peculiarly central evaluation for such an odd and garbled book," Jameson goes on, "immensely influential and yet a kind of literary white elephant, boring and fascinating all at once, and a perpetual question mark for the French and British traditions in which, as a text, it has played so small a role, yet which are incomprehensible without it."[30] Formally, *Wilhelm Meister* is loose, fluid, miscellaneous (a patchwork of genres and discourses, from lyric poetry to theatrical treatise and spiritual autobiography), as well as open, self-interrupting, unfinished (the story does not close but breaks off, with the hero about to set out on another journey): traits that will be exacerbated in the sequel, *Wilhelm Meister's Journeyman Years*.[31] Goethe's novel marks the debut in European fiction of the anthropological subject sketched in the early essays of Herder, a major intellectual presence in his early career.[32] The human being, wrote Herder, is "never the *whole human being*; always in development, in progression, in process of perfection," he is (in a resonant formulation) "nature's apprentice," "the apprentice of all the senses!, the apprentice of the whole world!"[33] In contrast to the hard-shelled protagonists of earlier picaresque, Goethe's Wilhelm is distinguished by his "easy adaptability" and "many-sided receptivity," according to Friedrich Schlegel, while his "inner dissatisfaction and mobility," according to Moretti, make him the representative creature of modernity, "[sharing] in the 'formlessness' of the new epoch, in its protean elusiveness."[34]

Critics promptly yoked this free-ranging form to a thematic emergence of human nature. "The work that appears to us in gentle radiance as the most general and comprehensive tendency of human *Bildung* is Goethe's *Wilhelm Meister*," declared Karl Morgenstern, coining the term *Bildungsroman* in a lecture in 1819: "no previous novel has to such a high degree and expansiveness attempted to represent and promote the harmonious formation of the purely human."[35] Giving the term wider currency nearly a century later, Wilhelm Dilthey distinguished the Bildungsroman from earlier biographical fictions (such as *Tom Jones*) on the grounds "that it intentionally and artistically depicts that which is universally human in [an individual] life-course": "A lawlike development is discerned in the individual's life; each of its levels has intrinsic value and is at the same time the basis for a higher level," that is, the development of personality as "a unified and permanent form of human existence."[36]

Subsequent accounts, amplifying these, characterize the Bildungsroman's goal or project as "the image of man in the process of becoming ... the novel of human emergence";[37] "a concern for the whole man growing organically in all his complexity and richness";[38] a "showing forth [of] the complex wholeness of human nature and the dynamic interplay between individual and organic natural order";[39] "the integration of a particular 'I' into the general subjectivity of a community, and thus, finally, into the universal subjectivity of humanity";[40] "a humanist process of transubstantiation by which the individual concretizes its abstract species image" and (hence) "a philosophy for positivizing human nature."[41] Bildungsroman commentary curates—even as it questions—a teleological thickening of Morgenstern's "harmonious formation of the purely human," sifting its readings of *Wilhelm Meister* through a set of contemporaneous disciplinary conversions of *Bildung* from Herder's conception of an organic developmental process, open to historical and geographical contingency and variability, to a philosophical program for the manufacture of a complex human totality: Schiller's aesthetic education, the state pedagogy of Wilhelm von Humboldt ("the true end of Man ... is the highest and most harmonious development of his powers to a complete and consistent whole"[42]), Hegel's world-historical evolution of Spirit.

Schiller's treatise, in particular, has shaped the reception of *Wilhelm Meister*, not least because of his close association with Goethe during the novel's composition (1794–96). Schiller read and criticized portions of the work in progress, and the simultaneous publication of *On the Aesthetic Education of Man* and the first part of *Wilhelm Meister*, in 1795, has encouraged programmatic applications of the former to the latter.[43] The aesthetic education is explicitly an anthropological project to reintegrate the "inner unity of human nature," fragmented by modernity's signal condition, the division of labor, under the aegis of Kantian reason. In the fourth letter, Schiller identifies the *state* as the objective form of human perfection:

> Every individual being, one may say, carries within him, potentially and prescriptively, an ideal man, the archetype of a human being [*einen reinen idealischen Menschen in sich*], and it is his life's task to be, through all his changing manifestations, in harmony with the unchanging unity of this ideal. This archetype, which is to be discerned more or less clearly in every individual, is represented by the *State*, the objective and, as it were, canonical form in which all the diversity of individual subjects strive to unite.[44]

The passage—endorsed by the explicitly civic prescriptions for *Bildung* of Humboldt and Hegel—has informed a robust vein of commentary for which the state or nation constitutes the Bildungsroman's pragmatic horizon. "Nationhood and adulthood [serve] as mutually reinforcing versions of stable identity," comprising the Bildungsroman's "soul-nation allegory of emergence," writes Jed Esty.[45] Readings of *Wilhelm Meister* in this vein identify the Society of the Tower, the crypto-masonic fellowship that claims tutelary authority over the hero's life story in the last two books, with an emergent or incipient nation-state formation: furnishing the "objective and canonical form" of Wilhelm's *Bildung*, retroactively ordering what has appeared up until now as a haphazard string of encounters.[46]

Such readings exaggerate the teleological confidence both of Goethe's novel and of Schiller's treatise. As the *Aesthetic Education* proceeds, the state—the political state—withers away.[47] If the program announced in the fourth letter implies a preformationist model of development, in which the individual bears already encased within him the "ideal man" his history will unfold, later passages veer toward the epigenetic conception of a form that discovers itself gradually, through time and circumstance—and may never achieve archetypal perfection. The *Aesthetic Education* itself follows an epigenetic logic of progression: as it moves through subsequent stages, Schiller's argument dialectically revises its informing categories and distinctions, so that what seem initially to be clear-cut oppositions merge into each other and breed further oppositions, and terms shift their relative positions.[48] The "aesthetic state" is at first posited as a "middle state" (123) or "middle disposition" (141), a bridge between the sense-bound state of nature and the moral state governed by reason. In the later letters, however, means becomes (or supplants) goal. The moral state recedes to the condition of a regulative principle, rather than a condition that can be achieved empirically, while the aesthetic state accommodates an actual convergence of individual and species being through the humanizing principle of the "play-drive" (*Spieltrieb*)—itself first posited as an intermediary between the sense-drive and the form-drive. "Man only plays when he is in the fullest sense of the word a human being, and *he is only fully a human being when he plays*" (107); "Beauty alone do we enjoy at once as individual and as genus, i.e., as *representatives* of the human genus [*Gattung*, species]" (217).[49] Schiller's closing paragraph separates the aesthetic state (now a "need," a longing) from any extant or possible political state:

> But does such a State of Aesthetic Semblance [*Staat des schönen Scheins*] really exist? And if so, where is it to be found? As a need, it exists in every finely attuned soul; as a realized fact, we are likely to find it, like the pure Church and the pure Republic, only in some few chosen circles, where conduct is governed ... by the aesthetic nature we have made our own. (219)

It is the esoteric property of a "few chosen circles," rather than a public or national resource.[50]

If this sounds like the Tower Society at the close of *Wilhelm Meister*, we should nevertheless be wary of readings of the novel that award the Tower a strong teleological function aligned with an incipient national destiny. Early in the "theatrical mission" (*Wilhelm Meisters theatralische Sendung*:

the original draft of the novel, revised into the first five books), Wilhelm and his troupe assume the mantle of national cultural revival: the drama will consolidate Germanness (*Deutschheit*) in the absence of a unified German state. Goethe's treatment of this "national fervor" (*Feuer des edelsten Nationalgeistes*) is unsparingly ironical.[51] The players adopt names from medieval German history or romantic drama, and the pregnant Madame Melina vows to christen her child Adelbert or Mechthilde (70). The project is literally aborted, in a heartless joke, with the stillbirth of Mechthilde (137, 140): so much for national allegory. Against the idea of a governmental (as opposed to cultural) supervisory framework for *Bildung*, the Tower's dynamic is centrifugal, characterized by a dispersal rather than concentration of forces. A cabal of aristocrats concerned with offshoring their assets to safeguard them against the coming wave of European revolutions (345), the Tower resembles a private corporation more than it does an incipient state formation.[52] Lothario advocates recognition of the state through payment of taxes—in order to guarantee nobles' property rights. Jarno's vision of the future Tower as a globalizing network reproduces, at the level of form, the jigsaw of petty duchies and electorates that make up the moribund Holy Roman Empire, saved from the forces of revolutionary nationalism by redistribution across the Atlantic settler-colonies.

Nor is the Tower's retrospective hold over the plot compelling. Despite its comprehensive archive and surveillance network, the society's guidance, like Wilhelm's progress, remains erratic, shivered in the quicksilver flow of Goethe's irony. In the closing chapters, characters refute or outright mock the Tower's pedagogical project. Wilhelm's bride-to-be Natalie, endowed with high moral clarity, disputes its educational program (while acknowledging its "tolerance in letting me go my own way," 323). Its busiest agent, Jarno, disavows its apparatus as "relics of a youthful enterprise that most initiates first took very seriously but will probably now just smile at" (335). The merry couple Friedrich and Philine set up their own private tower, complete with library: "We sat across from each other and read to each other, always bits and pieces, from one book and then from another." Reading fills the gap between erotic pleasure and boredom until it absorbs both: "We entertained ourselves day after day in this fashion, and thereby became so learned that we were astonished at each other.... We varied our means of instructing ourselves, sometimes reading against an hourglass that would run out in a few minutes, then be reversed by Philine as she began to read from another book, and when the sand ran out, I would begin my piece" (342). The lovers gleefully reduce Schiller's humanizing play-drive to the quotidian baseline aesthetic of the "interesting" (*das*

Interessante), theorized by Schlegel and quickly associated with the novel genre.[53]

This aesthetic antieducation counterpoints the irresolution or indirection of Wilhelm's *Bildung* (dignified in Schillerian commentary as "free play"), which finds him acknowledging a son and accepting a betrothed but still unsettled in a vocation, affirming—in the last words of the novel—a logic of "luck" and serendipity.[54] The marriage itself is deferred beyond the pages of the novel; indeed, Wilhelm will remain unmarried throughout *Wilhelm Meister's Journeyman Years*. The narrative is notoriously "tentative" and "oblique" in its account of the hero's development, withholding the epiphanic moment of "clear self-recognition" that characterizes later instances of the Bildungsroman.[55] The reader, rather than Wilhelm, comes to understand an antiteleological, "contingent and variational logic of play" as the novel's "comprehensive organizing structure."[56] Tentativeness and openness to contingency characterize Wilhelm's own purposive drive, which is just as intermittent, prone to distraction and hedonistic lapses, as the narrative that bears it.[57] These qualities provide for one of the innovations of *Wilhelm Meister*: the opening of narrative and character onto a provisional, indeterminate future, an ordinary and ongoing (as opposed to apocalyptic) time to come. The "interesting" (always on the verge, as Jameson suggests, of the boring) takes hold as the novel's governing aesthetic.

Among Goethe's contemporaries, Novalis criticized the "prosaic" issue of the work's poetic potential: despite (or rather because of) the merchant-class hero's adoption into a noble clique, *Wilhelm Meister* remains "a poeticized bourgeois and domestic story."[58] Hegel, sharpening the complaint, established the ironical reading of Wilhelm's story as an exemplary surrender of romantic aspiration to the banal constraints of everyday life:

> The end of such apprenticeship consists in this, that the subject sows its wild oats, builds himself with his wishes and opinions into harmony with subsisting relationships and their rationality, enters the concatenation of the world, and acquires for himself an appropriate attitude to it ... at last he gets his girl and some sort of position, marries her, and becomes as good a Philistine as others.[59]

Wilhelm's capitulation to things as they are (not yet the refined habitus of *Bildung* Hegel will call civil society) brings into relief a formal dialectic between "the poetry of the heart" and "the prose of the world"—the "world of finitude and mutability, of entanglement in the relative, of the pressure of necessity," of "external influences, laws, political institutions, civil relationships."⁶⁰ The prose of the world, the domain of novelistic realism, expresses the historic disintegration of the "world-situation" of ancient epic, in which actors and events occupy a spontaneous, organic relation to "the whole of [the] age and national circumstances."⁶¹ Hegel's critique, which treats the "apprenticeship" as symptomatic of the novel ("the modern popular epic"), laid the foundation for Lukács's *Theory of the Novel*, which contrasts the "organic" infinity of epic, its formal capacity to represent totality, with the "bad" infinity of the novel,⁶² a constitutive formlessness ("lack of limits") that necessitates an artificial imposition of form. "The novel overcomes its 'bad' infinity by recourse to the biographical form": archetypally in the Bildungsroman, which seeks "the reconciliation of the problematic individual, guided by his lived experience of the ideal, with concrete social reality"—a reconciliation in which, given modern historical conditions, the ideal is inevitably traduced.⁶³

Instead, in Lukács's account, the "dissonance special to the novel, the refusal of the immanence of being to enter into empirical life" (71), impels the novelist's resort to "creative irony." Goethe activates the "fantastic apparatus" of the last book (the Tower) while disclosing "its playful, arbitrary and ultimately inessential nature," deflating "the miraculous" into "a mystification without hidden meaning" (142)—a play of mere form, empty (to quote Hegel again) of "sensuously particularized" content.⁶⁴ Morgenstern had characterized a dialectical and reflexive operation of *Bildung* in the Bildungsroman: the formation of the hero forms the reader, in turn, through the act of reading—except that, now, the condition of the reader's formation is the absence or postponement of the hero's, in the formal mode of irony: "the normative mentality of the novel," writes Lukács, its only viable mode of "objectivity" (90). Irony, then, is the shadow of an unachieved humanity, visible to us if no longer to the hero, as he settles for worldly satisfaction (and the novel for being "merely interesting").

Infinity or Totality

Lukács's appeal to "creative irony" alludes to the best contemporary critic of *Wilhelm Meister*, Friedrich Schlegel,⁶⁵ whose powerful alternative to the Hegelian account of the Bildungsroman opens directly onto Staël's engagement with the form in *Corinne*. The leading Romantic theorist of the novel and of irony, Schlegel identified a tension between infinite process and formal containment as the structuring energy of the literary. Attuned to its mercurial aesthetic, Schlegel hailed Goethe's novel as one of three great tendencies of the age (alongside the French Revolution and the philosophy of Fichte), while his 1798 article "On Goethe's *Meister*" forgoes the symptomatic exegesis of (for instance) national allegory for a luminous essay in "poetic criticism."[66] *Wilhelm Meister* "turns out to be one of those books which carries its own judgment within it, and spares the critic his labour.... Indeed, not only does it judge itself; it also describes itself."[67] Schlegel practices an immanent mode of reading-along-with the novel that recreates its effects, such as the "divine nature of its cultivated randomness" (276) and "the irony which hovers over the whole work" (279), rather than translating them to another hermeneutic code or symbolic register.

Current Anglo-American criticism acknowledges Schlegel as a source of the Romantic irony developed, programmatically, by Paul de Man: irony, that is, not as a local rhetorical device but as the structural condition of *poesis*—the technique of what Lacoue-Labarthe and Nancy, in their exegesis of Schlegelian aesthetics, call "the literary absolute," literature "producing itself as it produces its own theory," through a reflexive turning upon its status and conditions.[68] Recent commentary has also recognized Schlegel as a key source for modern theories of the novel; his conception of the novel less as a form than as a force, a meta-genre breaking up, mixing, and recombining existing genres, prefigures Bakhtin, for example.[69] Yet Schlegel's theory of the novel has not been a strong resource in the theory or history of the novel in English. One reason may be that the emergent norms of novelistic practice in nineteenth-century Britain (at least as they are usually characterized) scarcely accommodate the heterogeneous, fragmentary, open-ended fictions—in Schlegel's phrase, the "mixture of storytelling, song, and other forms"[70]—produced in Romantic-period Germany: "the ironically self-reflexive, generically mongrel, philosophically abstruse tradition following on *Wilhelm Meisters Lehrjahre*," comprising works by Ludwig Tieck, E. T. A. Hoffmann, Jean Paul Richter, Novalis, and Schlegel himself.[71] These realize the ironic technique that Schlegel (adopting the term from Aristophanic comedy) called the "permanent parabasis," a literary work's insistent interruption of the mimesis in order to reflect on its own conditions and procedures.[72]

The English word "novel" is not, of course, an adequate translation of Schlegel's *der Roman*, etymologically intimate with "Romantic," just as *romantische Poesie* comprises dramatic and prose fictions as well as verse: since "every art and every science that works through discourse, when practiced as an art for its own sake, and when it achieves the highest peak, appears as poetry."[73] "A novel is a romantic book" ("Ein Roman ist ein romantisches Buch"), in Schlegel's famous phrase, equivalent to what he elsewhere calls "a universal, progressive poetry" ("Die romantische Poesie ist eine *progressive Universalpoesie*"): the literary work as medium of a recombinatory developmental energy that reconstitutes all genres and discourses in its drive toward an unrealized universal horizon.[74] "The romantic kind of poetry is still in the state of becoming; that, in fact, is its real essence: that it should forever be becoming and never be perfected."[75] It is a creative act homologous with the new, Romantic conception of life and with the Herderian conception of human nature—driving toward infinite differentiation, according to its radical principle, rather than an ultimate goal of totality or synthesis.[76] Schlegel's alignment of the novel (*Roman*), the medium of progressive universal poetry, with infinite emergence or becoming stands the Hegel-Lukács opposition between epic totality and novelistic "bad infinity" on its head: "Paradoxically," in Christoph Bode's summary, literature "*begins to resemble reality once it refuses to speak about its totality.*"[77] Anticipating Bakhtin, and rejecting (*avant la lettre*) the Hegelian view of the novel as a degraded epic form, Schlegel distinguishes the novel from the epic—the literary form of totality—and derives its mixed, variable, open mode, instead, from modern, romantic precedents, Shakespeare and Cervantes.

Lacoue-Labarthe and Nancy pose a key question: "From the moment that the novel, in the romantic sense, is always more than the novel, what happens to the novel itself, in the restricted sense?"[78] They are paraphrasing Schlegel's dialogic *Letter about the Novel*. "The Romantic [i.e., novelistic] is not so much a literary genre as an element of poetry which may be more or less dominant or recessive, but never entirely absent," declares Schlegel's main spokesman, who goes on: "I detest the novel as far as it wants to be a separate genre!" (*eine besondere Gattung*)[79]—as opposed to an emergent poetic energy, shaping all literary forms and settling in none, "a sort of beyond of literature itself."[80] Here Schlegel installs the hierarchical distinction of modern fiction between its exalted, authentic, "literary" iterations and the commodified repetitions of genre: the former modeling the boundless developmental potential of humanity (or life), the latter locked into a mechanical reproduction of form as formula, "bound to nothing and based on nothing."[81]

According to modern consensus, the novel's establishment as a "separate genre" takes place in Great Britain in the second decade of the nineteenth century, in the works of Scott and Jane Austen, which consolidate the techniques of realism and the topoi of domestic manners and national history that will define the novel in English for the next hundred years.[82] That settlement would seem to foreclose the permanent parabasis of

Romantic irony, which insistently undoes (in de Man's phrase) "the categories of the self, of history and of dialectic" that sustain the project of literary realism.[83] A Schlegelian idea of the novel seems better realized in those Germanizing works that occupy the experimental reaches of British Romanticism, outside or against an emergent realist mainstream: Mary Shelley's *Frankenstein*, Thomas De Quincey's *Confessions of an English Opium-Eater*, William Hazlitt's *Liber Amoris*, James Hogg's *Private Memoirs and Confessions of a Justified Sinner*, Thomas Carlyle's *Sartor Resartus*—the last of these, written to a "German" technical as well as theoretical prescription, systematically confounding national novelistic realism as a program of *Bildung*. (Carlyle abandoned his fledgling career as a novelist after translating *Wilhelm Meister* into English, in 1824, and *Sartor*—with its formal imitation of Jean-Paul and Hoffmann—recapitulates, with a distinctively Scottish Calvinist accent, the German idealist critique of the realist Bildungsroman's subjection to "the prose of the world."[84]) In these works, we find Schlegelian irony breaking up a realist mimesis and its philosophical foundations in the British common sense tradition, reopening the skeptical turn through which Hume had dismantled the metaphysical foundations of experience, undoing the stabilizing work of custom and sympathy. As for Germany itself: "graceful as are many of the German legends and fairy tales, fiction seems but little suited to the German genius, and novels of real life almost altogether beyond its range," G. H. Lewes commented as late as 1858.[85] A tradition of the realist novel would not take root until after the foundation of a united German national state. Dilthey found the social conditions for the early Bildungsroman's inward turn, to "the individual and his self-development," in the lack of a public sphere in "the small and middle-sized German states" and the consequent alienation of native intellectuals.[86] The diagnosis can be dialectically turned around: the absence of a political nation, far from impeding *Bildung*, kept open its horizon of universal potentiality.

Accordingly, in a taxonomic scandal that has exercised recent criticism, the Bildungsroman seems less to be a clear-cut, stable genre, categorically commensurate with other genres (such as the historical novel), than a principle that pervades the modern novel as such—its "conceptual horizon," its life force or formative drive.[87] Robert Musil identified the Bildungsroman with "the organic plasticity of man": "In this sense," he maintained, "every novel worthy of the name is a *Bildungsroman*."[88] Several commentators emphasize the reverse of this claim: everywhere in novelistic discourse, hence nowhere in particular, the Bildungsroman is a "phantom formation" conjured up by critics.[89] In this, again, it models its putative end or product, "humanity," that horizon of *Bildung* asymptotic with any actual human life. Schlegel discloses the division, constitutive of the Romantic novel, between a poetic energy or formative drive and a particular form or genre, materialized in a national history that will always fall short of its potential. And he helps us see that division as an anthropological predicament as well as a rhetorical one. Irony is not only the distinctive character or condition of the literary utterance: it is also the condition of being human. Schlegel summarizes the case in an early fragment (parsing Fichte): "humanity is *not wholly* present in the individual, but there only in part. The human being can never be present."[90] Schlegel, however, construes this human predicament of self-displacement as generative rather than disabling. The force of Schlegelian irony is "expansive and connective, not simply disruptive and corrosive," since, "though the subject may always be less than it is, it is also more"[91]—just as the novel is, or should be, "more than the novel."[92] The novel, like humanity, is always becoming, always in formation, reaching toward unrealized futures—infinity, not totality.

The Classical Form of the Historical Novel

Hegel's exaltation of epic as the poetic form of an organic integration of individual and world—realizing a full and original humanity—shadows subsequent accounts of the novel (such as Benedict Anderson's) that reclaim the form's epic status in identifying the nation as its horizon of subject-formation. Modern criticism has canonized the Bildungsroman, in particular, as the genre that aligns subject and nation in a comprehensive narrative of "the formation of the human"—the retroactive attachment of a national plot correcting the fault diagnosed by Hegel and Lukács. But more than the "soul-nation allegory" of critical projection, *Wilhelm Meister* itself resembles the dispersed "social network" novels of the early American republic analyzed by Nancy Armstrong and Leonard Tennenhouse: novels that precede national formation and, in their view, resist its ideological traction with an experimental battery of alternative techniques and tropes.[93] Granting the obvious, and drastic, differences between the former British settler colony and the patchwork of courts comprising the late Holy Roman Empire, both instantiate a historical condition in which the cultural forms of the modern, centralized nation-state have not yet taken hold.

In contrast to these cases, the Scottish historical novel is the flower of a strong modern state, which (at least to begin with) it makes its theme.

The defeat of the 1745 Jacobite rebellion, completing the political absorption of Scotland into Great Britain, supplies the historical topic of Scott's *Waverley*. If *Wilhelm Meister* looks forward, to an indeterminate (ahistorical) near future, *Waverley* looks back, regulating the formation of the present by the recent past, Lukács's "prehistory of the present."[94] Here, it would seem, the soul-nation allegory comes to fruition, as though the Bildungsroman finds its true form in the historical novel, or rather (to put the case less teleologically) the historical novel has conditioned subsequent interpretations of the Bildungsroman.[95] Lukács, in his later career, found in Scott's "classical form of the historical novel" the "purely epic character" that the Bildungsroman's biographical form failed to realize.[96] The plot of national formation directs the energy of *Bildung* and gives form to the novel's protagonist—or rather, it stabilizes his inner formlessness. The tendency of Goethe's hero to distraction and divagation has become chronic in Scott's, afflicted with a "wavering and unsettled habit of mind," stumbling into civil war through no stronger motive than "a curiosity to know something more of Scotland."[97] The end of the novel discloses this weakness as a secret strength, since Waverley's passive, receptive character, his adaptability to local climate and the looseness of his commitment to any political cause, all ensure his survival. Floating on the aesthetic surfaces of life, he drifts as lightly out of rebellion as he drifted into it, to emerge as the prototypical liberal subject of modern civil society.[98] Historical contingency, disguised as necessity—it appears to us now as an inevitable outcome, a destiny, because it happened—gives Waverley's life story a shape, a settlement, an end.

The shape of national history and hence of the hero's life is underwritten, in turn, by Scottish Enlightenment stadial history: a natural history of man that prescribes the progress of nations along a universal developmental axis of economic and social stages, from hunting tribes through pastoralism and agriculture to commercial modernity.[99] Its logic informs the closing "Postscript" to *Waverley*, charging the dissolution of residual feudal and clan communities in the north of Scotland, in the wake of the '45 rising, with an anthropological necessity. Its basis, a universal human nature, is invoked in Scott's introductory chapter, which reflects on his work's experimental character. Among the historical novel's innovations is its grasp of historical process as wholesale structural transformation, legible in the variable forms of social life—manners, customs, costumes—across time and space. Troping this potentially infinite field of formal difference as fashion ("fashion," writes Roland Barthes, "does not evolve; it changes"[100]), the novelist imposes order on it "by throwing the force of my narrative upon the characters and passions of the actors ... those passions common to men in all stages of society, and which have alike agitated the human heart, whether it throbbed under the steel corslet of the fifteenth century, the brocaded coat of the eighteenth, or the blue frock and white dimity waistcoat of the present day" (5). Scott's opening manifesto prepares us to read Waverley's unformed character as a figure for the chaotic potential of history as sheer change, indetermination, bad infinity—a condition that may infect the novel itself, launched amid the swirl of fashionable genres in the modern fiction market.[101]

Natural history, national history, and personal history regulate one another, with national history mediating between the other two. National history thus offers a solution to that anthropological crux, the disjunction between the life of the individual and the life of the species, by recasting it as an empirical discrepancy between temporal scales. It is the vital middle range that accommodates, indeed constitutes, the human.[102] *Waverley* installs "historical time," in Paul Ricoeur's formulation, as the common interface between "the time of the world," the long duration of cosmic history, and "the time of the soul," the fine-grained temporality of individual experience and feeling.[103] The action of Scott's new kind of novel, "neither a romance of chivalry, nor a tale of modern manners," takes place "Sixty Years since," on the hinge between past and present (5). The lifespan of human memory (sixty years) and the regional geography of a small nation (Scotland) define a gravitational field within which custom and sympathy may hold together a habitable domain of common life. The formula "Sixty Years since" curbs what might otherwise be a dislocating historical distance by shaping the difference between then and now with a human form, human proportions. In his Scottish novels, at least, Scott stays close to a familiar terrain, one where the collective mnemonics of custom and tradition maintain a vital human measure between past and present; and where the human, accordingly, is still knowable and readable, even (or especially) through the vague, dithering, malleable figure of the Waverley-hero.

Scotland sixty years since yields a temporary and provisional dispensation, not an absolute one, as the full title of *Waverley*'s closing chapter tells us. "A Postscript, which should have been a Preface" makes explicit the circular logic of historical retrospection, which views its own situation as the necessary outcome of a tangle of contending causes. *Waverley* repeatedly sounds the note of historical contingency—for example, in the set-piece description of the Highland chieftain, adapted from Thomas Blackwell's pioneering historicization (1735) of Homeric epic:[104]

> Had Fergus Mac-Ivor lived Sixty Years sooner than he did, he would, in all probability, have wanted the polished manner and knowledge of the world which he now possessed; and had he lived Sixty Years later, his ambition and love of rule would have lacked the fuel which his situation now afforded. (98)

The accidents of time and place shape character and its affordances of plot or destiny. This reflection raises questions about the integrity of the introductory distinction between the passions, "common to men in all stages of society," and manners, the historical forms in which those passions are expressed. Where can we draw the line between them? At what point might a change of form change its substance?[105]

"If you Saxon Duinhé-wassal (English gentleman) saw but the chief with his tail on!" boasts Evan Dhu Maccombich as he leads Waverley to Fergus (81). This startling glimpse of an evolutionist hypothesis—Lord Monboddo's[106]—is swiftly reduced to the English visitor's comic misprision:

> "With his tail on?" echoed Edward in some surprise.
> "Yes—that is, with all his usual followers, when he visits those of the same rank." (81)

The mild anachronism, an encounter with an earlier state of society, replaces the radical one, the eruption of a prehuman ancestor—and a breach of the universal nature that anchors the novel's historicism. Later, in fully anthropological mode, the narrator contrasts the fierce clan loyalty of the Highlanders with Waverley's civilized capacity for a diffusive, universal sympathy—"humanity"—which the Highlanders, their subjectivities articulated by local kinship networks, seem incapable of sharing:

> They would not have understood the general philanthropy, which rendered it almost impossible for Waverley to have past any person in such distress; but, as apprehending that the sufferer was one of his *following*, they unanimously allowed that Waverley's conduct was that of a kind and considerate chieftain, who merited the attachment of his people. (232)[107]

Both impulses express the social instinct that binds human nature to history. It is hard to say, however, whether the passion that constitutes "general philanthropy" is the same as the passion that constitutes tribal "attachment," since one will always translate the other into its own terms. The Highlanders' revision of Waverley's philanthropy downgrades it to just another local historical formation.

Scott's "Postscript" invokes a universal logic of stadial progress, drawn from Enlightenment historiography, at the same time as it acknowledges a melancholic loss of distinctive forms of human life. Civil war and its aftermath have accelerated two centuries of historical change into half a century, issuing in a notable cultural extinction:

> This race [of Jacobites] has now almost entirely vanished from the land, and with it, doubtless, much absurd political prejudice; but, also, many living examples of singular and disinterested attachment to the principles of loyalty . . . and of old Scottish faith, hospitality, worth, and honour. (363)

We glimpse, across the threshold of sixty years since, the afterglow of a more radical difference than of costumes and manners. Scott's later historical novels, from *Ivanhoe* onward, stretch the *Waverley* chronotope, the spatiotemporal continuum of the "prehistory of the present," beyond its breaking point—and with it the lifeline of national history and its corollary, a unified human nature. That disintegration of the nexus between national history and human nature in the late Romantic historical novel will be the topic of the following chapter.

The Dignity of the Human Race, the Glory of the World

> No more will she ride with you
> On horseback across the sky;
> The flower of her maidenhood will fade,
> Plucked by a husband—a masterful man
> She must henceforth obey.
> She'll sit and spin by the hearth,
> A target for scorn and mockery!
>
> —RICHARD WAGNER,
> *DIE WALKÜRE* (1870)[108]

"As I scan the great changes in the world and the succession of ages, I never divert my attention from one prime notion: the perfectibility of the human species," wrote Madame de Staël, echoing Condorcet, in her introduction to *Literature Considered in its Relation to Social Institutions* (1800).[109] Condorcet had included women in his vision of universal perfectibility, and he advocated their civil enfranchisement (*Sur l'admission des femmes au droits de la cité*, 1790). His immediate heirs, the Ideologues, were less progressive; even Cabanis, who attended female salons and corresponded with Mme. de Staël, declared women temperamentally unfit for citizenship.[110] The Napoleonic Code reaffirmed the legal subordination of women to their husbands and fathers in March 1804, a little over a year before Staël began writing *Corinne*. Defiantly, her novel's heroine affirms the right of "every woman, like every man, to make a way for herself according to her nature and her talents," and so to contribute to the general progress of humanity.[111] The defense of female *Bildung* is a rearguard protest against the gendered co-option of perfectibility by the revolutionary ideologues and their successors. *Corinne* constitutes an antithetical midpoint, a negative dialectical fulcrum, between *Wilhelm Meister* and *Waverley*. More polemically than they, Staël's novel acknowledges the claim of the nation-state and its ideological apparatus (history, domesticity, civil society) over human *Bildung*, and diagnoses its repressive force—despite its author's reputation as the leading proponent of Romantic nationalism after Herder. At the same time, Staël charges her protagonist with a more potent, purposeful subjective drive—a veritable *Bildungstrieb*—than either Goethe's hero or Scott's can lay claim to.

German reviewers, encouraged by Staël's visits to Germany and her association with August Wilhelm Schlegel, hailed *Corinne* as an offspring of German Romanticism. Dorothea Schlegel's translation, *Corinna oder Italien*, appeared in the same year, although (like her other published works) its title page bore the name of her husband, Friedrich.[112] In contrast to the immanent, ironical mode of *Wilhelm Meister*, *Corinne* embraces enthusiasm and allegory. It dramatizes the problematic we saw Schlegel bringing into view, in which national history precipitates a contradiction between the novel as an emergent force or universal potential and the novel as a finite, local, settled form, a genre among others. Staël's heroine personifies, instead, the Romantic ideals of the aesthetic education and universal progressive poetry. However, it is her misfortune to fall into a novelistic plot—and it kills her. Staël mounts a critique not only of domestic fiction and the marriage plot, as critics have recognized, but also of the universal claims of *Bildung* and the adequacy of national history as a medium for its realization in the emergent forms of Romantic fiction.

In her role as a rhapsodic performance-poet (*improvisatrice*), Corinne realizes the "mixture of storytelling, song, and other forms" Schlegel prescribed in his "Letter about the Novel." She combines them in voice and gesture, in a charismatic, total artwork that is authentic for being embodied and hence transient, ever varying, flowing through her rather than fixed.[113] Corinne's "cosmopolitan synthesis of improvisational styles"[114] embraces different forms, media, nations, and epochs: "It often happens that I depart from poetic rhythms and express my thought in prose; sometimes I quote the finest verses of the different languages I know," she affirms, with characteristic candor. "Sometimes, too, with chords and simple, national melodies, I complete on my lyre feelings and thoughts I cannot express in words" (46). Dancing the Neapolitan tarantella, Corinne evokes Indian "temple dancing girls" (*Bayadères*) and "dancing girls of Herculaneum," recreating "the poses depicted by the ancient painters and sculptors" (91).[115] In a triumphant fulfillment of aesthetic education, she reconstitutes an organic, universal artwork, before the arts were subject to the division of labor, delivering her audience to a transcendent state of unity:

> As she danced, Corinne made the spectators experience her own feelings, as if she had been improvising, or playing the lyre, or drawing portraits. Everything was language for her [*tout était langage pour elle*]. . . . An indefinable passionate joy, and imaginative sensitivity, stimulated [*électrisait*] all the spectators of this magical dance, transporting them into an ideal existence which was out of this world. (91)[116]

Because its medium is her body, rather than a closed graphic system, Corinne's art never exceeds the time—the eternally iterable "now"—of its performance; it is not abstracted and reified into the figure of a totality.

Correspondingly fluid and heterogeneous, Staël's "romantic book" interpolates a love story with extended descriptions of Italian scenery and monuments in the manner of a travelogue or guidebook, samples of the heroine's improvisations, her impromptu lectures on history and the arts, debates about politics, religion, and national character, frivolous salon *causerie*, and impassioned exchanges of letters and confessions. At first, Corinne plays the role of tutor to her lover Oswald—inverting the gendered ratio of authority in the Abelard-Héloïse plot that Rousseau imported into the modern novel in *Julie*. As it progresses, Staël makes the love story conventionally novelistic: desire is aroused only to be deferred, clogged by past secrets and family histories, thwarted by duty and circumstance, frozen into an anguished domestic triangle. The love story prevails, binding the heroine, mortifying her into another Italian tomb or ruin. "My talent no longer exists," she laments: "the source of everything is dried up" (356–57). She can only reclaim her art in the sacrificial performance of her own death. The formal ascendancy of the novel as a particular complex of conventions, subduing the book's other registers and discourses, extinguishes *Bildung* and eventually life itself.

This obliteration of the heroine's *Bildung* by the love story casts a bleak light back on what now becomes visible as the doubled structure of

Wilhelm Meister's Apprenticeship. Goethe's novel counterpoints Wilhelm's vocational plot, the theatrical mission and fellowship of the Tower, with a serial courtship plot. Mariane, Philine, the Countess, Aurelie, Therese, Natalie: the succession of fascinating women supplies the directional structure to Wilhelm's narrative that the vocational plot fails to deliver. Each of the women marks a stage in the hero's progress. Their assumption of masculine roles and costumes, a persistent "Amazonian" theme in the novel, highlights the receptive character of Wilhelm's sexuality, unformed or rather in formation, while their failure to accompany him to the next stage gives the narrative its progressive form, and casts the women, literally or figuratively, as sacrificial victims. The sacrificial role is reiterated on a formal level by the interpolation of the "Confessions of a Beautiful Soul" in book 6, dividing the Theatrical Mission from the Tower. The episode makes room for a closed, pietistic, autobiographical mode of feminine *Bildung* that, emphatically not Wilhelm's, is thus exorcised from the novel.[117]

The erotic plot of *Wilhelm Meister* makes explicit a disciplinary logic of substitution that structures the apparent free play of the hero's *Bildung*. As the Tower Society (itself theatrical, a masquerade) replaces the Theatrical Mission, so Wilhelm's accession to paternity and wedlock (deferred, however), paved with the series of spent women, replaces his vocation at the story's terminus. Where the first of these substitutions takes place diachronically, implying a dialectical succession, the second operates synchronically, nondialectically, that is, metaphorically or allegorically. The courtship plot displaces rather than subsumes the vocational plot, as the closing lines of the novel acknowledge. "You seem to me like Saul, the son of Kish, who went in search of his father's asses, and found a kingdom," Friedrich says to Wilhelm, who replies, "I don't know about kingdoms, but I do know that I found a treasure [*Glück*: fortune, happiness, luck] I never deserved. And I would not exchange it for anything in the world" (373). Luck, no plan or program, emancipates the merchant's son from his bourgeois destiny. An accidental exchange (finding a different object from the one he was seeking) yields the object that is beyond exchange, a wife—and one that (obeying the logic of narrative openness) will elude his possession even in the novel's sequel. Others, meanwhile, have paid for Wilhelm's fortune: the Countess, whose decline into pietism follows Wilhelm's impersonation of her husband; Aurelie, whose fatal inability to separate herself from her tragic roles liberates Wilhelm from the theatrical mission; Mariane, whose death in childbirth gives Wilhelm a son; and, above all, Mignon, whose death and apotheosis cast her as the novel's paradigmatic sacrificial victim. "The violence of her developing nature"

(156), wracking her with illicit passion, undoes her at last when she sees Wilhelm embracing his betrothed. (Again, it is typical of the narrative's movement that Theresa then turns out not to be Wilhelm's betrothed.) While Mignon's death makes it possible for the story to end, with Wilhelm's choice of a bride, her obsequies replace the marriage ceremony, postponing it beyond the novel's close. Turned into a beautiful corpse, Mignon becomes the centerpiece of a grotesquely ornate funeral pageant, set in the Hall of the Past, the site of a synthesis of the arts (architecture, painting, poetry, music) configured around mortality and memory.[118]

Corinne inherits Mignon's role as spectacular sacrificial victim, invested with pathetic sublimity, as well as the *Gesamtkunstwerk* aesthetic of the Hall of the Past. She takes over the show as author and manager, however, as though in fulfillment of Wilhelm's abandoned theatrical mission. Rallying her talents for a last performance, she directs the multimedia spectacle of her death, composes her own eulogy, and seizes posthumous control of the family romance.[119] "Since I must soon die, my only personal wish is that Oswald should find again in you and in his daughter some traces of my influence, and that at least he may never enjoy a feeling without recalling Corinne," she tells her successful rival (and half-sister), Lucile, colonizing the sacred interior space of the domestic affections (404).

Where Mignon submitted mutely to her fate, her lyric voice stifled, Corinne launches a sustained protest against the marriage plot, which her author codes as "English"—not just by making the Scottish peer Oswald choose fair, mild, English Lucile over dark, passionate, Italianate Corinne, but by making this outcome a reflex of national conditions, and setting it within the generic matrix of English domestic fiction. England (which apparently, if inconsistently, includes Scotland) is the novel's exemplary case of a historically ascendant, unified nation-state. A masculine public sphere of governmental and military institutions makes up the nation's carapace, entailing a gendered division of moral as well as economic labor: "in a country where political institutions give men honourable opportunities for action and public appearance, women must stay in the shade" (318). In England, "the strength and reality of [the] social order ... dominates all the more in that it is based on pure, noble ideas" (269)—ideas which, in the form of patriotic duty, justify Oswald's desertion of Corinne, and reciprocally produce his private character as melancholy and vacillating (he is a depressive Waverley *avant la lettre*).[120] And in England, Corinne's father warns her, "women have no occupation but domestic duties" (245). Corinne's narrative of her upbringing discloses the smothering psychic violence of this gendered ethos of duty in English provincial life: "I felt my talent slipping away. In spite of myself, my mind was occupied by petty things [*petitesses*], for in a society lacking all interest in science, literature, pictures, and music—in which, in short, no one is interested in the imagination—it is the little things [*petits faits*], the minute criticisms, which of necessity form the subject of conversations" (250).[121]

The English episodes of *Corinne* form a satiric counterpoint to Staël's encomia to the British constitution in her political writings (*Reflections on Peace Addressed to Mr. Pitt and to the French*, 1794; *Considerations on the French Revolution*, 1818), as well as her analysis of the role of women in English culture in *On Literature*. Thanks to their relegation to private life, "nowhere so much as in England have women enjoyed the happiness brought about by the domestic affections." This is the main reason, in turn, for the preeminence of the English in "one genre of imaginative works: fiction without marvels, without allegory, without historical allusions, based solely upon the invention of characters and events of private life"—in other words, the domestic realist novel developed by Richardson and Burney.[122] "These novels are created to be read by the people who have adopted the style of life portrayed in them—rural, domestic—in the time spared from regular pursuits and family affections." Despite longueurs that try the patience of French readers, "English novels hold our attention by a steady succession of accurate and moral comments on life's tender emotions."[123] But it is one thing to read the prose of the world and another to dwell in it. For Corinne, the "thousand little hurts" (*milles petites peines*) of domestic life, "like the bonds in which the pygmies wrapped Gulliver," provoke only "boredom, impatience, and loathing" (250)—the grimace of Romantic genius trapped in a Jane Austen novel.[124] In a further, ironic declension, the plot that captures Corinne once she succumbs to her love is a French (or Franco-Swiss) one: the blocked domestic triangle of dutiful married couple and anguished lover adapted from the close of Rousseau's *Julie*.

Corinne can thrive in Italy, conversely, because it is not a nation-state. (And in Italy "there is not a single novel," 94.) She redeems the pejorative association of Italian decadence and effeminacy made by northern travelers, such as Oswald: "Men's characters have the gentleness and flexibility of women's," he grumbles, "in a country where there are no military careers or free institutions" (97). For women, the case is otherwise. The absence of a state opens public life to Corinne. She assumes a national role, hymning "La gloire et le bonheur de l'Italie," crowned at the Capitol in the place of ancient emperors. The narrator notes her freedom from the gendered role-splitting dictated by the English system: "At one and the same time she gave the impression of a priestess of Apollo who approaches the sun-god's temple, and of a woman who is completely natural in the ordinary relationships of life" (23).

The heroine's role as "national metonymy," Corinne or Italy, is complicated by the historical condition of the country.[125] Corinne's performance convokes Italy as a realization of Schiller's "aesthetic state," in which the arts have replaced an extinct political sphere. Italy's (or rather Rome's) famed hospitality to the imagination derives from its condition as a former empire, enjoying "the last glory that is allowed to nations without military power or political independence, the glory of the arts and sciences" (99). The pastness of national power, fossilized in ruins, monuments, and artworks, inspires Corinne's poetry, which she dedicates to "the dignity of humankind and the glory of the world":

> I am a poet when I admire, when I despise, when I hate, not out of personal feelings, not for my own sake, but for the dignity of humankind and the glory of the world. (46) [*la dignité de l'espèce humaine et la gloire du monde*, 85]

Although at several points she joins Oswald in deprecating Italy's lack of republican virtue, Corinne issues no call for national political revival. Her coronation enacts the triumph of art over empire—whether Augustan, Napoleonic, or British.[126] The good cosmopolitanism of the aesthetic state reconstitutes Italy as a universal nation where all may find themselves at home because its political life lies in the past, not in the future.[127] "Is Rome not now the land of tombs?" she cries. "Perhaps one of Rome's secret charms is its reconciliation of the imagination with this long slumber" (32). Corinne's rhapsodic time, the time of an emergent humanity, is postnational and posthistorical.

My argument presses against the usual story of *Corinne*'s international reception, well summarized by Susan Tenenbaum:

> In *Corinne* Staël portrayed her heroine as an agent of social redemption. Corinne's inspired odes alluding to Italy's glorious past were intended to awaken its citizens' dormant civic consciousness: her poetic images of ancient Rome and the medieval Italian republics served as metaphors of lost dignity and inspiration to national rebirth.[128]

The story is encouraged, to be sure, by Staël's other writings on national questions (not least *Germany*); but here, as with her account of English domestic life, women, and the novel in *Literature*, we need to be wary of resorting to the author's essays and treatises for an explanation of her fiction, which develops its own internal logic. In contrast to the contemporaneous "Glorvina solution," the national-allegorical union of English milord and subaltern romantic heroine pioneered by Sydney Owenson in her 1806 novel *The Wild Irish Girl* and developed by Scott in *Waverley*,[129] *Corinne* undertakes a trenchant feminist critique of the nation-state as *Bildung*'s historical horizon, by separating a cosmopolitan aesthetic ideal of national life from its political reality. The topical deflection of that critique, from Napoleonic empire to the British example of a modern unified state founded on a gendered separation of spheres, also directs satirical energy at the ascendant English tradition of the domestic realist novel, which fixes the nation in a rigid ethos of duty and the asphyxiating habitus of provincial life. *Corinne* reroutes the relation between personal and universal histories—between individual *Bildung* and the perfectibility of the species—along the fault line of gender, split open rather than closed by the mediating term of national history. Staël's bold stroke is to make her heroine the charismatic claimant upon a universal humanity after "woman" had been cast as supplementary to the universal term of species being in Enlightenment philosophy, and excluded from political rights in the French Civil Constitution. If domestic fiction established "the modern individual [as] first and foremost a woman," in Armstrong's phrase, Corinne's defiance of a regime "in which women are invited to consider themselves as women or as human beings, but not as both at once," makes her "politically," as Moi contends, "the first modern woman in Western literature."[130]

Dark Unhappy Ones

True to the economy of sacrifice, Corinne, like Mignon, would transcend the novel that bore her and colonize the nineteenth-century imagination, in the phenomenon David Brewer calls "character migration": leading an independent afterlife in sequels and adaptations across a range of genres, media, and performance sites.[131] Terence Cave has traced Mignon's post-Goethean apotheosis in other novels, song settings, operas, plays, paintings, and even postcards. Corinne would claim more, and take on living flesh and blood; she would possess the person of her author, endowing Staël in turn with the uncanny immortality of a literary character.[132]

That afterlife—in which the heroine's sacrifice perpetuates her imaginary dominion—also looms at the close of Staël's novel. Barred from the marriage plot while alive, Corinne occupies it after her death, turning it into a haunted site of remembrance, repression, and regret. She will watch over Oswald from beyond the grave, she tells Lucile, "for you do not stop loving when the feeling of love is strong enough to destroy life" (403). The novel closes (as though Oswald's condition has infected the narrator) on a note of melancholy dubiety:

> Lord Nelvil was a model of the purest and most orderly domestic life. But did he forgive himself for his past behaviour? Was he consoled by society's approval? Was he content with the common lot after what he had lost? I do not know, and, on the matter, I want neither to blame nor to absolve him. (404)

Fifty years later, in *The Mill on the Floss* (1860), George Eliot's Maggie Tulliver declares her wish to avenge the wronged sisterhood of "the dark unhappy ones," the antiheroines of Staël and Scott.[133] If the secret of *Wilhelm Meister* and *Waverley* is that Goethe and Scott make their heroes *less* than the novels in which they appear, able to float lightly across the vicissitudes of plot, *Corinne* generates a modern tradition of female protagonists condemned to existential heaviness. Expelled from the novel because (in Schlegelian terms) they are "more than the novel," their darkness is pitched less against the douce blondes that replace them than against the lucky lightness of the Wilhelms and Waverleys. Their *Bildung* is deflected into a passionate intensity that, like Mignon's lyric pathos, eclipses all else around them. "Of all my gifts, the most powerful is my gift for suffering" (75): Corinne's promise, uttered early in her affair with Oswald, is confirmed in her last song: "Of all the faculties of soul given to me by nature, only suffering have I developed to perfection" (402).[134]

The "myth of Corinne" receives searching reassessment in two mid-Victorian romances of female artistic vocation, cosmopolitanism, and the aesthetic state. Elizabeth Barrett Browning's verse novel *Aurora Leigh* (1856) repairs the breach between *Bildung* and marriage for its Anglo-Italian poet-heroine, but at a high cost, since the combination still requires a sacrifice. That comes to bear on a substitute, Marian Erle, whose martyrdom allows Aurora to take her place in the marriage plot at the last minute, as though only by such drastic means can its fatality be averted.[135] Aurora must also renounce the purity of her earlier dedication to her art—in tandem, to be sure, with her betrothed's renunciation of his (failed) political career. Marriage as a mystical union with the divine nature sublimates both masculine and feminine paths of *Bildung* in the poem's closing rapture.

George Eliot's *Daniel Deronda* (1876)—to be revisited in chapter 5 of *Human Forms*—plays a more complicated set of variations on the figure of the performing heroine and the ideal of female *Bildung* through art. Early in the novel, the courtship plot of Julius Klesmer and Catherine Arrowpoint redeems the conjunction of aesthetic education and marriage by making the universal artistic genius a man and the heroine who weds him, in defiance of English domestic norms, his pupil: restoring the Abelard-and-Héloïse ratio of male tutor and female student capsized in *Corinne*. With the absorption of the English heiress Catherine into her husband's mixed-race cosmopolitan domain of art, their union also undoes the domestic national allegory. *Daniel Deronda* moves beyond this solution, marking it as exceptional but insufficient, to recombine the fable's key elements. Although antiheroine Gwendolen Harleth's lack of talent thwarts her attempt to escape from a destiny of English provincial marriage into art, Italy nevertheless affords her release from a soul-destroying fate into an ambiguous penitential afterlife (an equivalent of the convent to which Scott dispatches Flora Mac-Ivor at the end of *Waverley*).

A viable synthesis of *Bildung*, marriage, and national destiny coalesces, instead, around the novel's eponymous hero, who is the type of a more evolved humanity. The solution involves the dialectical repudiation of a provincial English ground via a succession of Jewish plots: first, the cosmopolitan cultural ideal of the aesthetic state claimed by Klesmer, and then, reinstating the national theme, the racial heritage and Zionist mission discovered and embraced by Daniel. Daniel's bride, Mirah, is a gifted musician, but she eschews a public forum for her art, and is absorbed into her husband's career. George Eliot's expansion of the Bildungsroman to a world-historical stage, as though in answer to Hegel's critique, decisively subordinates the marriage plot to ancillary status—a disposition reinforced by its catastrophic failure in Gwendolen's story.

Meanwhile, a vehement protest against the novel's resolution comes from Daniel's estranged mother, Princess Leonora Halm-Eberstein, "the Alcharisi," once an internationally celebrated opera singer and the most powerful avatar of Staël's heroine in *Daniel Deronda*. "Had I not a rightful claim to be something more than a mere daughter and mother?" she demands of her son. "Whatever else was wrong, acknowledge that I had a right to be an artist, though my father's will was against it. My nature gave me a charter."[136] Her devotion to her art has brought a double betrayal, in Daniel's eyes, of his paternal heritage and her maternal nature—a judgment that marks the limit of his hitherto all-capacious sympathy. Staël never endorses such a view of Corinne, who retains her "natural" womanliness even at the height of her public triumph. The Alcharisi's sacrifice of nature for art entails, in turn, the novel's compensatory sacrifice of her so as to achieve its resolution, solemnized with her lonely, painful death.

Where Corinne was able to outwit posterity, Leonora is forced to admit defeat. Nevertheless, her protest, amplified by her suffering, casts its bitter echo across the novel's close. The last-act apparition of the Princess Halm-Eberstein in *Daniel Deronda* admits—if only to exorcize—the specter of a female character too great for the novel that bears her.

CHAPTER THREE

Lamarckian Historical Romance

Monday, August 2, 1830.
The news of the Revolution of July, which had already commenced, reached Weimar today, and set every one in a commotion. I went in the course of the afternoon to Goethe's. "Now," exclaimed he to me, as I entered, "what do you think of this great event? The volcano has come to an eruption; everything is in flames, and we have no longer a transaction with closed doors!"

"A frightful story," returned I. "But what could be expected under such notoriously bad circumstances, and with such a ministry, otherwise than that the whole would end in the expulsion of the royal family?"

"We do not appear to understand each other, my good friend," returned Goethe. "I am not speaking of those people, but of something quite different. I am speaking of the contest, so important for science, between Cuvier and Geoffrey [sic] de Saint Hilaire, which has come to an open rupture in the academy."

—J. W. GOETHE, *CONVERSATIONS WITH ECKERMANN AND SORET* (1850)

C'était comme un nouveau monde, inconnu, inouï, difforme, reptile, fourmillant, fantastique.

—VICTOR HUGO, *NOTRE-DAME DE PARIS* (1831)

Of Paris

The last but one of Scott's Waverley Novels shatters the decorum of "the classical form of the historical novel."[1] Set in Constantinople at the end of the eleventh century—as far outside the developmental continuum of the "prehistory of the present"[2] as its author could reach—*Count Robert of Paris* explodes the taxonomy of national-historical types into a fantastic zoology. Its specimens include Greeks, Turks, Normans, Varangians, Africans, Scythians, a bluestocking princess, a warrior-countess, a seditious philosopher nicknamed "the Elephant," a real elephant, a tiger, a clockwork lion, and a giant orangutan named Sylvan. Sects, nations, races, and species ferment together in the postclassical world-city; distinctions between them, and between the cultural and biological, human and nonhuman, organic and mechanical, melt and fuse.[3] At the center of this flux, babbling an "unintelligible" language, gesticulates the orangutan, Romantic emblem of the dissolution of a unique and universal human nature. It seems the Author of *Waverley*, at the close of his career, has definitively abandoned the characterological, geographical, and genealogical apparatus of national historical romance he had put together seventeen years earlier.

Scott wrote *Count Robert of Paris* in fits and starts between December 1830 and September 1831, harassed by a series of strokes and by disagreements over the new novel with his publisher, Robert Cadell, and literary executor, John Gibson Lockhart. A little over a year later he was dead. Critics have not hesitated to diagnose the excesses and irregularities of *Count Robert of Paris*, published in 1832 as one of the *Tales of My Landlord, Fourth Series*, as symptoms of its author's encroaching apoplexy.[4] J. H. Alexander's recent restoration for the *Edinburgh Edition of the Waverley Novels* makes clear how much of the work's incoherence was due to Cadell and Lockhart, who cut and rewrote Scott's manuscript—dismayed by what they were reading—as it went to press.[5] Among the passages Alexander prints for the first time is an epilogue, in which the ailing Scott reflects on the experimental character of "the last of my fictitious compositions."[6] The quest for "novelty at whatever rate" has driven him outside the usual terrain of historical fiction, "domestic nature," to "lay his scene in distant countries, among stranger nations, whose manners are imagined for the purpose of the story—nay, whose powers are extended beyond those of human nature" (362). As examples of romances that feature powers beyond human nature, Scott cites Robert Paltock's *The Life and Adventures of Peter Wilkins* (1751), a fantastic tale of a sailor shipwrecked on an island of flying people, and "a late novel, also, by the name of Frankenstein, which turns upon a daring invention, . . . the discovery of a mode by which one human being is feigned to be capable of creating another" (363).

Of the two, *Frankenstein* is the prototype that gripped Scott's imagination. He had written one of the few appreciative reviews of it, for *Blackwood's Edinburgh Magazine* in 1818; his present reference may have been prompted by advertisements for the new edition in Bentley's *Standard Novels*, revised by Mary Shelley and published in October 1831, one month after he completed *Count Robert of Paris*. Scott, in effect, invites us to read *Count Robert of Paris* as a work of anthropological science fiction rather than as a historical novel. Here the link between genres, founded in British Romanticism with *Waverley* and *Frankenstein*, is revealed to be genetic as well as analogical.[7] The link has a name: the "unfortunate monster" Sylvan, standing eight feet tall, like Frankenstein's creature, exhibiting at once a "grotesque form" and "the form of a human being" (169–70)—disclosing, that is, the formation of the human *as* grotesque. In contrast to Shelley's gigantic artificial man, the Orang-outang embodies the most radical of late Enlightenment versions of natural man,[8] lately inserted by Jean-Baptiste Lamarck into the "animal series" as a transitional form between ape and human[9]—the scandalous figure of a not-yet-realized *Bildung der Humanität*.

A third anthropoid monster entered the European imagination in 1831, alongside Scott's sasquatch-sized orangutan and the second edition of Frankenstein's creature—and alongside the proclamation of a new science of monstrosity, a "teratology," in the philosophical anatomy of Etienne Geoffroy Saint-Hilaire and his son Isidore. Quasimodo, the "one-eyed, hunchbacked, crook-kneed" bell-ringer, is so compelling a presence in Victor Hugo's *Notre-Dame de Paris* that English translators retitled the novel after him.[10] Combining hideous deformity, malicious strength, and hidden depths of sensibility, Quasimodo is not the only monster in Hugo's novel, which unfolds, like Scott's, a metropolitan ecology of monstrous figures—from devils in human form to the stupendous cathedral of Notre-Dame, a stone sphinx or chimera that at one point transmutes into a visionary behemoth. Subsuming all of these—characters, cathedral, city, the novel, and our reading of it—there swarms across time and space "that myriapod monster, the human race" ("genre humain qui marche, monstre à mille pieds").[11] Like Sylvan in *Count Robert of Paris*, Quasimodo condenses up-to-date uncertainties about the philosophical status of human nature that roil across the novel's world and character system. "It's not a child," cry horrified ladies over the infant Quasimodo: "It's a monkey gone wrong" (156: "un singe manqué," 194)—a botched step in the Lamarckian evolutionary series, or, according to the new teratology, a case of arrested fetal development. "En effet, ce n'était pas un nouveau-né que 'ce petit monstre,'" shrugs the narrator: "(Nous serions fort empêché nous-même de le qualifier autrement)" (195). ("'The little monster' (we would be hard put to it ourselves to find another term for him) was indeed no new-born baby," 156.)

Hugo began writing *Notre-Dame de Paris*, which he had begun planning two years earlier, on July 25, 1830 (according to a manuscript note). He was interrupted two days later by the outbreak of the July Revolution; he resumed work on the novel at the beginning of September, and completed it in mid-January, 1831. *Notre-Dame de Paris* culminates, not coincidentally, in a fiery popular uprising, the truands' assault on the great cathedral of Notre-Dame.[12] "Of Paris": the titular genitive shared by Hugo's novel and Scott's names the Enlightenment capital of science and letters, the European world-city, in the epoch of a second French Revolution—and of impending Reform, which Scott and his party feared would unleash revolution, in Great Britain. 1830 brought a reiteration of the first Revolution's political redefinition of "man," its alignment of the term with the people, or national collective life embodied by the urban crowd—and the promise, or threat, of new forms of humanity, no longer given but remade. Now, however, that reiteration followed the disintegration of the claims of 1789 to regenerate human nature. Liberal and radical eyes viewed Revolution II: 1830 as a renewed opportunity to accomplish that regeneration, and realize the utopian goal of human perfectibility, while conservatives saw a deflation of the Enlightenment's progressive historiography and the world-striding figure of man. A counterrevolutionary rhetoric investing the insurgent populace with monstrosity, mobilized by Edmund Burke and others in the 1790s, returned in force.[13] 1830 revealed revolution as a recurrent rather than a singular event, a persistent, chronic condition of modern life rather than a once-and-for-all crisis, the onset of a new epoch. Revolution, dictatorship and empire, restoration, revolution again—it is as though history itself has become monstrous, according to the teratological definition: "the mixture of an old and a new order, the simultaneous presence of two states which, ordinarily, should succeed one another."[14]

Inevitably, the politics of the revolutionary era charged the intellectual debates and institutional rivalries that were agitating the emergent science of the forms of life, centered now in Paris. Arguing for the reform of knowledge as a necessary condition of political reform, scientific authors opposed to the Bourbon regime rallied around Lamarckism, and transformist natural history more broadly, throughout the 1820s. Lamarck's protégé Geoffroy Saint-Hilaire, emerging as a leading light of the liberal movement, made monstrosity a key research program of the new philosophical anatomy. Geoffroy collected his papers on monstrosity in the second volume of *Philosophie anatomique*, "Des monstruosités humaines," in 1822; ten years later, Isidore systematized his father's research in *Histoire générale et particulière des anomalies de l'organisation chez l'homme et les animaux*, the "Treatise on Teratology," volume 1 of which appeared on the heels of *Notre-Dame de Paris* and *Count of Robert of Paris* in 1832. Geoffroy and Isidore sought to reaffirm the orderliness of nature by insisting that monstrosities were natural phenomena, subject to natural law—deviations, on classifiable principles, from the archetypal regularity of the species, itself subject to the grand law of "unity of organic composition."[15] At the same time, monstrosity provided a mechanism for the transformation of species. In an 1821 essay, "Considerations from Which Rules Are Derived for the Observation and Classification of Monsters" ("Considérations d'où sont déduites des règles pour l'observation des monstres et pour leur classification": reprinted in *Philosophie anatomique*), Geoffroy approvingly cites the "prophetic" words of Julien-Joseph Virey: "The study of monsters will offer a research program, for the physiologist and the philosopher, into the processes by which nature effects the generation of species."[16] Attempting to induce anatomical deformities in unhatched chicks, Geoffroy argued for the susceptibility of the embryo to mutation under the direct influence of environmental pressures such as atmosphere and temperature. Embryonic mutability, together with other principles of Geoffroy's morphological synthesis, afforded a comprehensive apparatus for the evolution of living forms.[17] Although differing from it in certain key features, Geoffroy's synthesis built on and modified Lamarck's influential systematization of transformist ideas, such as the linked, progressive series of kinds, or "march of nature," and adaptive and heritable mutation. Lamarck's *Zoological Philosophy* (1809) had reinforced the association of transformism with materialist doctrines of a nature that was self-organizing and self-regulating, driven from below by innate progressive energies, rather than divinely ordained: a vision swiftly applied to the political aspirations of the age.

Controversial as it was, Lamarckism was in vogue in the decade preceding the July Revolution. Pietro Corsi has argued that Lamarck's writings, far from being sidelined and neglected (as the received account has it), aroused widespread interest across European scientific circles in the first third of the nineteenth century. That interest intensified with the politicization of medical and natural science in France in the mid-1820s, and the embrace of Lamarckian theory by critics and opponents of the regime. By the time of his death in 1829, Lamarck's reputation was at its height. Baron Georges Cuvier's notorious eulogy, in which the great comparative anatomist took the opportunity to demolish his late colleague's scientific standing, may have done his own reputation as much harm as it did Lamarck's, at least in the short term. Cuvier, Director of the Museum of Zoology and the ascendant power in French natural history in the early nineteenth century, had dismissed Lamarck's theory in the "Preliminary Discourse" to his *Recherches sur les ossements fossiles* (1812), lately reprinted in his most widely read work, *Discours sur les revolutions du globe* (1825); he remained implacably opposed to all tenets of transformism, maintaining a functional, not genetic, explanation of morphological variation within a taxonomy that rigorously separated the four main branches of the animal kingdom. Cuvier's tendency to lump together the various hypotheses of transmutationism, German *Naturphilosophie*, and philosophical anatomy in a polemical gesture of rebuttal contributed to the use of "Lamarckism" as a blanket term for a large, diversified field, whereas, Corsi shows, Lamarck's was one of many different, in some cases independent, versions of transformism proposed in the period.

In the mid-1820s, Geoffroy began to assimilate Lamarckian principles to his own morphological synthesis, even as it differed in crucial details from Lamarck's (notably, in locating fetal development as the stage where mutation occurred, and in ascribing causality to direct environmental pressure rather than heritable habit).[18] In an 1825 paper on the genealogical link between living and extinct species of crocodile, Geoffroy cited "the two laws posed by M. de Lamarck in his *Philosophie zoologique*" (the mutation of organs in response to environmental conditions and the heritability of those mutations), and recommended Lamarck's philosophical reflections to students.[19] Cuvier, Geoffroy's former friend and colleague, saw the emerging synthesis as a challenge to his own zoological system—a resurgence of the transformist heterodoxy he had spent his career opposing—and attacked Geoffroy's arguments in print. Their disagreement came to a head in the spring of 1830, in a public debate over morphological principles at the Académie des sciences that commanded the attention of scientific and philosophical circles across Europe. The ageing Goethe followed the debate, and wrote two articles on it, which were reprinted in major French scientific journals (*Revue encyclopédique*,

Annales des sciences naturelles) as well as a popular miscellany, *Paris, ou Le livre des cent-et-un*, to which Victor Hugo was a contributor. Goethe applauded Geoffroy for having introduced "the synthetic manner of treating nature" from Germany into France, thus ensuring "the universal victory of a subject to which I have devoted my life."[20] Cuvier, meanwhile, conflated Geoffroy's hypotheses with the "metaphysical, pantheistic, and ultimately nonscientific tendencies" of Lamarckism and *Naturphilosophie*; the principle of unitary composition was "a doctrine substantiated only in the imagination of some naturalist, more poet than observer."[21] (In his *Discourse on the Revolutions of the Surface of the Globe*, Cuvier had charged Lorenz Oken and the other *Naturphilosophen* with expounding "a philosophy which substitutes metaphors for reasoning."[22]) The redrawing of the methodological battle-lines of the debate between Kant and Herder, forty-five years earlier (discussed in chapter 1), is striking. Once more, the senior authority charges his adversary with forsaking properly scientific procedure and resorting to literary tricks of analogy and conjecture.

Cuvier's scornful treatment of Geoffroy's arguments discredited him in the eyes of liberal observers, who saw him as merely reactionary, attacking new ideas without taking the trouble to study them. Prospering under Napoleon and the Bourbon Restoration alike, Cuvier had come to personify the patronage-based corporate science of the old regime—an identification that solidified after the accession of Charles X and his ultraright government in 1824. Cuvier dodged the July Revolution, visiting the elite scientific institutions of London during the street fighting in Paris, and resumed high office on his return. Geoffroy, in contrast, managed to cast himself as the standard-bearer of "the ideals of political independence and free research championed by broad segments of the liberal movement," which rallied to his cause.[23] The contending savants helped bring Lamarck further into public view: Cuvier's polemical conflation of Geoffroy's thought with Lamarck's, no less than Geoffroy's advocacy of Lamarck and his adoption of key Lamarckian ideas, ensured that transformism was seen to be at stake in their debate, even though it was not the explicit theme of contention.[24] The publicity encouraged Lamarck's publisher to reissue *Zoological Philosophy* in 1830, passing off unsold copies as a new edition. In the limelight of Revolution, Lamarckism was visible as never before.

Across the channel, meanwhile, the ideas of Geoffroy and Lamarck were arousing intense, controversial interest in British medical circles, beyond the elite institutions of scientific patronage: above all in Edinburgh, the main center for the reception of Lamarckian thought in Great

Britain until the new University of London opened its doors in the late 1820s. In the decade after Waterloo, Scottish medical graduates flocked to Paris, where they studied Geoffroy's philosophical anatomy firsthand, imbibed Lamarckian theory, and encountered other proponents of the new morphology—and brought it all back home, along with its republican and democratic politics. The ideological stakes of the Cuvier-Geoffroy debate were played out around the London University curriculum, configured on the Scottish model and staffed by radical Edinburgh-trained professors like Robert Edmond Grant, the most zealous and authoritative advocate of Geoffroyan and Lamarckian views in Great Britain in the heady era of Reform legislation.

Despite the politicization of the new morphology (compellingly documented by Adrian Desmond), British interest in Lamarck was not confined to dissenting and democratic circles, as both Corsi and James A. Secord have argued.[25] The first favorable commentary on Lamarck in a British scientific periodical appeared in an anonymous article in the *Edinburgh New Philosophical Journal* in 1826. For a while, this was attributed to Grant, who, coincidentally, was introducing Lamarck's work to the young Charles Darwin, then a student at Edinburgh, two years before Grant went south to occupy the new chair of Natural History at London.[26] Secord makes a persuasive case, however, that Robert Jameson, Regius Professor of Natural History at Edinburgh, was the author of the 1826 article, not Grant. Jameson belonged to the moderate Tory establishment (Scott knew him); his even-handed comments on Lamarck elsewhere, and his journal's attentiveness to new developments in France and Germany, indicate the presence of "a wider circle of Scottish naturalists interested in evolution," less polarized than the field was becoming in London.[27] In his preface to the English translation of Cuvier's *Theory of the Earth*, also published in 1826, Jameson sounds an unmistakably evolutionist note (contra Cuvier's opposition to transformism) in his praise of "Geology, which discloses to us the history of the first origin of organic beings, and traces their gradual development from the monade to man himself, . . . and which even instructs us in the earliest history of the human species."[28]

Secord suggests that the tolerance for transformism in respectable Edinburgh, more than its vogue in radical London, provoked Charles Lyell to include an extensive, detailed refutation of Lamarck in the second volume of *Principles of Geology*, composed on his father's Forfarshire estate and published in January 1832, shortly after Scott's *Count Robert of Paris*. Lyell had visited Paris in 1828–29 and again in 1830, where he met allies and former collaborators of Lamarck and witnessed revolutionary street skirmishes. Back in London, Desmond suggests, he was alarmed to find the same intellectual doctrines fueling the agitation for Reform, and devised his own "extreme strategy" to counter Lamarckian science, "a fundamentally nonprogressionist life history" reliant on a "nondirectional fossil sequence."[29] Lyell also proposed the influential solution of a detour around the history of man (that minefield of infidel speculation) altogether, in a history of the earth from which human origins were excluded—a strategy Darwin would follow in *On the Origin of Species*.[30]

Above all, Desmond argues, "Lyell's science was an attempt to save man's 'dignity.'" In Paris in 1830, "after watching Jacobin sansculottes rampaging through the streets outside his house," he commented sarcastically on "the immense ages that would be required for 'Orang-Outangs to become men on Lamarckian principles.'"[31] The critique of Lamarck in *Principles of Geology* culminates in a vehement dismissal of the "progressive scheme" whereby "the orang-outang . . . is made slowly to attain the attributes and dignity of man"[32]—acknowledging the creature's persistent, scandalous centrality to debates about the formation and limits of human nature. Against Buffon's assertion that the orang-outang was "nothing but a real brute, endowed with the external mark of humanity, but deprived of thought, and of every faculty which properly constitutes the human species," Rousseau had speculated (with inscrutable irony) whether the mysterious great ape reported by travelers might not be the specimen of natural man required by his thesis.[33] Scottish philosopher Lord Monboddo, in a literal-minded elaboration of Rousseau's provocation, proposed that orangutans are "a barbarous nation, which has not yet learned the use of speech," endowed with "the sentiments and affections peculiar to our species."[34] Rousseau's irony and Monboddo's credulity served to exemplify a heterodox tendency of Enlightenment thought, of which Lamarckian transformism, according to Cuvier and Lyell, was the extravagant effusion. But the anthropomorphous orangutan declined to fade back into the woods. In 1827, Geoffroy's ally Bory de Saint-Vincent revived the taxonomic argument for a genetic link between apes and humans, fortified with detailed anatomical analysis; and the 1830 reissue of *Zoological Philosophy* brought back into circulation Lamarck's conjectural metamorphosis of a tree-dwelling quadrumanous brute into the bipedal, flat-faced lord of creation—the proximate target of Lyell's 1832 rebuttal. Richard Owen, Cuvier's British successor and nemesis of the British Geoffroy, Grant, would make demolition of the "Orang-Outang Hypothesis" a key objective in his campaign against materialist transformism in the following decade.[35]

Lamarckism was flying high, then, on the winds of the 1830 Revolution, and the Orang-Outang Hypothesis was back, disclosing human nature as the philosophical stake at hazard in the era's political turmoil. We can begin to see now why Scott might have highlighted Paris, not Constantinople, in the title of a romance of the late Byzantine Empire into which an orangutan keeps irrupting.

Paris: eighteenth-century capital of "international literary space," the forum—in conservative British eyes—of a once progressive, now discredited Enlightenment, brought down by a Europe-wide resistance to its universalist hubris in the form of local national mobilizations, Pascale Casanova's "Herder effect."[36] After 1815, according to Jon Klancher, a new, post-Enlightenment conception of the cosmopolis or world-city—imperial, modern, the ganglion of deterritorializing networks of finance as well as of corrosive new demographic and ideological forces—had superseded the old-regime cosmopolitanism of the Republic of Letters.[37] The Constantinople of *Count Robert of Paris* is the fantastic prototype of the multinational, heterodox world-city, decadent capital of the Enlightenment human sciences, of which the Orang-Outang is the monstrous mascot. Its bad cosmopolitanism makes it the opposite of Corinne's Rome, where imperial politics, embedded in the past, have fossilized into the monumental scenery of the "aesthetic state."

Scott's Toryism had hardened in his later years, chafed by the growing Reform agitation; never patient with radicals, he detested Robert Knox, the other leading Edinburgh Lamarckian after Robert Grant, although that was largely because of Knox's involvement in the 1827 Burke and Hare body-snatching scandal.[38] Yet Edinburgh accommodated a moderate openness toward Lamarckian views in the mid-to-late 1820s, and *Count Robert of Paris* does not offer anything so straightforward as a conservative rejection or even reactionary parody of the transformist speculations that run riot in it, as little as it does a clear-cut endorsement of them. The novel draws imaginative energy, rather, from the categorical confusion and instability, or radical reopening of philosophical and scientific certainty, marked by the orangutan in Romantic natural history.[39] Where various characters mistake Sylvan for the devil, or a knight transformed by witchcraft, the narrator is elaborately noncommittal as to his status: "The creature . . . it would have been rash to have termed it a man" (170); "the tremendous creature, so like, yet so very unlike to the human form . . . the creature in question, whose appearance seemed to the Count of Paris so very problematical, was a specimen of that gigantic species of ape—if it is not indeed some animal more nearly allied to ourselves—to which, I believe, naturalists have given the name of Ourang Outang" (171). Sylvan wears clothes, wields tools, understands orders spoken in Anglo-Saxon, and prattles in his own strange language. Like the blind prisoner Ursel (discussed later in this chapter), the orangutan's solitary, captive state raises fundamental questions about the terms that constitute humankind. Can the sounds he utters be considered a language, if there is no one with whom he can speak it? The presence of an interlocutor was a necessary condition for the production of language—an intrinsically social and dialogic event—according to Adam Smith and other Scottish Enlightenment philosophers. A larger question impends: whether Sylvan embodies the workings of perfectibility in nature, as an animal on its way to becoming human, or the opposite, the grotesque blockage of any such possibility—as in Herder's scenario of the ape's melancholy intuition of a rational horizon beyond its reach: "*it would perfect itself.* But this it cannot ... the door of human reason is shut against the poor creature, leaving it perhaps an obscure perception, that it is so near, yet cannot enter."[40]

While Scott recoiled from the glare of Reform at the end of his life, the 1830 Revolution offered the twenty-eight-year-old Victor Hugo a fulcrum around which he could swerve from fanatical youthful royalism to a no less fervent liberalism. After breaking off *Notre-Dame de Paris*, he wrote a newspaper ode, "À la Jeune France," in which he praised the revolutionary students as heirs of Napoleon; an explanatory note by Saint-Beuve ("like a publicity manager at a press conference"), published along with the poem, sought "to pilot it through the still narrow straits of triumphant liberalism."[41] The clamorous reception of *Hernani* (1829–30) had already cast Hugo as the muse of revolutionary youth; his identification with the liberal cause seemed unambiguous over in Britain, where local nuances were invisible. Thus William Hazlitt *fils* could flaunt the author's republican tendencies in the preface to his English translation of *Notre-Dame*, provoking Tory reviewers to attack both the novel and its translator.[42] It is far from easy to extract a coherent set of political attitudes (let alone principles) from *Notre-Dame de Paris*. Hugo grants one character, the artisan Flemish envoy Jacques Coppenole, the prophetic gift of forecasting a revolutionary future. When Louis XI (a terrifying enhancement of the Machiavellian monarch of Scott's *Quentin Durward*) dismisses the truands' uprising as merely "a mutiny[,] which I shall put down whenever it pleases me to wear a frown," Copenolle replies, "In that case, it will mean that the people's hour has not yet come" (444). The scene takes place in the Bastille:

> Louis's face became sombre and thoughtful. He remained for a moment in silence, then patted the thick wall of the keep gently with his hand,

as one might stroke the cruppers of a war-horse. "Never!" he said. "You won't fall so easily, will you, my good Bastille?" (445)

But Hugo's narrative drives against this rather facile dramatic irony: not so much because, indeed, the people's time has not yet come (the insurgents are crushed by the royal guard), but because it is hard to imagine their representatives in the novel, the denizens of the Court of Miracles, as harbingers of any new social order—they are the feral rabble, eager to pillage and burn, of counterrevolutionary propaganda. Here, as elsewhere in *Notre-Dame de Paris*, the populace is as monstrous as the king—but many where he is one, hot where he is cold.[43]

Hugo's novel, like Scott's, quivers with the tremors of current scientific debate, in which human nature and organic form convulse, change shape, dissolve. Once again, the imaginative resources of metamorphosis, rather than a determinate ideological stance or scientific paradigm, energize the romance.[44] The foundling Quasimodo is a monkey *manqué*, a misstep on the Lamarckian progression of forms—as his name suggests, "incomplete and half-finished . . . hardly more than an 'approximation'" (163: "n'était guère qu'un *à peu près*," 203). He develops into something that resembles a botched orangutan, not so much ill grown as badly made—a hybrid of Scott's and Shelley's monstrous conceptions:

> A huge head sprouting red hair; between the two shoulders an enormous hump . . . a system of thighs and legs so strangely warped that they met only at the knees and looked, from the front, like two scytheblades joined at the handle; broad feet and monstrous hands; and somehow, along with such deformity, an appearance of formidable energy, agility and courage. . . . He was like a giant broken in pieces and badly reassembled. (70–71)

Swarming up the cathedral towers, swinging on bells, he recapitulates ancestral simian habits. Quasimodo's condition as "defective ape" conforms (as William Graham Clubb has argued) to the Geoffroyan diagnosis of monstrosity as an environmentally induced retardation of fetal development.[45] Another observer offers a different hypothesis. The wen over the child's eye, according to Maître Robert Mistricolle, is "an egg containing another devil, just the same, and that contains another little egg with another devil inside, and so on" (157). This diagnosis reproduces—and equates with medieval superstition—a recognizably preformationist account of organic development, in which the ancestral ovum or sperm (it was debated as to which) contains within it the miniature form of the grown creature as well as all subsequent generations. Echoes of the late Enlightenment controversy between preformationist and epigenetic theories reverberated into the 1820s: Geoffroy dismissed the "pre-existence of germs" as an "unintelligible" doctrine, founded on a "metaphysical explanation" rather than empirical evidence, in the peroration to *Philosophical Anatomy*.[46] Victorian histories of science would cite Geoffroy's "great law of 'arrested development'" as a decisive rebuttal of the preformation theory.[47]

As Quasimodo grows up, the epigenetic account prevails, blossoming into a bizarre idyll of the coadaptation of an organism with its environment:

> [He] looked, with his human face and his animals' limbs, like the native reptile of these damp and somber flagstones.... So it was that, little by little, developing always in harmony with the cathedral, living in it, sleeping in it, hardly ever leaving it, subject day in and day out to its mysterious pressure, he came to resemble it, to be incrusted on it, as it were, to form an integral part of it. His salients fitted into the building's re-entrants, if you will allow the metaphor, and he seemed to be not only its inhabitant but also its natural content. One might almost say that he had taken on its shape, just as the snail takes on the shape of its shell.... So deep was the instinctive sympathy between the old church and himself, so numerous the magnetic and material affinities, that he somehow adhered to it like the tortoise to its shell. The gnarled cathedral was his carapace. (163–64)[48]

Notre-Dame affords him, outcast from human moral and aesthetic norms, the consolation of symbiosis—with a hint of the development of a new kind of Gothic-cybernetic entity in the "magnetic and material" conjunction of man and building. Ecstatically riding the giant bells, Quasimodo becomes "a strange centaur, half man, half bell" (168): Hugo's rhapsodic rhetoric outstrips the figure of the orangutan, and organic form itself, for something weirder—as though language, rather than life, is the vehicle (deliriously clanging) of an uncontainable metamorphic drive.

Uncharacteristically, in the passage just quoted, Hugo draws attention to the figurality of his description by apologizing for it: "His salients fitted into the building's re-entrants, if you will allow the metaphor" (164) ("Ses angles saillants s'emboîtaient, qu'on nous passé cette figure, aux angles rentrants de l'édifice," 204). Once again, the obsolete biology of preformation is evoked, this time through the technical term "s'emboîter." While its primary sense here comes from carpentry, Hugo's apology points to its rhetorical usage in the old embryology, where "emboîtement" ("encasement") refers to the preformation of germs in the ancestral womb.[49] Here the term underscores a radically epigenetic scenario, at once Lamarckian and Geoffroyan, of the formation of Quasimodo—an arrested fetus discarded by his biological mother—within the archetypal matrix of Notre-Dame. Hugo's technical trope works as a philological jest, a provocation: it prompts the question, raised throughout the novel, of the fatality (ANÁΓKH) of which Quasimodo represents an extreme case. (Citing St. Augustine, Geoffroy rebukes preformation for being a fatalistic doctrine: "*Est quod futurus est, traduisible par, C'est déjà ce que cela deviendra plus tard.*"[50]) Deformation bears the burden of preformation, in that Quasimodo's destiny is seemingly fixed (like that of Frankenstein's monster) by his repulsive physical appearance. His assimilation to the sublime body of the cathedral, turning him into the building's genius, completes his exile from the human community.

The narrator goes on to describe Quasimodo as imprisoned inside the monstrous carapace of his biological form, unable to evolve spiritually:

> If now we were to try and penetrate to Quasimodo's soul through that hard, thick bark; if we could sound out the depths of that malformed constitution; if it were given to us to shine a torch behind those untransparent organs, to explore the murky interior of that opaque creature, to elucidate its dark corners and absurd cul-de-sacs, and to shine a bright light suddenly on the unhappy psyche chained in the depths of that lair, we should surely find it in the wretchedly stunted and rachitic posture of those prisoners in the leads of Venice, who grew old bent double in a stone box that was too low and too short for them. The spirit must atrophy in a misbegotten body. Quasimodo could just sense, stirring blindly within him, a soul made in his image. The impressions of objects underwent a considerable refraction before they reached his intelligence. His brain was a peculiar medium: the ideas that passed through it emerged all twisted. The reflections which resulted from this refraction were necessarily warped and divergent. (165–66)

Quasimodo's encounter with Esmeralda awakens a latent capacity for chivalry and sentiment, conditioned by his own awareness of his deformity, which acts as a moral shield between them, allowing him to address her as though they are beings of different orders: "When I compare myself to you, I feel very sorry for myself, poor unhappy monster that I am! To you I must seem like an animal. . . . I am something frightful, neither man nor animal, but something else, harder, more downtrodden, more misshapen than a stone!" (368). This fragile possibility—the opening of a path of *Bildung* through a pure, exalted love—is snuffed out in the novel's horrible closing scenes. From the height of the cathedral, Quasimodo weeps as he watches the only beings he has loved (Esmeralda and Frollo) die gruesome, agonizing deaths. In recapitulating the pathos of Herder's orangutan, aware of an ontological horizon he will never reach, Quasimodo achieves a more intense humanity, as a quality of feeling, than anyone else in *Notre-Dame de Paris*.

Retrograde Evolution

The Constantinople of *Count Robert of Paris* inaugurates a new setting for a new kind of historical romance: the world-city as conjectural arena for a post-Enlightenment world history, the natural history of man. Byzantium is doubly cut off from the path to modernity, first by the schism between the Greek and Roman churches, which made it the languid shadow of a more vigorous Western civilization, and then, decisively for early nineteenth-century British readers, by the Ottoman Conquest of 1453. The novel's opening paragraphs evoke the neoclassical schema of the *translatio imperii*, according to which Constantine's city on the Golden Horn would succeed Rome (and be succeeded in turn by Charlemagne's Paris) as the "seat of universal empire," only to reflect upon the decadence of artificially restored empires: Constantinople is like "a new graft . . . taken from an old tree," bound by an organic fatality to resume an internal chronology of decline (3–4). The world-city pullulates with heterogeneous creeds and races, while alien hosts—Frankish Crusaders, Moors, Turks, Scythians—besiege it from without. Contending monotheisms jostle alongside strange heresies, including Manichaeism (in a canceled episode, 367–77), and Pagan survivals (the "brutal worship of Apis and Cybele," decayed from a state religion into vulgar superstition, 89). More than once Scott compares the Byzantine court ceremony with the "court of Pekin" (7, 147), the Enlightenment antitype of a progressive nation, drawn into the geopolitical horizon of British interest through the embassies of Lord Macartney (1792–94) and Lord Amherst (1816).[51]

One effect of this imaginary distention of world space is a diffusion and fragmentation of historical time, which scatters any sense of a unitary direction along which history might unfold. *Count Robert of Paris* thwarts readers' expectations that the western Crusaders, for instance, might represent a romantic futurity, distinguished by individualist virtues of honor and courage, destined to supersede orientalized Byzantine decadence, according to the "clash of civilizations" and "nature versus art" schemata interpreted by some critics of the novel.[52] While informed readers would know about the sack of Constantinople in the Fourth Crusade, as well as the Ottoman conquest, the narrator refrains from harping on these future catastrophes—in contrast to Hugo, nudging us about the fate of the Bastille. The eponymous Count Robert and his warrior-bride Brenhilda, formidable types of Crusader valor on their first appearance, prove worse than useless in the novel's plot. Brenhilda's valor culminates in a grotesque Amazonian duel with the historian Anna Comnena, in which she endangers her own pregnancy.[53] Count Robert, despite a mid-story outburst of action-hero alacrity, accomplishes nothing toward the resolution of the various predicaments he crashes into. He does not even keep his promise to liberate the noble captive Ursel; that task is taken care of by the Emperor, in one of Scott's virtuoso essays in anticlimax. Byzantine policy trumps Crusader prowess, despite the Western narrator's professions of contempt for the former and admiration for the latter, as the Emperor Alexis successfully manages and diverts the Western intruders. The Emperor's disavowed virtuosity consists above all in rhetoric, a crafty linguistic excess, aligned with the supervening practice of the Author of *Waverley*. The narrator's mockery of the "fair historian" Anna Comnena, sealed with a stylistic parody of her *Alexiad*, frames the open secret of their affinity. "Scott moves to become what he beholds," notes Jerome McGann; since "his own style has over the years grown increasingly elaborate and formulaic[,] [t]o pastiche Comnena's prose . . . is to fashion a critical measure of his own"—ornate, murky, Byzantine indeed.[54]

Any sense of a governing historical progress or developmental direction is exploded into a multitude of competing paths. The world-city opens for the imagination alternative modes and directions to the progressive model of Enlightenment conjectural history, supposed to structure particular national destinies, as well as its premodern precursor, the *translatio imperii*. We learn that the Anglo-Saxon insurgents who sought refuge in the greenwood after the Norman Conquest—avatars of those mythic figures from *Ivanhoe*, Robin Hood and his band—"made a step backwards in civilisation, and became more like their remote ancestors of German descent, than they were to their more immediate and civilized predecessors." "Old superstitions had begun to revive among them," drained however "of the sincere belief which was entertained by their heathen ancestors" (209). At the same time, these "Foresters," natural Malthusians, regulate their population by chaste practices of "moderation and self-denial" (210). Those who regress from civilization to the woods, as a consequence of world-historical defeat, are able to sustain a virtuous organic community—whereas the original "man of the woods," Sylvan, is a deracinated, melancholy figure, one among the novel's gallery of captives and exiles.

Volume 2 of *Count Robert of Paris* closes with a flamboyant display of what the narrator calls "retrograde evolution" (255). Having solemnly sworn "never to turn back upon the sacred journey" to Jerusalem, the Crusaders cross the Bosphorus to the Asian shore. There, upon learning that Count Robert and his bride are in trouble back in Constantinople, they must figure out a way to help them without breaking their oath. "Are we such bad horsemen, or are our steeds so awkward, that we cannot rein them back from this to the landing-place at Scutari?" one suggests. "We can get them on shipboard in the same retrograde manner, and when we arrive in Europe, where our vow binds us no longer, the Count and Countess of Paris are rescued" (253). The inhabitants of Scutari are accordingly treated to the spectacle of a column of knights riding their horses backwards onto the transport barges.

The Crusaders' retrograde evolution enacts, with grotesquely literal insistence, a developmental dynamic or potential that pulls at the various histories, destinies, and identities entangled in *Count Robert of Paris*. History is not bound to move forward, whatever forward might mean; it might fall backward, slide sideways, or go off in some other, unheard-of direction. It is not only progress that is at stake—enmeshed by Lamarckian science with an evolutionary natural history. Scott's Constantinople frames a disintegration of the monogenetic figure of origins that has underpinned the national history addressed in the great sequence of Scottish Waverley Novels. Monogenesis designates the orthodox principle that all humans descend from a common (Adamite) ancestor and thus comprise a single race or species. Powerfully secularized in the Enlightenment theme of a universal human nature, the principle governs the ideology of unified development, a shared single history, which a particular people or nation, such as Scotland, must join on the path to modernity.

In those key reflections on his art, the introductory chapter to *Waverley* and the dedicatory epistle to *Ivanhoe*, Scott cites a universal human nature, constant across the differences of time and place, as the philosophical basis for the kind of novel he is writing:

> Considering the disadvantages inseparable from this part of my subject, I must be understood to have resolved to avoid them as much as possible, by throwing the force of my narrative upon the characters and passions of the actors;—those passions common to men in all stages of society, and which have alike agitated the human heart, whether it throbbed under

the steel corslet of the fifteenth century, the brocaded coat of the eighteenth, or the blue frock and white dimity waistcoat of the present day.[55]

Scott asserts the immanence of this universal principle in his novel: "It is from the great book of Nature, the same through a thousand editions, whether of black letter or wire-wove and hot-pressed, that I have venturously essayed to read a chapter to the public" (6). The historical novel typifies the "book of Nature" by virtue rather than in spite of its attention to the variability of local forms through which that nature is transmitted.[56] At the same time, Scott's renovation of an ancient metaphor reopens the difference it seeks to absorb. The book of nature is at once a virtual entity, a text, and a material one, an edition; it consists of a universal code and a historically contingent mode of production—paper-manufacture and printing, "black letter or wire-wove and hot-pressed."[57] Where can we draw the line between "manners" and "passions," between the historical and geographical variables of culture and a universal nature? What if the distinction itself is an effect of historical perspective? How far away or how far back do we go before it disappears?

We do not have to read much further in Scott's novels to find the conservative medium of national history and its corollary, a unified human nature, breaking apart. The figure of man becomes as elusive as it was in the conjectural anthropologies of Rousseau and Ferguson. Historical difference dilates into anthropological difference in Scott's revision of the romantic historicism of *Waverley* in *Rob Roy*. Perplexed by a mysterious Highlander who keeps appearing and disappearing in his way, the novel's protagonist, Frank Osbaldistone, demands to know the stranger's name and purpose. "I am a man," comes the reply: "He that is without name, without friends, without coin, without country, is still at least a man; and he that has all these is no more."[58] Rob Roy is the last fully human being, the complete yet always-escaping embodiment of what it means to be "a man," in the uncertain dawn of commercial modernity. Scott (or rather Frank) repeats an allusion made by Rousseau, in the *Discourse on the Origins of Inequality*, to an anecdote of the Cynic philosopher Diogenes, searching the city streets in vain to find "a man."[59] Yet Frank's descriptions exhibit Rob Roy to the reader as a weird, prehistoric, "half-goblin half-human" creature, with a shaggy red pelt, broad shoulders, and long sinewy arms:

> [His] shoulders were so broad in proportion to his height, as, notwithstanding the lean and lathy appearance of his frame, gave him something the air of being too square in respect to his stature; and his arms,

though round, sinewy, and strong, were so very long as to be rather a deformity.... This want of symmetry ... gave something wild, irregular, and, as it were, unearthly, to his appearance, and reminded me involuntarily of the tales which Mabel used to tell of the old Picts who ravaged Northumberland in ancient times, who, according to her tradition, were a sort of half-goblin half-human beings, distinguished, like this man, for courage, cunning, ferocity, the length of their arms, and the squareness of their shoulders. (273)

Rob Roy resembles the orangutan that Rousseau, Monboddo, and Lamarck had proposed as a type or ancestor of "natural man"—and at the same time he bears an aura of the supernatural, since the aboriginal Picts are "half-goblin half-human beings." (David MacRitchie, a late Victorian exponent of polygenetic racial theory, did not hesitate to cite Scott's novel for ethnological evidence of a prehistoric British population of subhuman aborigines.[60]) In an uncanny inversion of Rousseau's argument, man appears *less natural* the closer he is to his original nature—a distortion, presumably, of civilized man's alienated perspective. Meanwhile, unlike the Mac-Ivors in *Waverley*, Rob Roy and his clan do not die out at the end of the adventure. It is as though the wild Highlander is still out there somewhere, paying the occasional visit to civil society and escaping with a growl of contempt, like Adam Ferguson's savage.

Scott's novels keep reopening the join between temporal scales that provided the historical novel with its scientific premise. The historical and geographical bounds of human nature are stretched more drastically in *Ivanhoe*, set six hundred years since, at a time when England—crucially— is not yet a nation. In the "Dedicatory Epistle," the author reaffirms his anthropological principle:

> The passions, the sources from which [sentiments and manners] must spring in all their modifications, are generally the same in all ranks and conditions, all countries and ages; and it follows, as a matter of course, that the opinions, habits of thinking, and actions, however influenced by the peculiar state of society, must still, upon the whole, bear a strong resemblance to each other. Our ancestors were not more distinct from us, surely, than Jews are from Christians.[61]

The story that follows turns this last proposition into a profoundly vexed question: how distinct are Jews from Christians? The equivalent differences Scott invokes—between our ancestors and ourselves, between Jews and Christians—are resolved in the universal history authorized by scripture, according to which Christianity digests its Judaic heritage. This scriptural version of universal history, with its anthropological subtext, underwrites the story of national *Bildung* prefigured in *Ivanhoe*, in which conquering Normans and colonized Saxons are assimilated into a future English homeland.

That is not quite what happens in the novel. The Jews, European history's perpetual, tragic strangers, disrupt the official story of national foundation based on a reconciliation of alien peoples into a common humanity.[62] Isaac and Rebecca, in the teeth of Christian anti-Semitism, invoke "the great Father who made both Jew and Gentile"—an ecumenical appeal that implicitly reverses the Christian assimilation of Judaism. While Isaac is routinely degraded to subhuman status by the other characters (who spurn him with such epithets as "dog of a Jew"), Rebecca, the most admirable figure in *Ivanhoe*, embodies Christian and chivalric virtues of honor, fidelity, charity, and courage—the official virtues of the English nation-in-waiting—more convincingly than any of the novel's knights and clerics. A rhapsodic heroine à la Corinne, she pleads ardently for what modern readers are trained to recognize as humane values.[63] She is, in short, more human than "our ancestors."

Historical necessity (what happens to have happened) excludes Rebecca from the protonational community invoked in the novel's closing chapters. She and her father prepare for an exile that will transport them not just in space but in time: three centuries into the future, to the court of "Mohammed Boabdil, King of Grenada" (499). Abu Abdullah Muhammad XII ("Boabdil"), the last Nasrid Sultan of Granada, surrendered his kingdom to the Catholic Monarchs Ferdinand and Isabella in 1492; the *Reconquista* issued in the final expulsion of the Jews from Spain, along with their Muslim hosts. The anachronism reminds us that further cycles of dispossession await Rebecca and her people—a fate Scott's readers should have recognized as reaching into their own historical present.[64] Through Rebecca we glimpse a categorically different history from the official English destiny of compromise, reconciliation, and settlement that Scott's novel is usually taken to be promoting. Scott's heroine gestures toward a radically unfinished, unsettled history of worldwide dispossession and wandering, which is at the same time one of utopian aspiration and humanist hope—the byproducts, it seems, of the condition of dispossession and wandering. Like Staël's Corinne, Rebecca personifies the ideal of an evolved universal humanity—a Herderian *Bildung der Humanität*—that cannot be accommodated within a merely national history, since the bonds of custom and sympathy that secure the imagined community are also, as her case reveals, bonds of prejudice and xenophobia. The domestic union that closes the story—Ivanhoe's marriage to Rowena, portending the English nation to come—seems smaller in every sense, lacking spiritual grandeur, than the sublime horizon that widens around Rebecca.

The enlargement of historical distance in *Ivanhoe* and the novel's disintegration of England into a welter of alien castes anticipate the more radical experiment of *Count Robert of Paris*, which opens up a polygenetic potentiality of evolutionary trajectories and forms, enabled by its setting of a world-city severed from Western religious orthodoxy and the developmental path to modernity. Although the terms "monogenesis" and "polygenesis" would not be formulated until the 1860s, arguments over the unity or multiplicity of human nature were brewing among philosophical anthropologists in the latter decades of the eighteenth century, heated by the scientific encounter with hitherto unknown populations in the South Pacific and the intensifying debate over slavery. Kant disputed the claims of George Forster, who had accompanied the second Cook expedition, that humanity consisted of several distinct species.[65] Johann Friedrich Blumenbach, author of the *Bildungstrieb*, inaugurated the dubious nineteenth-century discipline of anthropometry by collecting, measuring, and grading human skulls, arguing for the existence of four and then five distinct races of man; at the same time, he affirmed the fundamental unity of the species—the zoological premise for which, he insisted (after Buffon), was the "profound interval, without connexion, without passage," that separated apes from human beings. Blumenbach's English disciple William Lawrence enlisted the principle in a liberal defense of the unity of mankind against polygenetic justifications for slavery.[66] Meanwhile, pro-slavery apologist Edward Long had marshaled early evolutionist speculation to argue "that [negroes] are a different species of the same *genus*" as Europeans and "that the oran-outang and some races of black men are very nearly allied" as early as 1774 in his *History of Jamaica*.[67] That same year, Scottish philosopher Lord Kames gave scientific credence to the heterodox doctrine of separate human species in his *Sketches of the History of Man*. Addressing the question "whether there are different races of men, or whether all men are of one race without any difference but what proceeds from climate or other external cause," Kames came to the conclusion that "there are different species of men as well as of dogs."[68] Species differentiation occurred, along with linguistic differentiation, as "an immediate change of bodily constitution" after the fall of the Tower of Babel.[69] Polygenetic speculation was becoming increasingly current by the late 1820s, undermining what had been a monogenetic orthodoxy, until (inflamed by anti-abolitionist polemic in the United States and British reactions to colonial uprisings in India and Jamaica) it hardened into a racial science in the mid-nineteenth century. Robert Knox, the Edinburgh Lamarckian disgraced by the Burke and Hare scandal, would publish a polemical polygenetic thesis, *The Races of Men: A Fragment*, forecasting wars of racial extermination accompanying the global spread of European colonialism, in 1850.[70]

Conjectural tentacles of "the polygenetic imagination" extend across *Count Robert of Paris*.[71] By the logic of retrograde evolution, cultural difference seems always to be on the point of falling into racial difference, racial difference into species difference—since, as Lamarck had argued (after Buffon and against Cuvier), species is a nominal, not a real, entity, and variable over time: all nature knows is individuals.[72] Race is a problem everywhere in the novel, its status and boundaries the objects of constant interrogation. Count Robert, Hereward, and Anna Comnena argue about whether or not the Normans and Franks are the same people (142–43). Alexius lectures Brenhilda on the clash of civilizations memorialized in the *Iliad*: "the offences of Paris were those of a dissolute Asiatic; the courage which avenged them was that of the Greek Empire" (195)—a distinction belied by the dissolute and Asiatic character of his own Greek Empire.

Most striking is the novel's set of limit cases of human racial difference. The first of these is instantiated (predictably) by Africans, reduced to slavery and subject to a not-always-specified physical deformation. (Mutes and eunuchs appear to share the same kind of disability in *Count Robert of Paris*.) The philosopher Agelastes hypocritically holds forth on the separate status of "the race of Ham" as justification for their enslavement (128). His slave Diogenes rebukes Hereward's "childish" suspicion that he might be the devil with an ironical reflection on the marks of racial difference:

> "Thou objectest sorely to my complexion," said the negro; "how knowest thou that it is, in fact, a thing to be counted and acted upon as a matter of reality? Thy eyes daily apprise thee, that the colour of the sky nightly changes from bright to black, yet thou knowest that this is by no means owing to any habitual colour of the heavens themselves. . . . How canst thou tell, but what the difference of my colour from thine own may be owing to some circumstance of a similar nature—not real of itself, but only creating an apparent reality?" (90)

Well might he ask, in a city where the confessional difference between Eastern and Western Christianity has acquired a biological, embodied cast, so that heterodoxy is expressed as a physical deformity. Whereas the

Greek cross decently exhibits "limbs of the same length," an "irregular and most damnable error prolongs the nether limb of that most holy emblem" in the Latin cross, the Orthodox Patriarch complains (84).

Disproportion of limbs characterizes the most outlandish of the novel's human limit cases. In Constantinople, "the race of the Greeks was no longer to be seen, even in its native country, unmixed, or in absolute purity; on the contrary, there were features which argued a different descent" (125). The theme takes a nightmarish turn:

> A party of heathen Scythians, presented the deformed features of the daemons whom they were said to worship—that is, having flat noses with expanded nostrils, which seemed to admit the sight to their very brain; faces which extended rather in breadth than length, with strange unintellectual eyes placed in the extremity; figures short and dwarfish, yet garnished with legs and arms of astonishing sinewy strength, disproportioned to their body. (125)

Scott alludes to a legend, mentioned in Gibbon's *Decline and Fall*, according to which "the witches of Scythia ... had copulated in the desert with infernal spirits, and the Huns were the offspring of this execrable conjunction" (525).[73] The conjecture of demonic descent infects the novel's anthropology. The combination of "strange unintellectual eyes" with nostrils that admit "sight to [the] very brain" suggests, meanwhile, a weird short-circuiting of a human physiology of vision-based cognition—a recurrent topic in *Count Robert of Paris*.

These mutations and metamorphoses condense around the most outrageous of the novel's figures of category implosion, the Orang-Outang, a being also endowed with limbs "much larger than humanity" (170). Sylvan poses a more up-to-date challenge to the monogenetic basis of philosophical history than fables of diabolical miscegenation. Along with his other human traits, he expresses the same passions—sorrow, lust, fidelity, resentment, fear—as the people in the novel. The threat to the idea of a universal human nature, then, is not so much that it might vary across space and time but that it *might not*, or not enough, if its core qualities are held in common by other creatures. Forty years later, in *The Expression of the Emotions in Man and Animals* (1872), Charles Darwin would argue that the passions are indeed constant for men in all stages of society—and constant for nonhuman beings too. (David Hume had intimated as much, in book 2 of his *Treatise of Human Nature*.) The passions express not human nature but something larger and more fundamental, the life force itself, the developmental drive that endlessly modifies all natural forms.

It turns out that the appearance of the Scythians foreshadows uncanny crossings between the human and demonic after all. In the course of a diatribe against Christian "superstition," Agelastes is startled by the intrusion of a figure resembling "the Satan of Christian mythology, or a satyr of the heathen age" (270), which turns out to be Sylvan. Monboddo had accepted satyrs, along with mermaids, men with tails, dog-headed men, and men with eyes in their breasts, as legitimate varieties of humankind alongside the orangutan, although Scott is probably alluding to a more modest conjecture that ancient accounts of satyrs might have been based on travelers' tales of great apes.[74] Sylvan's role as satyr will be confirmed at his last appearance in the novel, as closing act to the tragicomical sequence of combats. More consequential is his role here (not for the first time) as the devil. Agelastes has just been scoffing at "the Christian Satan," whose "goatish figure and limbs, with grotesque features," in violation of the scriptural principle of monogenesis, represent a bad theology as well as bad biology and bad aesthetics (270). Sylvan proceeds to throttle Agelastes, an event interpreted by the onlookers as "the judgment of Heaven" (271). Certainly this is a startlingly abrupt explosion of poetic justice. Agelastes is the worst of the book's villains, a seditious freethinker, just the sort of rogue who would be promoting transformist, polygenetic, and other materialist speculations in Scott's day. Yet Sylvan, conjured up by his victim's denunciation of the Christian Satan, realizes a euhemeristic demystification of that improbable demon, in keeping with the demystification of the heathen satyr: he is himself the grotesque hybrid Agelastes ridicules. Scott reasserts the heterodoxy that is ostensibly being punished in the form of a complicated joke. The orangutan might not be the devil— but the devil may be nothing more than an orangutan.

Reading in the Dark

Whenever he shows up in *Count Robert of Paris*, Sylvan provokes strange textual disturbances. His incursions set off figural mutations and slippages which trace a lateral or retrograde motion, an associative logic not of science, of an empirically verifiable order of cause and effect, but of something like dreamwork, of metaphor and metamorphosis: a logic of romance. His second irruption into the tale leads to the recasting of Hereward, a Varangian soldier in the imperial guard, as a "Forester," a virtuous version of the "man of the woods." Enter a beautiful young lady, pursued by the Orang-Outang. After chasing him off, Hereward recognizes his long-lost love, Bertha, who recognizes Hereward in turn by the scar of a boar's tusk on his brow. We find ourselves rapt into a suddenly intensified zone or atmosphere of romance, charged with allusions not just to the *Odyssey* but (as Clare Simmons points out) to *The Faerie Queene*, in which the chivalrous forester Sir Satyrane defends virgins from giants and wild men. "Have I but dreamed of that monstrous ogre?" Bertha asks. Hereward's reassurance, "That hideous thing exists," gives the cue for the recognition scene and—true to the Odyssean analogue—a retrograde narrative plunge to his youthful encounter with another "hideous animal" or "monster," the wild boar, decades earlier (208). The recognition scene generates, in turn, the back-story of the Foresters, Saxon insurgents who "made a step backwards in civilization" and returned to the woods.

The regressive, oneiric, metamorphic logic of romance is sustained most powerfully in the weird sequence, midway through the novel, in which Sylvan makes his first appearance. On this occasion, the orangutan closes rather than initiates the sequence of figural mutations. Invited to dinner at the imperial palace, Count Robert is startled by the ramping of the Emperor's mechanical lion, and smashes its skull with a blow of his fist; clockwork cogs and springs litter the floor. After the banquet, Robert wakes up (he has been drugged) to find himself in an underground dungeon, menaced by a tiger—a real tiger, this time, not a mechanical one. Once again he reacts by smashing its skull. Scott's description renders the tiger, despite its reality, more like an effect in a magic-lantern show— "two balls of red light"—than a flesh-and-blood animal: "he gazed eagerly around, but could discern nothing, except two balls of red light which shone from the darkness with a self-emitted brilliancy, like the eyes of a wild animal while it glares upon its prey" (161). This apparition then generates the voice of a fellow prisoner, invisible in the darkness, who informs Robert that *his* eyeballs have been put out with red-hot irons. The mutilation of eyes or tongues (or, less explicitly, genitals), in a metonymic chain of disfigurements recurrent across the novel, marks the victim's removal from fully human status. Deprived of the organs of speech or vision, he is reduced to a mere body, to "bare life," the condition of a beast or slave.[75] (The prisoner's name, Ursel, also recalls a bear, the anthropomorphic beast of the woods that historically precedes the theriomorphic man of the woods, the Enlightenment orangutan.) So at last the grotesque quasi-human captive Sylvan makes his appearance, murmuring strange sounds that—in the absence of his tribe—may or may not constitute a language.

What logic moves this strange narrative sequence, with its delirious transitions and transformations, its cryptic sequence of antitheses? Mechanical animal versus living animal; animal eyes without a body versus a human body without eyes; the man bereft of the endowment of humanity (vision) versus the animal that possesses it (language). These symbolic oppositions recapitulate the historical set of conjectures about the essential distinction between humans and animals that informed the Enlightenment project of the science of man: from the Cartesian account of the animal as machine, through the empiricist abstraction of a vision-based cognition, to the Romantic investment in language as uniquely human faculty.[76] Implicit in the Orang-Outang himself, as a scandalous crux, is the property or quality that had come to subsume all these—progress or perfectibility: which Scott's narrative, declining to settle the issue of the creature's status, leaves in suspense.

The set of conjectures resurfaces, recombined, near the end of the novel. The captive Ursel turns out not to have been blinded after all. Released from prison, high on a terrace overlooking the city, he experiences a return of vision that is scarcely less traumatic than an actual blinding:

> His eyeballs had been long untrained by that daily exercise, which teaches us the habit of correcting the scenes as they appear to our sight, by our actual knowledge of the truth as it exists in nature. His idea of distance was so confused, that it seemed as if all the spires, turrets and minarets which he beheld, were crowded forward upon his eyeballs, and almost touching them. With a shriek of horror, Ursel turned himself to the further side. (295)

This instance of retrograde evolution—from blindness to a sight that bears the violent effect of blindness—recalls the Platonic allegory of enlightenment as emergence from a cave. Scott's formulation however is radically empiricist rather than Platonic. More particularly, the episode recalls a long-running Enlightenment debate as to whether sensory cognition is learned or innate, focused on the case of a blind man cured of cataracts by William Cheselden in 1728, and recently resumed in the Edinburgh press around the treatment of the deaf-blind James Mitchell (b. 1795).[77] The most famous intervention in the debate, Denis Diderot's *Letter on the Blind for the Use of Those Who See* (1749), endorsed the empiricist argument of John Locke and William Molyneux:

> We see nothing the first time we use our eyes; and during the first moments of sight we only receive a mass of confused sensations, which are only disentangled after a time and by a process of reflection. It is by experience alone that we learn to compare our sensations with what

occasions them; that sensations having no essential resemblance with their objects, it is from experience that we are to inform ourselves concerning analogies which seem to be merely positive.[78]

Diderot's treatise includes an early articulation of the evolutionist (Lucretian) hypothesis, subsequently developed in *D'Alembert's Dream* (1769), that would be given scientific substance by Lamarck: "if we went back to the origin of things and scenes and perceived matter in motion and the evolution from chaos, we should meet with a number of shapeless creatures, instead of a few creatures highly organized."[79] Ursel's accession of vision allusively unlocks (in short) the larger tradition of heterodox natural philosophy that culminates in Lamarckian transformism.

Shortly afterwards, the narrator recurs to this episode, relating the reaction of the conspirator Nicephorus Briennius to the upsetting of his plans:

> The pardoned Caesar . . . found it as difficult to reconcile himself to the reality of his situation as Ursel to the face of nature, after having been long deprived of enjoying it; so much do the dizziness and confusion of ideas, occasioned by moral and physical causes of surprise and terror, resemble each other in their effects on the understanding. (322)

Nicephorus suffers what is more explicitly a narrative vertigo, as he discovers himself to be a character in a plot when he had imagined himself to be the author of it. His "dizziness and confusion" mirror the dizziness and confusion that perturb readers of *Count Robert of Paris*. Only, as readers of a novel, we are apt to experience that perturbation—the regression from a conditioned blindness to a vision that is itself blinding—as a mediated effect of aesthetic enjoyment rather than as sensory distress.

Scott's romance does not attempt, any more than Victor Hugo's, to pacify the turmoil afflicting the science of man circa 1830 into fixed knowledge. We are not to look in these pages for a determination, even an allegorical one, of the taxonomic status of man as a race or species. "In our culture," writes Agamben, "man has always been thought of as the articulation and conjunction of a body and soul, of a living thing and a *logos*, of a natural (or animal) element and a supernatural or social or divine element. We must learn instead to think of man as what results from the incongruity of these two elements, and investigate not the metaphysical mystery of conjunction, but rather the practical and political mystery of separation."[80] Scott's novel, through the techniques peculiar to fiction, begins that work of disarticulation and opening. The underground apparition of "two balls of red light," at once juxtaposed with and detached from the prisoner who believes he has been blinded, constitutes the reflective, metatextual core of *Count Robert of Paris*—the figure for vision, metonymically dislodged, as a mark of the reader's gaze. Here, disoriented from a daylight logic of cause and effect, flung deepest into a delirium of romantic adventure and weird science, we catch a glimpse of ourselves: neither as human countenances, nor organic bodies, but as dislocated perceptual fragments embedded in a meaning-generating apparatus—a work of literature. It is the reflection of our own vision, bloodshot with passionate amazement and with the sheer effort to see, reading in the dark.

Le grotesque au revers du sublime

Count Robert of Paris articulates a categorical crisis that is as much aesthetic as it is scientific—one that bears on the representation of states of feeling as well as of forms of knowledge, and in either case on the integrity of form as such. Ursel's blinding access of vision, echoed in Nicephorus's narrative vertigo, dramatizes the empiricist rerouting of the aesthetic from ideal form (as in the proportional ratios of Vitruvian neoclassicism) to physiological sensation, developed as a scientific theme by the Scottish philosophical physician William Cullen and applied to the perception of natural and artificial forms by Edmund Burke in *A Philosophical Enquiry into the Origin of Our Ideas of the Sublime and Beautiful* (1757).[81] Along with the allusion to Diderot's essay on the blind, Scott's scene offers a gloss on Burke's account of the sublime. As the urban panorama, with its heterogeneous architecture of "spires, turrets and minarets," is "crowded forward upon his eyeballs . . . almost touching them," Ursel suffers an implosion of the aesthetic distance, encoded in the structure of vision, that should sustain the experience of the sublime: bringing "horror" instead of "delight" (according to Burke, "the sensation which accompanies the removal of pain or danger").[82] The novel's reiteration of scenarios of sensory and physical deprivation or mutilation combined with a painful crowding of sensation—a pressure of mixed forms that eradicates the distance across which one sorts and orders them—enacts a collapse of the sublime and, in turn, of aesthetic distinction itself.[83] A new aesthetic mode, one that accommodates the dynamic mixture and metamorphosis of form, here associated with the urban phenomenology of the crowd, emerges: not the sublime but the grotesque. The grotesque, in Friedrich Schlegel's definition, "plays with the wonderful permutations of form and matter, loves the illusion of the random and the strange and, as it were, coquettes with infinite arbitrariness."[84] Its representative in Scott's novel is (once again) the Orang-Outang. Endowed with the "goatish figure and limbs, with grotesque features" of the "Christian Satan" and the "grotesque form" of an animal "so like, yet so very unlike to the human form," Sylvan embodies the aesthetic regime that governs the new natural history of which he is the type.

In the preface to his play *Cromwell* (1827), received as a manifesto for French Romanticism, Victor Hugo designates the grotesque as the mode of representation appropriate to modern culture. Hugo sketches a conjectural history of poetic expression across three epochs or stages: the primitive, typified by Genesis, the ancient, typified by Homeric epic, and the modern, typified by Shakespearean drama. Christianity, the ideological regime of modernity, "leads poetry to the truth," meaning scientific as well as religious truth:

> Like it, the modern muse will see things in a higher and broader light. It will realize that everything in creation is not humanly beautiful, that the ugly exists beside the beautiful, the unshapely beside the graceful, the grotesque on the reverse of the sublime, evil with good, darkness with light.... It will set about doing as nature does, mingling in its creations—but without confounding them—darkness and light, the grotesque and the sublime; in other words, the body and the soul, the beast and the intellect; for the starting-point of religion is always the starting-point of poetry. All things are connected.[85]

The type of this truth-to-nature is man himself, whom Christianity has designated the quintessentially mixed being:

> First of all, as a fundamental truth, it teaches man that he has two lives to live, one ephemeral, the other immortal; one on earth, the other in heaven. It shows him that he, like his destiny, is twofold: that there is in him an animal and an intellect, a body and a soul; in a word, that he is the point of intersection, the common link of the two chains of beings which embrace all creation. (8)

Hugo resuscitates (again) Buffon's *Homo duplex*, brandishing the figure's theological stemma—but to insist that man's hybrid nature makes him exemplary of the system of nature rather than an anomaly within it. Man is "the point of intersection, the common link" with a nature that is itself double, compounded from contrary principles, and in which "all things are connected." Nature is a network: man the hub. The "new form of art" that expresses this truth to nature, distinguishing "modern from ancient art, the present form from the defunct form, . . . *romantic* literature from *classical* literature" (12), is the grotesque. Hugo at once opposes the grotesque to the canonical modes of the sublime and the beautiful, and—since the mixture of forms includes all forms—subsumes them within it:

> The beautiful has but one type, the ugly has a thousand. The fact is that the beautiful, humanly speaking, is merely form considered in its simplest aspect, in its most perfect symmetry, in its most entire harmony with our make-up. Thus the ensemble that it offers us is always complete, but restricted like ourselves. What we call the ugly, on the contrary, is a detail of a great whole which eludes us, and which is in harmony, not with man but with all creation. That is why it constantly presents itself to us in new but incomplete aspects. (16)

The grotesque governs an emergent vision of the mixture and diversity of nature that exceeds the bias of human aesthetic preference; "what we call the ugly" harmonizes with an order—a natural history rather than a taxonomy—in which all form is deformation, in which variation is the principle of reality and the ideal type an illusion, the hypostasis of a local quirk of taste.

In the opening scene of *Notre-Dame de Paris*, Quasimodo wins the face-pulling contest that inaugurates the popular saturnalia, the Feast of Fools, thereby establishing (or so it seems) his singular, exceptional ugliness. The course of the narrative, discovering not only the abominable inhabitants of the Court of Miracles but the moral and spiritual deformities of the novel's outwardly fair or respectable characters (Frollo, Phoebus, Jehan, the king), establishes "what we call the ugly" as a characterological norm: indeed, Quasimodo's latent sensibility, his capacity for chivalry and pathos, sets him above all others when it comes to his treatment of Esmeralda. Esmeralda herself exemplifies the subjection of the beautiful to the code of the grotesque. "The beautiful strikes us as much by its novelty as the deformed itself," as Burke puts it (93): her beauty is no less freakish, no less an index of an estrangement from ordinary human life, than Quasimodo's ugliness. Far from inculcating disinterested admiration of a manifestation of ideal form, the one provokes an automatic physiological response (erotic desire: this is what Hugo means, presumably, by beauty's "most entire harmony with our make-up") as reliably as does the other (phobic revulsion). And Esmeralda too conforms to the novel's pervasive condition of formal hybridity: as Quasimodo becomes "a strange centaur, half man, half bell" (168), she is "a regular hornet," "a sort of bee-woman" (263).[86] She too belongs to the novel's menagerie of monsters.

In contemporaneously establishing a "teratology," a science of monsters, Geoffroy Saint-Hilaire and his son Isidore decisively relocated monstrosity within the order of nature. "Monstrousness is no longer a random disorder, but another order, equally regular and equally subject to laws," declared Isidore: "it is the mixture of an old and a new order, the simultaneous presence of two states that ordinarily succeed one another."[87] It is, in aesthetic terms, grotesque.[88] Both Hugo's tale and Scott's, placing anthropomorphic or theriomorphic monsters at the center of their labyrinths, give this hypothesis a dialectical shove: the presence of monsters in nature reveals nature itself to be monstrous—a dynamic and evolutionary succession of states, rather than a fixed scale, in which earlier forms are continually changing and merging into later ones. "By revealing the precariousness of the stability to which life has habituated us . . . the monster bestows upon the repetition of species, upon morphological regularity, and upon successful structuration a value all the more eminent in that we can now grasp their contingency," writes Georges Canguilhem: "Monstrosity is the accidental and conditional threat of incompleteness or distortion in the formation of the form; it is the limitation from within, the negation of the living by the nonviable."[89] Like rocks lobbed into a pond, Sylvan and Quasimodo generate concentric ripples of monstrosity across the character system of each novel's respective "Paris."

The dynamic of monstrous metamorphosis surges most powerfully through the urban populace in *Notre-Dame de Paris*, especially when viewed in its nocturnal, underworld aspect. The city's coiled depths enclose a nightmare of species drift.[90] Reprising the role of the unwitting Scott hero who conducts the reader into a new, symbolically freighted anthropological space (as Franco Moretti has noted),[91] Pierre Gringoire blunders into the Court of Miracles:

> The hands and heads of the crowd stood out black against the background of light, gesturing weirdly. Now and then a dog that looked like a man or a man that looked like a dog would go past along the ground, where the flickering light from the fires merged with the great indeterminate shadows. In this city, the boundaries between race and species seemed to have been abolished, as in a pandemonium. Amongst this population, men, women, animals, sex, age, health, sickness, all seemed communal; everything fitted together, was merged, mingled and superimposed; everyone was part of everything. . . . It was like some new world, unknown, unprecedented, shapeless, reptilian, teeming, fantastic. (100–101)[92]

Through Gringoire's hallucinated gaze we glimpse not just an anthropological new world but—"inconnu, inouï, difforme, reptile, fourmillant, fantastique"—a new order of nature. This vision radicalizes for nineteenth-century literature the Romantic intuition that the city, the paradigmatically total, constructed human environment, is the arena where humanity disintegrates, its singularity overridden by a general, irresistible mutational drive, a total economy of change that invests all organic and inorganic forms: *fourmillante cité*. At first the reader hesitates between opposing views of this grotesque mass life form, even as it appears monstrous in the eyes of the stranger. Will it unfold a fuller realization of the human, in a collective mobilization of sensation, affect, intelligence, agency—or its decomposition, into something base and abhorrent? If the latter potential quickly asserts itself, with the increasingly hellish turn of Gringoire's adventure, the former is latent, perhaps, in the Truands' insurgency toward the end of the novel, when they erupt from their dens to rescue Esmeralda from Notre-Dame.[93]

However, as Jacques Coppenolle remarks, the people's hour has not yet come: that is to say, this lumpen horde of beggars and thieves is not yet a *people*, regimented by a revolutionary consciousness. As the mob improvises siege-engines to storm the cathedral, it morphs into "a monstrous millipede" (410), "a steel-scaled serpent" (418).[94] The narrator dramatizes the assault as (Godzilla-style) a combat between monsters:

> There was no resisting this rising tide of terrifying faces. Their savage features glowed red with fury; their grimy foreheads streamed with sweat; their eyes flashed fire. All these grimaces, all this ugliness was besieging Quasimodo. It was as if some other church had despatched its gorgons, its mastiffs, its drees, its demons, its most fantastic sculptures, to the assault of Notre-Dame. They were like a layer of living monsters on top of the stone monsters of the façade. (420)[95]

They are not, however, equivalent monsters. The siege unleashes the sheerly destructive potential of the crowd, as it devotes its immense collective force to the bliss of looting, smashing, and burning. Notre-Dame—quickened by its resident genius, Quasimodo—is something else altogether. The novel has consistently referred to the cathedral as the greatest of the *kaiju* of Paris: "a sort of chimera among the old churches of Paris," with "the head of one, the limbs of another, the cruppers of a third; something of all of them" (128); "an enormous two-headed sphinx sitting in the midst of the town" (187).[96] In Claude Frollo's delirium, it turns "animate and alive" and assumes the form of a "prodigious elephant, breathing and walking, with pillars for its feet, the two towers for trunks and the immense black hangings for a housing" (361).[97]

The Great Book of Mankind

Delirium is the intensification of a subjective mental state, locked into the feverish agitation of the body; it is within the individual perspective, confined in a local place and historical moment, that the cathedral appears monstrous. Recall, however, Hugo's essay on the grotesque: "What we call the ugly... is a detail of a great whole which eludes us, and which is in harmony, not with man but with all creation." The grotesque is manifest, in other words, in two aspects or modalities: as "what we call the ugly," viewed from an individual, time-and-place-bound human perspective; and as what we call the sublime, viewed from the transhistorical, transhuman perspective of "all creation," the dynamic order of nature extensive throughout time and space.

The reader is granted access to this sublime-grotesque perspective in the series of philosophical meditations on architecture that punctuates the narrative of *Notre-Dame de Paris*. In them, Hugo views the cathedral as the authentic realization of human nature, a collective expression of the creative energy of the race or species, by virtue of its enlargement beyond personal scales of space and time. Notre-Dame becomes a more faithful emblem of humanity than the human body in this novel, where the hideous hunchback is a nobler being than the handsome, preening, vacuous Captain of the Guards, where Esmeralda's beauty subjects her to relentless sexual persecution, and where the body as such—in Gringoire's adventure in the Court of Miracles—becomes an unstable, warping apparition. In contrast:

> Architecture's greatest products are less individual than social creations; the offspring of nations in labour rather than the outpouring of men of genius; the deposit left behind by a nation; the accumulation of the centuries; the residue from the successive evaporations of human society; in short, a kind of formation. Each wave of time lays down its alluvium, each race deposits its own stratum on the monument, each individual contributes his stone. Thus do the beavers, and the bees; and thus does man.... The man, the individual and the artist are erased from these great piles, which bear no author's name; they are the summary and summation of human intelligence. Time is the architect, the nation the builder. (128)[98]

The cathedral materializes human creative labor as a slow aggregation, a sedimentary accumulation, of local and infinitesimal concrete acts.

Humanity comes into itself—its species being is manifest—in gradual secretions of collective instinct, like the beehive or beavers' dam, that are more like geological processes than acts of purposeful expression. Enlightenment conjectural anthropology had applied an architectural metaphor to its account of human natural history. "In the human kind, the species has a progress as well as the individual," wrote Ferguson: "[men] build in every subsequent age on foundations formerly laid; and, in a succession of years, tend to a perfection in the application of their faculties."[99] But in literalizing the conjunction between architecture and natural process, Hugo blurs the claim of human distinctiveness it was meant to uphold. The significant individual life is an invisible detail, an ornamental curlicue, upon the vast edifice of the history of the race or species.

In a precocious essay in comparative media history added to the 1832 edition of *Notre-Dame de Paris*,[100] Hugo identifies architecture as "the great book of mankind" (189), "the great script of the human race" (191), "the chief, the universal form of writing" (194).[101] It is so, that is, "up until Gutenberg" (194; *jusqu'à Gutenberg*, 244). After 1500, print succeeds architecture as the expressive medium of human species being. Hugo's sublime style erupts, once more, in a pyroclastic flow of epithets: the printing press is "the ant-hill of the intellect," "edifice of a thousand storeys," "metropolis of the universal mind" (201). Hugo represents the succession as catastrophic—as an exterminating revolution. In his famous formulation, "This will kill that" (*Ceci tuera celà*): "the book will kill the building" (188), "printing will kill architecture" (189). The breach between them is constituted by a material difference that bears a formal and semiotic difference:

> In its printed form, thought is more imperishable than ever; it is volatile, elusive, indestructible. It mingles with the air. In the days of architecture, thought had turned into a mountain and taken powerful hold of a century and of a place. Now it turned into a flock of birds and was scattered on the four winds, occupying every point of air and space simultaneously. We repeat: who cannot see that in this guise it is far more indelible? Before, it was solid, now it is alive. It has passed from duration to immortality. You can demolish a great building, but how do you root out ubiquity? (196)

The difference can be summed up in the terms articulated in the previous chapter, in the contrasting aesthetics of Hegel and Schlegel: architecture achieves totality, print infinity. Architecture expresses an ancient, sacred, and organic (Hegel would say epic) wholeness of human labor—a conjunction of idea, word, and force—in relation to its local materials and scene, expressed in the synecdochic relation between architecture in general, considered as a universal human project, and the particular buildings in which it is realized:

> While Daedalus, who is force, measured, and Orpheus, who is intelligence, sang, the pillar which is a letter, the arcade which is a syllable, the pyramid which is a word, simultaneously set in motion both by a law of geometry and a law of poetry, formed groups, they combined and amalgamated, they rose and fell, they were juxtaposed on the ground, and superimposed in the sky, until, at the dictate of the general idea of an epoch, they had written those marvellous books which were also marvellous buildings: the pagoda of Eklinga, the Ramesseum of Egypt, the Temple of Solomon.
>
> The idea that engendered them, the word, was not only the foundation of all these buildings, it was also in their form. The Temple of Solomon, for instance, was not merely the binding of the sacred book, it was the sacred book itself. From each of its concentric ring-walls, the priests could read the word translated and made manifest to the eye, and could thus follow its transformations from sanctuary to sanctuary until, in its ultimate tabernacle, they could grasp it in its most concrete yet still architectural form: the ark. . . . And not only the form of buildings but also the site chosen for them revealed the idea which they represented. (190–91)

Hugo evokes the printing press with an architectural image: "when we try to compose in our minds a total picture of the sum of the products of the printing-press up till our own day, does the whole not appear to us as a vast construction, with the entire world as its base, at which mankind has been working without respite and whose monstrous head is lost in the profound mists of the future?" (201).

But the synecdochic relation between process and product is lost in the turn to print. Printed works remain particular, atomistic, resisting totalization, because they are replicable, multitudinous:

> For the rest, this prodigious edifice remains perpetually unfinished. The printing-press, that giant machine, tirelessly pumping the whole intellectual sap of society, is constantly spewing out fresh materials for its erection. . . . This indeed is a construction which grows and mounts in spirals without end; here is a confusion of tongues, ceaseless activity, indefatigable labour, fierce rivalry between all of mankind, the

intellect's promised refuge against a second deluge, against submersion by the barbarians. This is the human race's second Tower of Babel. (201–2)

Hugo shifts the combination of organic and inorganic forces that had characterized architecture (biology and geology together) from the ground of nature to artifice. The shift is ominous: the "giant machine, tirelessly pumping the whole intellectual sap of society," is a sort of industrial-robotic replicant of the tree of knowledge. It is a refuge against barbarism and darkness—and a second Tower of Babel, doomed to ruin and dispersal. "The human race has two books, two registers, two testaments: masonry and printing, the bible of stone and the bible of paper" (200). The bible of stone is more fully a book, singular, entire, and self-contained, than the bible of paper, the type of which is rather the potentially interminable series of a feuilleton.

Where does the novel we are reading fit into this sublime vision? What is the status of the individually authored book in an era of industrial mass production, and in the *longue durée* of the history of the species? Bearing the title *Notre-Dame de Paris*, Hugo's work evokes not only the cathedral itself but the vast productive forces of which it is the realization, as though aspiring to the totality that architecture once had.[102] At the same time, it is just one more brick in the perpetually unfinished edifice of print rising into the mists of time—hard though it may be to imagine Victor Hugo (of all authors) accepting such heroic anonymity. "The incompleteness of the book's edifice is for Hugo the very condition of its dynamic power," writes Victor Brombert: "Like all acts of language, like all collective effort, [books] participate in a ceaseless becoming."[103] At the same time, "Hugo's ambition is to 'totalize' himself in a book [*se totaliser dans un livre complet*]": his writing "brings into conflict the demands of the poem with those of poetry, embraces 'totalitarian' and 'unrealizable' projects, and strains towards a poetics of perpetual movement, a textuality without boundaries."[104] It aims to realize human species-work even as it breaks apart in the force of that aspiration.

Hugo's paeans to print and architecture invert the emergent aesthetic reckoning with geological time as an inhuman sublime, a "dark abyss of time" in which the middle scale of human endeavor and human meaning is lost from view.[105] Instead, this is the dimension in which the human comes fully into its own; this is the authentic time of human nature, bodied forth in a centuries-long accretion of multitudinous acts of creative labor. As for our daily, lived time, the ethical time of the will and appetite, choice and action, it renders us not just as ugly but as inhuman, as the sentimental rhetoric of *Notre-Dame de Paris* brings home with unrelenting panache. Hugo develops melodramatic techniques that refuse realism for a collectively scaled representation: formulaic plots of lost children and foundlings that are a kind of communal literary property, larger-than-life sensations and emotions fitted to the crowd. These occupy the present tense of reading, bound to the vital, sensate body. Hugo renders that simultaneously individual and collective feeling as a sadistic voyeurism, excited by prospects of torture, execution, and riot, and solicits our sympathetic participation in its scenes of cruelty and anguish—the flogging of Quasimodo, the torture of Esmeralda, the attack on the cathedral—with irresistible force. In the unforgettable final scene, Quasimodo weeps as he watches Esmeralda and Frollo perish in agony beneath him. We, the reader, watch all three: with what blend of horror, relish, and dismay?

It is not quite the very end of *Notre-Dame de Paris*. The novel's closing paragraph yields the last of its grotesque exhibitions: the discovery, eighteen months later, of the skeletons of Quasimodo and Esmeralda entangled together in the vile charnel house of Montfaucon. Convergent scientific associations cloud the scene: Buffon's criterion of procreation as the term that marks the boundary between species; the philosophical anatomists' biological grounding of sexual difference in the human skeleton;[106] Cuvier's virtuosity in parsing the relics of extinct creatures. Quasimodo's necrophiliac embrace—the last in a series of sexual assaults aimed at Esmeralda—yields a final, superbly horrible emblem of human impossibility: an aesthetic exaggeration of sexual dimorphism so extreme that it constitutes the pair as remains of separate species, joined in a sterile copulation of skeletons.

CHAPTER FOUR

Dickens

TRANSFORMIST

The word humanity *strikes us as strangely discordant, in the midst of these pages; for, let us boldly declare it, there is no humanity here.... [Humanity] is in what men have in common with each other, and not in what they have in distinction. [Dickens's characters] have nothing in common with each other, except the fact that they have nothing in common with mankind at large.... Who represents nature?*

—HENRY JAMES, ON DICKENS'S *OUR MUTUAL FRIEND* (1865)[1]

Domestic races of the same species, also, often have a somewhat monstrous character.

—CHARLES DARWIN, *ON THE ORIGIN OF SPECIES* (1859)

No Humanity Here

Anthropoid monsters, botched men, Sylvan and Quasimodo embody the *genius loci* of the world-city in *Count Robert of Paris* and *Notre-Dame de Paris*. Almost or not quite human, transfixed on a broken history, they manifest the crisis of Enlightenment knowledge and progressive ideology circa 1830. Their deformation (but not obliteration) of the figure of "man" is legible in the aesthetic medium of the grotesque, the signature of fragmented, hybrid, mutant form.

Two decades later, in great novels of mid-century, monsters rise from the depths of geological time opened by the new natural history, to overwhelm—like the deluge that is their element—the form and scale of man. The "Himmalehan, salt-sea Mastodon" looms in Herman Melville's *The Whale* (the novel's London title, 1851) long before his spectacular breach in the book's final chapters: "the mightiest animated mass that has survived the flood; most monstrous and most mountainous!"[2] Ranging across the "unshored, harbourless immensities" of the world, the whale precedes human history and will outlast it. "I am, by a flood, borne back to that wondrous period, ere time itself can be said to have begun; for time began with man," muses Ishmael, in vatic submersion. "I am horror-struck at this antemosaic, unsourced existence of the unspeakable terrors of the whale, which, having been before all time, must needs exist after all humane ages are over" (498). The abyss sounded by Leviathan is temporal, stirring metaphysical turbulence, as well as oceanic and global. Like God before Moses, he cannot be seen:

> Dissect him how I may, then, I go but skin deep; I know him not, and never will. But if I know not even the tail of this whale, how understand his head? much more, how comprehend his face, when face he has none? Thou shalt see my back parts, my tail, he seems to say, but my face shall not be seen. But I cannot completely make out his back parts; and hint what he will about his face, I say again he has no face. (414)

No face—no prosopopoeia: a nature that does not return our gaze.[3] Instead, the whale presents a massive *brow*—a feature of daunting power and inscrutable intelligence: "you see no one point precisely; not one distinct feature is revealed; no nose, eyes, ears, or mouth; no face; he has none, proper; nothing but that one broad firmament of a forehead, pleated with riddles; dumbly lowering with the doom of boats, and ships, and men" (179).

If we cannot know the whale, we may achieve an authentic intuition of him—a frisson of his sublimity—by seeking him in his marine environment: "Only in the heart of quickest perils; only when within the eddyings of his angry flukes; only on the profound unbounded sea, can the fully invested whale be truly and livingly found out" (495). Such a revelation the narrative of *Moby-Dick* transmits, more faithfully than the cetological dissertations it parodies, to land-bound readers. Truly and livingly found out: Melville registers dynamic nature, the whale in its ecosystem, as the phenomenon that exceeds human life and yet (hence) supplies the condition for its fullness. In *Moby-Dick*, the indomitable otherness of living nature, incarnate in the whale, recharges human experience with epic grandeur, for perhaps the last time in Western literature (since nature's fall darkens the historical horizon).[4]

Charles Dickens's *Bleak House*, the first serial installment of which appeared six months after *Moby-Dick*,[5] opens with a vision of London in November:

> As much mud in the streets, as if the waters had but newly retired from the face of the earth, and it would not be wonderful to meet a Megalosaurus, forty feet long or so, waddling like an elephantine lizard up Holborn-hill. Smoke lowering down from chimney-pots, making a soft black drizzle, with flakes of soot in it as big as full-grown snow-flakes—gone into mourning, one might imagine, for the death of the sun.[6]

Deluge, dinosaur, and death of the sun evoke cosmological timespans stretching before and after the brief glimmer of human history—temporal gulfs that swallow both individual lives and the life of the species.[7] "All discoveries which extend indefinitely the bounds of time must cause the generations of man to shrink into insignificance and to appear, even when all combined, as ephemeral in duration as the insects which live but from the rising to the setting of the sun," wrote Charles Lyell in 1827.[8] Nonhuman monsters that confound the measure of man: but where the living whale commands the final catastrophe of *Moby-Dick*, megalosaurus and solar extinction are visionary conceits, gags or punch-lines, in the opening paragraph of *Bleak House*. Their comic ostentation marks them not as sublime but as grotesque: "waddling like an elephantine lizard up Holborn-hill," "snow-flakes—gone into mourning, one might imagine." The apocalypse of Victorian London, with its collapse of time, is vouchsafed to the reader ironically, foreshadowing the judgment ("On the Day of Judgment," 21) that will arrive near the end of the story, when the great cause is consumed by its own costs, leaving behind waste paper and empty mirth.[9]

If the implacable alien vitality of the Whale shores up human meaning, the megalosaurus heralds its dissolution:

> Dogs, undistinguishable in mire. Horses, scarcely better; splashed to their very blinkers. Foot passengers, jostling one another's umbrellas, in a general infection of ill-temper, and losing their foot-hold at street-corners, where tens of thousands of other foot passengers have been slipping and sliding since the day broke (if the day ever broke), adding new deposits to the crust upon crust of mud, sticking at those points tenaciously to the pavement, and accumulating at compound interest. (13)

In a parody of evolutionary history, the epoch of the megalosaurus yields to an indiscriminate swarm of biological populations, seething up from and sinking back into the mud and murk, in the no less inhuman time of capital.[10] Dickens brings to a head the Romantic intuition about urban life developed by Scott and Hugo in their novels of Paris: the world-city, the total human habitat, is where human nature comes undone. Monsters belong here, as modern cinema confirms. For all both novels' shared vision of a scale of natural history overwhelming human life, Dickens's leviathan affords an insight the reverse of Melville's. Where the whale is a living creature, the embodiment of a planetary ecosystem, Dickens's dinosaur is a phantasmatic emanation of the Victorian metropolis—an allegorical figure for the "Dickens World." (Its avatar coalesces later in this opening chapter: "Jarndyce and Jarndyce still drags its dreary length before the Court," 17.) Against *Moby-Dick*'s sublime vision of the world as a nonhuman natural order, upon which humanity imprints its violent signature of epic striving, the world of *Bleak House* is unnatural, man-made, an "artificial nature," which reconstitutes its human origins in the aesthetic mode of the grotesque, according to a genetic logic of monstrosity, and is legible through the techniques of allegory.[11]

Early reviews of *Bleak House* fixed a standard complaint: Dickens "[makes] the unnatural in character predominate over the natural," he populates his story with "monstrosities" rather than recognizable human beings, and "the grotesque and the contemptible [take] the place of the humorous."[12] The complaint, fastening on the representation of character, would justify Dickens's exclusion from twentieth-century canons of realist fiction, from Georg Lukács's *Studies in European Realism* to F. R. Leavis's *The Great Tradition*. Recent criticism has been more willing to appreciate "the paradoxically central role of minor characters—comic, grotesque, eccentric—in Dickens's novels,"[13] and the force of the Dickensian grotesque in unsettling the distinctions that sustain realist representation: "living people thought to be dead, inanimate objects made animate, human beings who become things before our eyes," writes John Bowen, who asks, "What if the human itself was also at stake in these novels, not a given but a question?"[14] Where criticism has largely addressed this question in Marxist or Freudian terms (reification, fetishism, the uncanny), the present chapter takes it literally, as a zoological question. What if, indeed, Dickens's characters are not human, according to the canons of (mainstream, mimetic) Victorian realism—what if the Dickensian character system maps something other than the taxonomy of a singular, unified human nature?

Hugo proposed a positive theory of the grotesque in his preface to *Cromwell*, where he identified it as the aesthetic mode typical of modernity, characterized by an unstable mixture of forms, registers, and moods. The grotesque expresses the restless diversity and fluidity of natural forms, exceeding the bias of human preference, such that all form is deformation, variation and recombination are principles of reality, and the ideal type the projection of a local quirk of taste. *Notre-Dame de Paris*, a crucial work for Dickens and other authors of "urban mysteries" fiction in the 1830s and '40s,[15] attaches this aesthetic to the evolutionist natural history of Lamarck and the philosophical anatomists Geoffroy and Isidore Saint-Hilaire. Broadcast to a British public in Robert Chambers's *Vestiges of the Natural History of Creation* (1844), the new evolutionism also supplied a scientific basis for Dickens's art. "Organic beings are, then, bound together in development, and in a system of both affinities and analogies": the developmental hypothesis summarized by Chambers, with its recourse to an allegorical formal logic, provides a rationale for the well-noted turn from improvisation to formal design in Dickens's composition practice in the late 1840s.[16] The present chapter argues that the world of *Bleak House* and other major Dickens novels is that of a transformist natural history; and that Dickens's transformism is congruent with the Lamarckian-Geoffroyan synthesis popularized in *Vestiges*, more than it is with Charles Darwin's canonical account (not yet published) of "variation with modification through natural selection."[17]

Rather than marking a negative deviation from realist norms in Dickens's art, monstrosity and the grotesque are aesthetic signatures of an alternative realism, allegorical rather than mimetic in method, and an alternative scientific account of the reality generally taken to be the basis of Victorian representation.[18] *Bleak House* restructures the formal systems of narration and character that sustain realism's premise: a world held in common on a universal ground of human nature, constituted by a convergence of subjective intuition and objective reality, mediated through social existence. "I have purposely dwelt on the romantic side of familiar things," Dickens declares in his preface to the volume edition (7), anticipating George Eliot's identification of "common things" and "commonplace things" as the stuff of realism six years later in *Adam Bede*,[19] but without specifying the angle of variance of his art's "romantic side." The romantic, we infer, defamiliarizes what we take to be real, the "things" that make up our shared environment. *Bleak House* brings into collision the two major, oddly incommensurate, categories that have organized modern accounts of the nineteenth-century literary field: Realism (a formal category associated with the novel and with the Victorian period) and Romanticism (an aesthetic ideology, associated with lyric poetry, disguised as a period category). The second half of this chapter considers Dickens's invocation of a device from Romantic lyric, the visionary prospect, within the breach *Bleak House* makes—literally, splitting the narrative—in the apparatus of Victorian realism. The lyric insertion signals the disruption of a key technique with which realism gives back a human form to the world: an impersonal omniscient narration regulated by free indirect discourse. Casting its gaze into the opening of a failed correlation between subject and world, lyric vision cedes to lyric audition, and the registration of an inhuman vibration, a resonance that is not ours, the noise of the world.

Bleak House perfects a distinct aesthetic of the novel, scaffolded by early transformist natural history, much as George Eliot's *Middlemarch* will perfect the high realist tradition of the novel in English two decades later—an achievement Gillian Beer has linked to Darwinist canons of the complex, dynamic interdependency of natural systems and a gradual rather than catastrophic pace of organic change.[20] Pre-Darwinian transformism makes sense of those qualities of *Bleak House* that Dickens's critics have found most eccentric or idiosyncratic, "Dickensian," least amenable to a realist consensus. Nor were Dickens and Hugo the only major novelists of the age to draw on the intuitions of transformist science. Taking stock in 1842 of the great sequence of novels he had begun a dozen years earlier, Balzac appealed to Geoffroy's vindication of Lamarckian principles (in the latter's debate with Cuvier, discussed in the previous chapter) to make the case for *La Comédie humaine* as a scientifically coherent project—a natural history of society:

> "The Animal" is elementary, and takes its external form, or, to be accurate, the differences in its form, from the environment in which it is obliged to develop. Zoological species are the result of these differences.... Does not society modify Man, according to the conditions in which he lives and acts, into men as manifold as the species in Zoology? ... If Buffon could produce a magnificent work by attempting to represent in a book the whole realm of zoology, was there not room for a work of the same kind on society? But the limits set by nature to the variations of animals have no existence in society.[21]

Social environments exert a differentiating pressure on human life that is equivalent to, indeed exceeds, the pressure of natural environments on the production of animal species.

Modern criticism withheld from Hugo and Dickens the realist credentials it allowed Balzac. Nevertheless, we may read *Bleak House* as the strong growth of a divergent branch of nineteenth-century realism, rising through Balzac and Hugo, in which a catastrophist and saltationist evolutionism yields a true account of reality and human nature. Steampunk, so to speak, *avant la lettre*: in this Victorian metropolis, it would not be wonderful to glimpse a megalosaurus (recently described by Professor Owen, its life-size effigy under construction at Sydenham Park) waddling up Holborn Hill. A marine-store proprietor perishes of spontaneous combustion, none other of all the deaths that can be died, adding his aliquot of greasy particulate matter to the London fog. That fog, Jesse Oak Taylor points out,[22] is an artificial production, a toxic blend of organic waste miasma and fossil fuel combustion, and an early symptom—we now know—of the wholesale anthropogenic transformation of natural systems. It was "on or about 1848," write Nathan K. Hensley and John Patrick James, "that human action was confirmed in scientific terms to have definitively shaped the course of biotic evolution, collapsing any still-lingering divide, ontological or conceptual, between nature and culture. This discovery installed into the scientific record the fact that human action could now redraw the very blueprint of life on earth."[23] Man has abolished nature and replaced it with an unnatural order. The present chapter looks beyond these recent ecocritical reckonings to see how that unnatural order, in a virulent recursive logic, is remaking man.

The Poetry of Science

The opening of *Bleak House* sinks Victorian London in the temporal abyss opened by the new astronomy and geology. The massively expanded timescale of earth history proposed in the Romantic-era "geohistorical revolution" was well established by the early 1850s, although its extent remained controversial.[24] Its popular diffusion was accelerated by Robert Chambers's controversial bestseller *Vestiges of the Natural History of Creation*, a grand-narrative synthesis of modern evolutionist science, published anonymously in 1844. Chambers opens *Vestiges* with an evocation of the tremendous magnitudes of the solar system and the greater "astral system," as well as their temporal correlative, "the sublime chronology to which we are directing our inquiries."[25] The argument that follows combines Laplace's nebular hypothesis, the astronomical discoveries of the Herschels, Lyellian geology, the transcendental anatomists' and developmental morphologists' hypotheses of embryonic evolution and unity of plan, Lamarckian and Geoffroyan arguments for the transmutation of species, and other ingredients, both providential (unity of structure and the law of development are evidence of ultimate design) and heterodox (speculations on the spontaneous "chemico-electric" generation of life). Updating the progressive scheme mooted by Herder in *Ideas for a Philosophy of the History of Mankind*, Chambers sketches an evolutionary natural history according to the universal "principle of development": from the coalescence of the sun and planets from a fiery cosmic dust-cloud to the emergence of "the whole train of animated beings, from the simplest and oldest up to the highest and most recent," culminating in the ascent of man (203).[26]

Although *Vestiges* adopts the Lamarckian (and Herderian) conception of a progressive scale of organic life, Chambers rejects Lamarck's account of the mechanism of transmutation by the organism's "wants and the exercise of faculties" in response to changing environmental conditions (230–32). Instead, he proposes a progressive "law of development in the generative system," which overrides the genetic imperative of organic forms to make identical copies of themselves:

> *The simplest and most primitive type, under a law to which that of like-production is subordinate, gave birth to the type next above it, that this again produced the next higher, and so on to the very highest*, the stages of advance being in all cases very small—namely, from one species only to another. (222)

What safeguards the process from chaotic random mutation, Chamber insists, is the regulative force of unity of structure, manifest in the law of progressive embryonic development of organic forms, which subdues all modifications to an overarching providential design. "Nor is man himself exempt from this law. His first [embryonic] form is that which is permanent in the animalcule. His organization gradually passes through conditions generally resembling a fish, a reptile, a bird, and the lower mammalia, before it attains its specific maturity" (199). In the case of man, this morphological evolution achieves an archetypal and teleological synthesis of lower forms. "In [man] only is to be found that concentration of qualities from all the other groups of his order.... Man, then, considered zoologically, and without regard to the distinct character assigned to him by theology, simply takes his place as the type of all types of the animal kingdom, the true and unmistakable head of animated nature upon this earth" (272–73).[27]

Vying with Dickens's novels in readership (it had reached a tenth, revised edition by 1853), *Vestiges of the Natural History of Creation* familiarized the Victorian public with the main tenets of transformist natural history in the fifteen years preceding *On the Origin of Species*. Its enormous popular appeal, spanning the social ranks from Mechanics' Institutes to the royal household, defied the contempt of the British scientific establishment. The book's naturalization of universal progress and upward mobility, on the one hand, and its insistence on the stabilizing force of providential design, on the other, made it ideologically attractive to all but the most conservative readers. An early, enthusiastic review of *Vestiges* appeared in John Forster's liberal weekly *The Examiner*, to which Dickens was a frequent contributor.[28] Reviewing another work of popular science in the *Examiner* a few years later, Robert Hunt's *The Poetry of Science, or Studies of the Physical Phenomena of Nature* (1848), Dickens himself paid tribute to *Vestiges* as "that remarkable and well-abused book." It was one of the scientific texts he owned.[29]

Recent scholarship has largely dispelled the old complaint that Dickens was "indifferent or hostile to the scientific development of his age."[30] As well as *Vestiges* and works by Buffon, Cuvier, and Lyell, Dickens owned the second (1860) edition of *On the Origin of Species*; he befriended Richard Owen, the leading comparative anatomist of the day (and taxonomist of the megalosaurus), who warily endorsed an archetypally regulated evolutionism. Dickens commissioned a brace of articles on Darwin's theory, "Natural Selection" and "Species," for *All the Year Round*, which appeared in June and July 1860; these were followed by a third essay, "Transmutation of Species" (on the longer genealogy of transformist speculation), in March 1861.[31] The articles offer a reasonably detailed, accurate, and broadly sympathetic account of the argument advanced in *On the Origin of Species*. (The main review, "Natural Selection," cautions readers about the discrepancy between evidence and conclusions in Darwin's argument, while giving a fair summary of it.) Given his close editorial supervision of work published in his own journals, Dickens would have been well informed about Darwin's argument by these articles alone, however closely he may or may not have read the *Origin* itself. As George Levine suggests, the appearance of "essays so generously indulgent of the development theory in a journal as tightly controlled as *All the Year Round* [seems] very unlikely unless Dickens were ready to endorse the idea himself."[32] Dickens, in other words, was already disposed toward transformist science before the publication of the *On the Origin of Species*.

Studies of Dickens's work in the light of Darwin, pioneered by Levine and focused on the major post-*Origin* novel *Our Mutual Friend*,[33] have turned more recently to Dickens's engagement with pre-Darwinian geology and paleontology. Emphasizing the two men's friendship, Gowan Dawson argues that Owen's providentialist natural history informs Dickens's concern with the governing design of his own, formally "monstrous" serial fictions.[34] Katharina Boehm and Adelene Buckland argue that Dickens's interest in the imaginative potential of scientific knowledge was nourished less by the publications of accredited experts than by the Victorian media of popular and commercial science: "mesmeric trials and popular scientific shows, the movement to establish hospital care for children, popular medical manuals, germinating pediatric and psychological debates in the periodical press, anatomical museums and scientific lectures,"[35] dioramas, panoramas, melodramas, and other spectacles—including the life-size concrete models of the megalosaurus and iguanodon, designed by Waterhouse Hawkins to specifications by Owen, that were mounted in the grounds of the Sydenham Crystal Palace while *Bleak House* was in serialization.[36] Bernard Lightman aligns *Vestiges of the Natural History of Creation* with these popular shows and spectacles.[37] Yet scholars have largely neglected its impress on Dickens's work, with the notable exception of Boehm, who cites *Vestiges* as a primary source of the pre-Darwinian evolutionary theory that informs the later novels.[38] Not only was *Vestiges* the most influential of popular scientific syntheses of the kind that Dickens himself advocated, but its account of organic development came close to views held by Owen in the 1840s. Indeed, Owen would cite Chambers's teratogenic account of the mechanism of transformation in preference to Darwin's theory of natural selection in his 1860 *Edinburgh Review* article on the *Origin of Species*.[39]

"The science of *Bleak House* remains outside of any religious orthodoxy and is unconcerned with systematic coherence," Levine has remarked,[40] while Buckland issues a caution against "reading theoretical geological positions into Dickens's work," which was responsive to the poetic potential rather than the strict logic of scientific ideas.[41] The natural history of *Bleak House* may be more coherent than these accounts allow; at any rate, poetic potential and scientific coherence need not be at odds, for Dickens as for any other writer. *Vestiges* was important for Dickens, to be sure, because of its character as a popular synthesis—capacious and suggestive rather than theoretically rigorous, a mingled yarn of hypothesis, argument, illustration, and conjecture spun from miscellaneous sources. The charges leveled against it for being more like a "historical tale" or "fictitious narrative" than real science (a recurrent motif in the story we are tracing) point up its affinity with Dickens's own literary projects.[42] Dickens put his finger on Chambers's achievement in his *Examiner* review of *The Poetry of Science*: "by rendering the general subject popular, and awakening an interest and a spirit of inquiry in many minds, where these had previously lain dormant, [the author of *Vestiges*] has created a reading public—not exclusively scientific or philosophical—to whom such offerings can hopefully be addressed."[43] Dickens expresses an appreciation of the common environment of these works of popular science and his own journalism and fiction, the expanding liberal public sphere of the age of Reform, in which he and Chambers both forged their early careers.[44]

Dickens's Teratology

"Radical science," crowned by Lamarckian-Geoffroyan transformism, loomed large in that public sphere, the turbulent forum of popular journalism and political debate sustained by the new industrial print media of periodicals and miscellanies in the 1830s and 1840s.[45] As well as works of popular science, one of its characteristic genres was serial fiction. Long before Henry James's denigration of nineteenth-century serial novels as "large, loose, baggy monsters, with . . . queer elements of the accidental and the arbitrary," reviewers drew an analogy between the forms of periodical fiction and the prehistoric beasts reconstructed by Victorian paleontologists.[46] Serial publication, with its promise (or threat) of a potentially infinite extension of the time of reading, exacerbated the charge of formlessness to which novels had always been susceptible. Formally homologous with its mass-public urban setting—sprawling, overcrowded, polyglot—a new kind of monster-fiction arose in the 1830s and 1840s, attuned to collective life, to transpersonal surges of energy and agency, to the affective flow of crowds, flouting what Nancy Armstrong has described as the realist novel's ideological task, the production of the liberal individual.[47] Cheap popular serials and feuilletons in Great Britain and France spawned radical, sensational mutations of "urban mysteries" fiction in the wake of Hugo's original metro-Gothic monsterpiece, *Notre-Dame de Paris*.[48] As the most successful English novelist to command the format of part-issue serialization, Dickens was the frequent target of reviewers' comparisons of his fiction to the "ungainly and gigantic antediluvian creatures" that fascinated the public in the second quarter of the nineteenth century.[49] His placement of the megalosaurus at the very opening of *Bleak House* awards it, in this light, the defiantly jaunty role of figurehead for his most ambitious work to date.

Monstrosity is traditionally an effect of formal heterogeneity—the grotesque—as well as of huge size. It affronts classical norms of unity, scale, and proportion that have roots, Dawson reminds us, in Aristotle's prescription of a "certain amplitude" and an "orderly arrangement of parts" for poetic works, based on an analogy with living bodies—paradigmatically, the human body. An enormous form, whether natural or artificial, cannot be beautiful, "since our view of it is not simultaneous, so that we lose the sense of its unity and wholeness as we look it over; imagine, for instance, an animal a thousand miles long. Animate and inanimate bodies, then, must have amplitude, but no more than can be taken in at one view; and similarly a plot must have extension, but no more than can be easily remembered."[50] *Bleak House* features a teeming multitude of bizarre characters, a bewildering tangle of plots, moods, styles, and voices, and a tectonic fracture between third-person present-tense and first-person past-tense narrations across the storytelling apparatus. Aristotle's strictures on aesthetic magnitude are notable for correlating temporal and spatial dimensions. The vast size of Dickens's novel comprises not just hundreds of pages but the time required to read it: nineteen months for the first issue, published in twenty installments (culminating in a final double number) from March 1852 to September 1853. Serialization, with its production of the story in parts, precluded the "tell-tale compression of the pages" that would allow readers of a novel in book form, even one as big as *Moby-Dick*, to keep its dimensions in view—to grasp it as a whole. Unable to discern the novel's form while immersed in it, contemporary reviewers complained of an "absolute want of construction."[51] To its first readers, at least until September 1853, *Bleak House* literally appeared shapeless, boundless—an animal a thousand miles long.[52]

It was not size as such that moved James's complaint about "large loose baggy monsters." Huge, sprawling serial novels fail to subdue their animating energy, by force of "composition," to "organic form."[53] Organic form, a regulative, teleological principle of design, prescribes a synecdochic equivalence between art and life, mediated by the human scale—a knowable, accountable totality. That equivalence is lost from view in the new natural history, which recasts humanity—the history of the species as well as the individual—as transient, epiphenomenal, barely constitutive of an event. Against astronomical, geological, and microbiological measures of the infinitely large and infinitely small, established across the domains of natural science in the first half of the nineteenth century, realism codifies a "middle scale"—a spatial, temporal, psychological, ethical, ideological *via media*—as the proper range of human life: "an *anthropocentric* scale, where readers are truly 'the measure of things,'" writes Franco Moretti.[54] Realism's middle scale is secured by a spatiotemporal "grammar of perspective," analogous to pictorial vanishing-point perspective, that confirms "the uniformity at the base of human experience and solidarity of human nature," according to Elizabeth Deeds Ermarth.[55] It comprises the stylistic ethos of "the middle, the intermediate," analyzed by Moretti, and the "middling condition" George Levine identifies as realism's proper field;[56] the "narrative middle," which is the forum of the novel's social engagements, according to Caroline Levine and Mario Ortiz-Robles;[57] the affective "mediocrity," a zone of "social abstraction," that sustains moral agency, in Rae Greiner's account of "sympathetic realism;"[58]

the geography of "expanded border zones" (Devin Griffiths) or provincial life (as in "Middlemarch") favored by so much Victorian fiction;[59] and the "more or less mediocre, average" protagonists, introduced in the Romantic Bildungsroman and historical novel, given scientific substance in mathematician Adolphe Quetelet's statistical fiction of *l'homme moyen*, "the average man."[60]

Dickensian characterology elides the middle scale of a realist human psychology, as Victorian reviewers noted in their complaints that the "odd and eccentric" crowd out the "ordinary" in *Bleak House*.[61] "Mr. Dickens has committed a grave error in bringing together such a number of extraordinary personages, as are to be found huddled *en masse* in this romance," groused *Bentley's Miscellany*: "is it, we ask, within the rightful domain of true art to make the unnatural in character thus predominate over the natural? In *Bleak House*, for every natural character we could name half a dozen unnatural ones."[62] The troubling consequence was a wholesale takeover of the character system by this "unnatural" aesthetic: "the reader might be excused for feeling as though he belonged to some orb where eccentrics, Bedlamites, ill-directed and disproportioned people were the only inhabitants. Esther Summerson, the narrator, is, in her surpassingly sweet way, little less like ordinary persons than are Krook and Skimpole."[63] The mention of Esther is shrewd: the ostensible protagonists and patterns of virtue are not exempt from the criterion of the grotesque. No one in *Bleak House* is ordinary, whole, "round"; all are distorted, fragmented, "flat"—in Alex Woloch's compelling analysis, "[conscripted] into minorness."[64] Henry James, reviewing *Our Mutual Friend*, summarizes the offense to realism, and its premise of a world held in common, posed by Dickensian character:

> Among the grotesque creatures who occupy the pages before us, there is not one whom we can refer to as an existing type.... Like all Dickens's pathetic characters, [Jenny Wren] is a little monster.... The word *humanity* strikes us as strangely discordant, in the midst of these pages; for, let us boldly declare it, there is no humanity here. Humanity is nearer home than the Boffins, and the Lammles, and the Wilfers, and the Veneerings. It is in what men have in common with each other, and not in what they have in distinction. The people just named have nothing in common with each other, except the fact that they have nothing in common with mankind at large.[65]

Everyone in the Dickens world is an extreme or singular case. Everyone is a monster.

Instead of a realist character system, shoring up a normative conception of human nature, Dickens writes a *teratology*, in the sense postulated by Geoffroy and Isidore Saint Hilaire, popularized by Chambers in *Vestiges of the Natural History of Creation*, and adopted as a transmutationist hypothesis by Richard Owen. With evolutionist conceptions of "the production of new beings, normalized by their unpremeditated adaptation to new conditions of existence," writes Georges Canguilhem, "monstrosity [becomes] the rule and originality a temporary banality."[66] In her study of Dickens and the "sciences of childhood," Katharina Boehm argues that a pre-Darwinian vision of species transformation by abrupt mutation, blended in *Vestiges* from the rival embryological theories of Geoffroy Saint-Hilaire and Karl Ernst von Baer, informs the figure of the "monstrous child" (a portent of radical social change) in *Our Mutual Friend*.[67] A dozen years earlier, this transformist vision governs *Bleak House*. Chambers proposes a "law of development in the generative system," regulated by length of gestation, by which organisms give birth to successively "higher" (more complex) forms, "the stages of advance being in all cases very small—namely, from one species only to another" (222). The stages might be very small, but they represent jumps, nonetheless, from one distinct form to another: in other words, in contrast to the gradualist theory outlined fifteen years later in the *Origin of Species* (with its adherence to "that old canon in natural history of 'Natura non facit saltum'"[68]), this is a saltationist model of biological transformation.

Its direction is not necessarily progressive: "We see nature alike willing to go back and to go forward. Both effects are simply the result of the operation of the law of development in the generative system. Give good conditions, it advances; bad ones, it recedes" (218). Indeed, Chambers is unable to offer any positive evidence for the progressive tendency, since no one has observed or at any rate recorded the production of one species from another. Instead, he routes his argument through the negative case of Geoffroyan teratology, which demonstrates the law of development "in the production of certain classes of monstrosities" (218)—regressive rather than progressive mutations, forms arrested at lower stages of fetal development:

> Here we have apparently a realization of the converse of those conditions which carry on species to species, so far, at least, as one organ is concerned. Seeing a complete specific retrogression in this one point, how easy it is to imagine an access of favourable conditions sufficient to reverse the phenomenon, and make a fish mother develop a reptile

heart, or a reptile mother develop a mammal one. It is no great boldness to surmise that a super-adequacy in the measure of this under-adequacy (and the one thing seems as natural an occurrence as the other) would suffice in a goose to give its progeny the body of a rat, and produce the ornithorynchus [duck-billed platypus], or might give the progeny of an ornithorynchus the mouth and feet of a true rodent, and thus complete at two stages the passage from the aves to the mammalia. (219)

The evolutionary process only becomes visible through its "converse." This passage sparked a brusque comment from Charles Darwin: "The idea of a Fish passing into a Reptile (his idea) monstrous."[69]

At the beginning of his own research into the species question, in 1837, Darwin recorded Owen's suggesting to him "that the production of monsters (which Hunter says owe their origin to very early stage and which follow certain laws according to species), present an analogy to production of species."[70] Reviewing the *Origin of Species* two decades later, Owen would accept the principle of organic descent but reject the mechanism of natural selection, and appeal instead to Chambers's hypothesis: "[The Author of *Vestiges*] cites the results of embryological studies, to show how such 'monster,' either by excess or defect, by arrest or prolongation of development, might be no monster in fact, but one of the preordained exceptions in the long series of natural operations, giving rise to the introduction of a new species." He adds, self-servingly, "Owen has not failed to apply the more recent discoveries of Parthenogenesis to the same mysterious problem."[71] Owen would incorporate the hypothesis into his own system:

> I deem an innate tendency to deviate from parental type, operating through periods of adequate duration, to be the most probable nature, by way of operation, of the secondary law, whereby species have been derived one from the other.... [N]o explanation presents itself for such transitional changes, save the fact of anomalous, monstrous births.[72]

Teratology is the funhouse mirror in which we view the production of species as a production of monsters—a sudden, spontaneous appearance of inchoate forms, halfway between the familiar parent and something new, on which we are bound to fix the aesthetic label of the grotesque.

Bleak House exploits the paradox of an organic developmental drive that is manifest in, as, malformation. The novel's character system maps a saltationist, catastrophic transformism unregulated (in contrast to Owen's archetypalist account) by a progressive teleology, in which mutations are as likely (or likelier) to be retrogressions, or grotesque lateral deformations, as upward steps. It is visible in patterns of resemblance and affinity across the bizarrely differentiated figures that populate the novel's world, as well as in the hereditary relation between parent and child. The novel swarms with cases of blocked, retarded, or premature development (children who have grown up too fast or who never had a childhood, adults who have regressed to childhood or who have never grown up or who affect childishness), as well as with not-fully-human beings, endowed with weird metabolisms, that we read both metaphorically and literally: allegorical-biological hybrids. The mock-Lord Chancellor, Mr. Krook, is a salamander, "the breath issuing in visible smoke from his mouth, as if he were on fire within" (68); his death from spontaneous combustion makes for the novel's most flagrant breach of scientific probability, as G. H. Lewes complained, provoking a public exchange with Dickens, who cited medical authorities in his defense. Krook spreads his hands "like a vampire's wings" (166), but it is the lawyer Vholes, his "long thin shadow . . . chilling the seed in the ground as it glided along" (698), who has "something of the Vampire in him" (924). The Reverend Chadband, "a large yellow man, with a fat smile, and a general appearance of having a good deal of train-oil [whale-oil] in his system," is "not unlike a bear who has been taught to walk upright. He is very much embarrassed about the arms, as if they were inconvenient to him, and he wanted to grovel" (304–5).[73] The regressive reversal of the drive to stand erect, a Lamarckian signature of human becoming, informs an absurd appeal to the natural theologians' argument from design: "Why can we not fly? Is it because we are calculated to walk? . . . What should we do without strength, my friends? Our legs would refuse to bear us, our knees would double up, our ankles would turn over, and we should come to the ground" (306–7). (In the same vein, Tony Jobling's announcement of a developmental ascent from birth to maturity with the successive courses of his dinner offers a neat burlesque of recapitulation theory, 321–22.) The noxious Smallweeds, infesting the novel's core, have by horrible parthenogenesis managed to constitute themselves as a separate species. Young Bart is "a weird changeling, to whom years are nothing . . . a fossil Imp," who "drinks, and smokes, in a monkeyish way" (318). The Smallweed patriarch is "a horny-skinned, two-legged, money-getting species of spider" (332); his great-granddaughter "seemed like an animal of another species" among her schoolfellows (335); the family "has had no child born to it, and . . . the complete little men and women whom it has produced, have been observed to bear a likeness to old monkeys with something depressing on their minds" (333). The evolutionary leap from bipedal spider to dejected ape, mysteriously bypassing natural childbirth, parodies Chambers's law of development.[74]

A more poignant teratological case is Caddy's and Prince's afflicted baby: "It had curious little dark veins in its face and curious little dark marks under its eyes like faint remembrances of poor Caddy's inky days" (768).[75] The pathos, characteristic of Esther's narration, tempers what might register elsewhere as a grim joke. Also on a somber note, the impersonal narrator reflects on the potential of neglect and abuse to induce retrogression. Jo the crossing-sweeper and a drover's dog stop to listen to musicians in the street, "probably with much the same amount of animal satisfaction":

> Likewise, as to awakened association, aspiration or regret, melancholy or joyful reference to things beyond the senses, they are probably upon a par. But, otherwise, how far above the human listener is the brute!
>
> Turn that dog's descendants wild, like Jo, and in a very few years they will so degenerate that they will lose even their bark—but not their bite. (259)

Chambers aligns his evolutionary scheme with Lamarck's in allowing for an environmental role in shaping organic development—especially in human societies: "Man's mind becomes subdued, like the dyer's hand, to that it works in. In rude and difficult circumstances we unavoidably become rude, because then only the inferior and harsher faculties of our nature are called into existence" (303). Environmental conditions may exert a degenerative as well as a progressive force.[76] The passage cited by Chambers, from Shakespeare's Sonnet 111, also shows up in Dickens's preface to the volume edition of *Bleak House* (August 1853), where it underlines a key satirical principle of the environmental deformation of character. Environment, in *Bleak House*, is man-made; the Dickens World is the Dickens City, an anthropogenic ecology of climatic, architectural, political-economic, and symbolic systems, of which Chancery is the dominant (but not the only) manifestation.[77] Its distorting force bears down, dismayingly, on the innocent and virtuous as well as the malignant and corrupt. It turns persons into cases: Gridley, the "man from Shropshire," warped by a Chancery suit; Richard Carstone, the novel's main study (typified by "indecision of character," 197) in thwarted and perverted *Bildung*; and John Jarndyce, who manages to survive by rehearsing his deformation as a protective ritual, externalizing it in a battery of eccentric quirks and tics.

"The system! I am told, on all hands, it's the system," cries Gridley, indicting those human agents who disclaim ethical liability for their part in an evil apparatus (231). At the same time, he ventriloquizes a fearful truth. The institutions and economies that comprise the world of *Bleak House*—the imaginary totality James Buzard calls "the national anticulture"[78]—function autopoetically, with baleful effects on their human origins. Nowhere in the novel is "character" what the realist novel projects, a relatively autonomous, self-regulating phenomenon. Instead it is contaminated, penetrated, and restructured by impersonal, systemic forces—human inventions that, achieving a sufficient level of complexity and abstraction, have become self-organizing and self-perpetuating.[79] The true horror of Chancery is not simply that it destroys people, spreading ruin, insanity, and disease—but that they adapt to it. It breeds Smallweeds and Vholeses. A human invention, transcending its origins, rewrites the script of human nature.

And it has no outside.[80] There is no original nature, or none that is accessible, any more. The scarred and blasted waif Phil Squod relates a dream of the country, which he has never visited:

"How did you know it was the country?"
"On accounts of the grass, I think. And the swans upon it," says Phil, after further consideration.
"What were the swans doing on the grass?"
"They was a eating of it, I expect," says Phil. (418)

Chancery reconstitutes pastoral as industrial raw material: "these pleasant fields, where the sheep are all made into parchment, the goats into wigs, and the pasture into chaff" (661). Toward the end of the novel, Mr. Jarndyce gifts a replica of Bleak House, miniaturized to a cosy rustic cottage, to Esther, right before he gifts her to Allan Woodcourt. The Yorkshire scenery, with its Dingley-Dell vista of a distant cricket match, is conspicuously artificial, an idyllic retreat, an embroidered pocket within the novel's world rather than a heterotopic outside: the last best work of Jarndyce's patronage.[81] Instead, the novel cherishes a fragile affective residue of authentic humanity in the battered lives of orphans, fugitives, and other castaways: Jo the crossing-sweeper, Charley the bailiff's daughter, Miss Flyte, Guster, Phil Squod, Caddy Jellyby, Prince Turveydrop and their stricken child, the wards in Chancery, Esther Summerson herself, her lost father, Nemo, and, eventually, her desolate mother. These register deformation as damage, rather than monstrosity, in a sentimental rather than satiric modality of the grotesque. Innocents bearing distortions and impairments of which they are not the cause, refugees from an original, now broken nature, they distil a pathos that is the last essence of the human.

The Prose of the World

The sentimental division of the character system refracts the "bold and perennially fascinating technique," without precedent in the English novel, for which *Bleak House* is famous: Dickens's formal division of the story between Esther's first-person narrative, written in the historic past tense, and a present-tense third-person narration that emanates from no identifiable diegetic source.[82] Esther's narrative—feminine, local, perspectivally bounded—provides a reservoir for humane pathos and affection;[83] while the impersonal narrative—implicitly masculine, universal, unrestricted—commands the satiric function of knowledge as it tracks the flow of transhuman, systemic energies across the novel's world, via "multiple distributed networks."[84] This narrative flaunts its power over space and time, scrutinizing as though from "an eternal present"[85] the story's inner motives and connective links, while keeping them hidden from the reader until the appointed time. It swoops in and out of the conjectured viewpoints of "lower animals" (horses, dogs, poultry: 103–5), trees ("frowning woods, sullen to see how trees are sacrificed," 446), and even gas-lamps ("as the gas seems to know," 14), as well as of the novel's vast human population. It transmits a multitudinous clamor of registers, styles, jargons, and idiolects, managing a ventriloquism of local voices that sometimes resembles (although it is not the same as) free indirect discourse. This ventriloquism amplifies the narrator's relentlessly external representation of character, reading and decoding outward appearances for the reader (Hortense, 187; Mr. George, 341), teasing us with withheld knowledge of the thoughts of Mr. Tulkinghorn and Lady Dedlock as they fight their stealthy duel of mutual suspicion (23–25, 195–96, 459).

It is tempting to call this an omniscient narration, and indeed, it is one of nineteenth-century fiction's most exuberant performances of the "fantasy of unlimited knowledge and mobility" that, in Audrey Jaffe's account, invests the realist project at its most ambitious—enhanced, here, by the recessive, introverted tonality of Esther's utterance.[86] As Jaffe goes on to argue, however, the formal coexistence of Esther's narrative, "constituting as it does a boundary omniscience cannot cross," disables any totalizing claim that might be made for the impersonal narrative. "Omniscience in *Bleak House*," she writes, "is paradoxically proscribed, limited to one half of the novel," which cannot then, by definition, be all-knowing.[87] Developing this argument, I suggest that Dickens's division of the narrative breaks apart the functions that, joined together, come to constitute the normative technology of Victorian realism: an omniscient narration based on the strategic fusion of subjective and impersonal viewpoints in free indirect discourse.

Free indirect discourse comes to the forefront of fictional practice in nineteenth-century realism, within a general formal constellation F. K. Stanzel has called the "figural narrative situation," in which an impersonal third-person narration is mediated through a diegetically embedded character (a "reflector").[88] Names for related, constituent, and possibly synonymous techniques include *style indirect libre* in French, *erlebte Rede* in German, and in English (variously) represented speech and thought, psychonarration, and narrated monologue.[89] Narrative theorists have debated their terminology, taxonomy, and function, and whether they amount to a unified or convergent practice or different, disparate ones. For convenience's sake, I shall use the common English term, free indirect discourse, to designate a third-person narration that represents the mental and somatic states—the thoughts and feelings—of novelistic characters without formal separation from its own utterance: absorbing a first-person perspective, in other words, into a general narrative medium. Roy Pascal defines free indirect discourse as "a dual voice, which, through vocabulary, sentence structure, and intonation, subtly fuses the two voices of the character and the narrator";[90] for Gérard Genette, "the narrator takes on the voice of the character, or, if one prefers, the character takes on the voice of the narrator, and the two instances are *merged*."[91] Ann Banfield stresses the transaction's negative valence: free indirect discourse sublimates a first-person voice into language that belongs to no one, "unspeakable sentences, the sentence of narration and the sentence representing consciousness."[92] Other critics (following Stanzel) describe a graduated scale or spectrum of intersections between an impersonal narrative and characters' voices and thoughts.[93]

Many critics consider free indirect discourse constitutive of realism, in the full flower of its development, if not, more generally, constitutive of fiction or the literary as such.[94] Ermarth's "grammar of perspective," confirming "the uniformity at the base of human experience and solidarity of human nature," relies upon the technical capacity of a third-person narration to sustain a "consensus among multiple viewpoints" through the management of "modulations from one mind to another: from character to character, from narrator to character, from reader to narrator."[95] Free indirect discourse secures these subjective modulations (and a range of affective stances, from ironical to sympathetic) within its larger mediation between individual consciousness and an impersonal, putatively objective representation of the world.[96] A critical consensus sees the technique fully emerging as a coherent practice in the Romantic period, to become the advanced narrative medium of realist fiction in nineteenth-century Europe. In the English tradition, free indirect discourse achieves early formal perfection in the novels of Jane Austen. Victorian authors, foremost among them George Eliot, amplify it into a flexible technology of omniscient narration, staking a claim over the totality of its world on its capacity to represent the inner lives of characters in correlation with what Eliot calls (after Auguste Comte and Herbert Spencer) the "social organism."[97]

The world realism holds in common by means of free indirect discourse is thus, in the first place, a social world. (Hannah Arendt calls the novel "the only entirely social art form."[98]) Developing the definitions of Pascal and Genette, Moretti argues that the technique's blending of individual character and impersonal narration synthesizes "a *third voice*, intermediate and almost neutral among them: the slightly abstract voice of the achieved social contract."[99] For Greiner, it is realist fiction's quintessential technique of the "middle ground," establishing "the social basis of private subjectivity."[100] Social life constitutes, and is phenomenologically identical with, the world as such in Austen's novels—an effect secured by her focus on a restricted social and topographical range. *Emma* mounts a virtuoso set piece of free indirect discourse as medium of a liberal immanence of being-in-the-world,[101] as the heroine steps aside from the business at hand (a shopping expedition) to look at the view:

> Emma went to the door for amusement.—Much could not be hoped from the traffic of even the busiest part of Highbury... when her eyes fell only on the butcher with his tray, a tidy old woman travelling homewards from shop with her full basket, two curs quarrelling over a dirty bone, and a string of dawdling children round the baker's little bow-window eyeing the gingerbread, she knew she had no reason to complain, and was amused enough; quite enough still to stand at the door. A mind lively and at ease, can do with seeing nothing, and can see nothing that does not answer.[102]

Here is the technique at its most subtly compelling. The narration raises Emma's perspective to an abstract affirmation beyond her own capacity to articulate it (one of Banfield's "unspeakable sentences"), even as she fully inhabits it: to "a mind lively and at ease," the world is enough. The reiterated "nothing" of that last sentence ("seeing nothing... nothing that does not answer") bespeaks an existential plenitude in the rhythm of exchange—of the economic and social contract that binds this world into a habitable whole. The cycle of exchange ("traffic") integrates the mind lively and at ease with the outward scene, supplying, in other words, the material conditions for liveliness and ease to those minds that may, in turn, animate the scene they look upon. For the handsome, clever, and rich heroine, and the discerning reader, the sensible world yields all that there is to imagine. The passage affords a bracing contrast with Flaubert's later elaboration of the technique in *Madame Bovary*, likewise trained on the "nothing" of provincial life but to ironically caustic effect, measuring "the gap between the world as experienced and the world as it actually is."[103]

Reliance on free indirect discourse, with its naturalization of the claim on the world as a totality, differentiates Victorian omniscient narration from the earlier omniscient mode attached to a strongly personified, self-consciously performative, masculine narrator, typified in the novels of Henry Fielding. Rhetorically exhibitionist third-person narration in the mode of Fielding enjoys a strong revival in the nineteenth century at the hands of Scott, Thackeray, and Dickens. Hence, two of the major theorists of free indirect discourse (Pascal and Dorrit Cohn) agree that it is not characteristically Dickensian: the typical Dickens narrator is stylistically too extroverted for the discreet merging of personal and impersonal registers the technique prescribes.[104] This presents us with the paradox that what is supposed to be the nineteenth-century novel's major formal innovation was not particularly important to one of its major novelists. Nevertheless, the Victorian expansion of free indirect discourse into a complex, modified mode of omniscient narration is crucial to Dickens's art in *Bleak House*.

The division of the narration in *Bleak House* does not merely juxtapose the distinct narrative techniques, first person and impersonal, that are merged in free indirect discourse. It splits them, in an act of aesthetic violence that splits the figure of human fitness, of an immanent being-in-the-world, that free indirect discourse composes.[105] (The split between present and past-historic tenses also destroys the grammatical medium for *style indirect libre*, perfected by Flaubert in contemporary French realism, the *passé imparfait*.[106]) Moretti's "third voice," "the slightly abstract voice of the achieved social contract," has no purchase here; the world of *Bleak House* is not a world held in common, as in Ermarth's and Greiner's accounts, by realism's narrative technology. Whence James's complaint: humanity consists "in what men have in common with each other, and not in what they have in distinction," whereas Dickens's characters "have nothing in common with each other, except the fact that they have nothing in common with mankind at large." To draw again on Jaffe's insight, that *Bleak House* disables the omniscient narration by hobbling it with a first-person memoir: far from making two halves that add up to a whole, the novel's two narratives coexist in a broken, maimed, incommensurate relation to one another. To see how this works, I turn now to Dickens's distinctive use of the *prospect*, the heroine's gaze that opens the world, an example of which we have just been reading in *Emma*. The remainder of this chapter seeks to recover the aesthetic radicalism charging Dickens's modest announcement, in the volume preface to *Bleak House*, that he has "purposely dwelt on the romantic side of familiar things."

Visionary Dreariness

In the tenth number of *Bleak House*—midway through the story—Esther Summerson looks at the view. Gazing at the night sky over Saint Albans, on the brink of a series of catastrophic transformations in her life, she experiences a mysterious access of feeling:

> In the north and north-west, where the sun had set three hours before, there was a pale dead light both beautiful and awful; and into it long sullen lines of cloud waved up, like a sea stricken immoveable as it was heaving. Towards London, a lurid glare overhung the whole dark waste, and the contrast between these two lights, and the fancy which the redder light engendered of an unearthly fire, gleaming on all the unseen buildings of the city, and on all the faces of its many thousands of wondering inhabitants, was as solemn as might be.
>
> I had no thought that night—none, I am quite sure—of what was soon to happen to me. But I have always remembered since, that when we had stopped at the garden gate to look up at the sky, and when we went upon our way, I had for a moment an undefinable impression of myself as being something different from what I then was. I know it was then, and there, that I had it. I have ever since connected the feeling with that spot and time, and with everything associated with that spot and time, to the distant voices in the town, the barking of a dog, and the sound of wheels coming down the miry hill. (488–89)

In his study of *Bleak House*, John Jordan calls this "one of the most puzzling passages in the book," and remarks that, of the several critics who have discussed it, "none has been able to identify or explain the strange experience it describes."[107] It is as if readers can only reiterate the character's sense of a revelation withheld, just beyond the threshold of understanding.

The mystery is deepened by Dickens's quiet but insistent echo of William Wordsworth's phrase for remembered moments of visionary illumination, "spots of time," in his autobiographical poem *The Prelude*, published two and a half years before the appearance of this passage in the December 1852 number of *Bleak House*.[108] "I have ever since connected the feeling with that spot and time, and with everything associated with that spot and time," Esther remarks. "There are in our existence spots of time," writes Wordsworth: reservoirs of a "renovating virtue" with which our minds, depressed by the burden of "trivial occupations" and "ordinary intercourse," may later be "nourished and invisibly repaired."[109] The poet recalls one of them:

> It was, in truth,
> An ordinary sight; but I should need
> Colours and words that are unknown to man,
> To paint the visionary dreariness
> Which, while I looked all round for my lost guide,
> Invested moorland waste, and naked pool,
> The beacon crowning the lone eminence,
> The female and her garments vexed and tossed
> By the strong wind.[110]

An ordinary sight, not a sublime or beautiful one, affords the indescribable mood or atmosphere Wordsworth calls "visionary dreariness." Dickens opens his description in a more conventionally visionary style. The dead sea and fire over London call to mind the spectacular biblical canvases of John Martin, completed while *Bleak House* was in serialization (*The Destruction of Sodom and Gomorrah*, 1852; *The Great Day of His Wrath*, 1853), returning to its scriptural matrix the apocalyptic motif framed in the novel's opening paragraph as a scientific hypothesis (the death of the sun). But then the description stoops, from the apocalyptic to the ordinary, from heaven to earth, from sight to sound: distant voices, a dog barking, wheels coming down the hill. Commonplace, everyday noise, rather than a vision of the world's end, mediates Esther's surge of feeling.

Both episodes, Wordsworth's and Dickens's, have in common the speaker's intuition of something beyond the phenomenal impressions of sight and sound: "Colours and words that are unknown to man," "an undefinable impression of myself as being something different from what I then was." Sense stalls against a cognitive limit. "I had no thought that night," Esther insists, "of what was soon to happen to me." Readers of *Bleak House* are able to assign a content to her premonition when, soon afterwards, she is disfigured by smallpox: "I had never been a beauty, and had never thought myself one; but I had been very different from this," she says, of her first glimpse of her scarred face in the mirror (572). Later in the same chapter, Esther encounters her lost mother and learns the dark secret of her birth. For now, though, that content is unavailable to her, locked in the historic past tense of the scene and moment. All she apprehends is an "undefinable impression," a purely formal and negative sensation of "being something different from what I then was."

Dickens bought a copy of *The Prelude* in July 1850, just a few weeks after its posthumous publication.[111] Evidently the poem made an impression on him: scholars have detected traces of it in his mature fiction, from *David Copperfield* and *Bleak House* to *Great Expectations* and *The Mystery of Edwin Drood*.[112] Nor is this the only echo of Wordsworth in *Bleak House*, a novel exceptionally rich in allusions to and recollections of British Romanticism.[113] Dickens's echo of *The Prelude*, at his story's critical midpoint, makes audible the poetic sources of a topos adopted by the British novel in the Romantic period: the visionary or lyric prospect. A character looks at a view and feels the intimation of a reality beyond what is available to the senses: the aura of an elsewhere, a hidden past or latent future, a numinous presence behind the scene, a totality investing it, an otherness that haunts the perceiving self. In the long eighteenth century, a georgic tradition of loco-descriptive poetry (Denham, Pope, Thomson, Dyer) infused the view of British landscape with a glamour of national destiny. Late Enlightenment discourses of the sublime and picturesque, applying ecphrasis to the forms of nature, brought aesthetic refinement. The "greater Romantic lyric" enhanced the prospect with metaphysical and cosmological resonances: opening it for an intuition of the imagination's sovereignty (the "glory" of Wordsworth's and Coleridge's odes of spiritual crisis), or, conversely, an external power aloof from human cognition (Shelley's summit of Mont Blanc). The prospect is the lyric realization of the organic tropism to stand upright "and command a large and distant view," to own the world in knowing it, which enacts a uniquely human becoming in the anthropology of Herder and Lamarck.[114]

The lyric prospect provided a rich resource for the novel as it expanded its formal repertoire in the Romantic period. In the English tradition, in a notable development, it became the property of a feminine viewing subject: decisively, in the novels of Ann Radcliffe, which cite a "Romantic" poetic canon (Shakespeare, Milton, Thomson, Collins, Mason, Gray) in epigraphs, embedded quotations, and formal imitations, condensed around the sensibility of an endangered heroine. (Both *The Romance of the Forest* and *The Mysteries of Udolpho* carry the subtitle, "interspersed with some pieces of poetry."[115]) These poetic citations signal a blending of narrative and lyric modes that achieves set-piece expression in the visionary prospect, in which the heroine finds solace in looking out at the view (archetypally nocturnal, moonlit) from her Gothic confinement. The forward motion of the narrative, brimming with threat, is paused for a suffusion of the scene with lyric atmosphere. Gazing at the night sky, Emily (in *The Mysteries of Udolpho*) hears mysterious distant music, reverberations of a lost idyllic past and a salvific future.

Writing in the 1790s, Radcliffe makes the prospect the generic topos of romance. (She is the first author to use the term systematically to differentiate her works from "the novel.")[116] The lyric prospect is also the emergent topos of what Fredric Jameson calls "affect," one of his "antinomies of realism": an aesthetic mood or climate to which the sensitive subject is receptive, but which is not exclusively hers, since it comes partially from outside, the emanation of a "scenic impulse."[117] As such, it is a providential sign of the heroine's privilege—a secular grace. Emily's attunement (*Stimmung*) to the scene and moment marks her special sensitivity and her occult power over the story: however besieged, apparently helpless, she will endure and flourish, because she stands at the center of reality. Scott adopts the device—with a concomitant feminization of the male protagonist—in *Waverley*. "He had now time to give himself up to the full romance of his situation," Scott writes of Waverley's entry into the Highlands, remarking his imaginative receptivity to the scenic impulse as he contemplates the view (also nocturnal, moonlit).[118] As in the case of Radcliffe's Emily, the promise of this romantic susceptibility is redeemed with the hero's safe delivery at the end of the story.

The greater Romantic lyric makes the prospect a vehicle for what M. H. Abrams called the "illuminated moment" or (simply) "the Moment": "a deeply significant experience in which an instant of consciousness, or else an ordinary object or event, suddenly blazes into revelation."[119] Abrams tracks its genealogy from Wordsworth, Hölderlin, and Shelley to its resurgence in early twentieth-century fiction, when it becomes a signature device of literary Modernism: Joyce's "epiphany," Conrad's and Woolf's "moments of vision."[120] The nineteenth-century novel is conspicuous by its absence from this genealogy. The radiant promise of Romanticism, overcast in the long dull afternoon of Victorian realism, breaks out again in Modernism's radical aesthetic experiments—which emancipate the novel, at last, from prose into poetry. Following Henri Bergson's philosophical reaffirmation of Romantic vitalism, the Modernists identify "life" as the mystic content, the revelation, of the illuminated moment. "Life is a luminous halo, a semi-transparent envelope surrounding us from the beginning of consciousness to the end," writes Virginia Woolf in 1919, pitting the "spiritual" art of the high Modernists, in quest of that vital aura, against the "materialists" (Bennett, Wells, Galsworthy), mechanical reproducers of a defunct Victorian realism, who "write of unimportant things [and] spend immense skill and immense industry making the trivial and the transitory appear the true and the enduring."[121] Realism has outworn its mediating function.

So we encounter, again, the Hegelian antithesis between "the poetry of the heart" and the "prose of the world," the banal praxis of novelistic realism, discussed in chapter 2. Where Woolf reaffirms the antithesis pragmatically, as an artistic credo, Abrams installs it as an aesthetic ideology, swallowing the conception of Romanticism as a literary-historical period, in one of the influential projects of Cold War-era humanities scholarship. The novels of Radcliffe, Austen, Scott, and others are written out of Romanticism—indeed, the "Romantic novel" as a category ceases to exist (save always the exception that proves the rule, *Frankenstein*). Its practitioners either perfect an eighteenth-century form (Austen) or rough out a Victorian prototype (Scott). The succession of periods, Romantic to Victorian, marks an aesthetic watershed or paradigm shift, realized in their respective typical forms, lyric and novel. Victorian realism, the prose of the world writ large, does not just chronologically follow the revolutionary opening of Romantic poetry: it closes it, covers it up, in a historical dialectic that yields, in turn, Modernism's reiteration of the Romantic rupture at an advanced technical level.[122]

The naturalization of the lyric prospect in nineteenth-century realism supplies the missing antithetical turn in this Romanticism-Modernism dialectic. Realist fiction replaces the prospect's transcendental intuition with an immanence of individual being in the world, in which the imagination inhabits the larger, implicitly total, field of social life and its natural and material environments. Its refined narrative technique is free indirect discourse, brought to early perfection by Austen, as we saw in the previous section of this chapter. In the scene from *Emma*, the heroine's view of village life, subordinating local detail to the abstract gravitational field of social and commercial traffic, triumphantly overcomes the Hegelian antagonism between novelistic realism and the poetry of spiritual fulfillment. Half a century later, free indirect discourse becomes the instrument of an omniscient narration encompassing a totality still greater than social life in the major novels of George Eliot. This enlargement of scope entails not only the physical enlargement of the novel (*Middlemarch* is as long as *Bleak House*, twice as long as *Emma*) but also a qualitative enlargement of what constitutes its world. Toward the close of *Middlemarch*, a spectacular realist transmutation of the lyric prospect signals a resolution of the heroine's spiritual crisis:

> She opened her curtains, and looked out towards the bit of road that lay in view, with fields beyond outside the entrance-gates. On the road there was a man with a bundle on his back and a woman carrying her baby; in the field she could see figures moving—perhaps the shepherd with his dog. Far off in the bending sky was the pearly light; and she felt the largeness of the world and the manifold wakings of men to labor and endurance. She was a part of that involuntary, palpitating life, and could neither look out on it from her luxurious shelter as a mere spectator, nor hide her eyes in selfish complaining.[123]

Eliot reclaims the prospect's visionary charge by a more radical naturalization. The intuition of a supersensory presence is that of "life" itself, organic matter and its formative drive. (I revisit this passage in the next chapter.) Thus *Middlemarch*, masterwork of Victorian realism, looks forward to the Modernist exaltation of life as mystic principle, as though glimpsing realism's horizon or limit: at which the novel, more than "a personal impression of life," becomes itself "a living thing, all one and continuous, like every other organism," in the words of Eliot's avowed successor Henry James.[124] At the same time, the prospect's centering of a human content—stabilized in the archetypal figures of the man with the bundle and the woman with her baby—guarantees the middle scale of human experience and meaning promised in the novel's title ("A Study of Provincial Life"): even as it infuses that content with a faint glimmer of obsolescence, that of pastoral, in the vague "figures moving" (at the vanishing point?) that might or might not be "the shepherd and his dog." (Shepherds would be unlikely in Middlemarch, an agricultural and commercial district.[125])

The lyric prospect and free indirect discourse, ascendant in Romantic poetry and fiction, invest from different angles a common project in nineteenth-century realism: giving the world a human form. Commensurate with what Quentin Meillassoux has called "correlationism," systematized in the critical philosophy of Kant, they are formal devices designed to fit human perception and experience to a world the natural sciences are disclosing as radically inhuman in scale and meaning.[126] Both devices shape the external world around human subjectivity—consciousness and feeling—so as to restore it to the center of the world: whether "the world" is posited as a phenomenological field, a totality of social relations, a geopolitical arena, or a network of vital forces. Thus Wordsworth announces his neo-epic project, in the "Preface" to *The Excursion*:

How exquisitely the individual Mind
(And the progressive powers perhaps no less
Of the whole species) to the external World
Is fitted:—and how exquisitely, too—
Theme this but little heard of among Men—
The external World is fitted to the Mind.[127]

The prospect tenders a promise of the world's response to the call of human presence in it, naturalized as an exquisite equivalence between imagination and world, lyrically apprehended. Where the great odes of Wordsworth and Coleridge rehearse the faltering of this equivalence, in a crisis of the poetic imagination, realism becomes the literary technology of the immanence of the human. Recent work in evolutionary literary studies has pressed the account of realism's anthropomorphic agenda to literal conclusions. Building on Lisa Zunshine's analysis of literary techniques of "mind reading," Blakey Vermeule argues that free indirect discourse, codifying a sophisticated ability to imagine other subjects' mental states, expresses a quintessentially human cognitive adaptation.[128]

Twenty years before *Middlemarch*, *Bleak House* unfolds its critique of what we may call (without condescension) the liberal-humanist biopolitics of Victorian realism, wrought by George Eliot to technical and ideological perfection. Dickens's recollection of Wordsworthian lyric to mark the blockage of the heroine's visionary illumination participates in a general derangement of realist narrative's tropism toward omniscience, the command of a fully humanized world. Rekindling the prospect's Romantic aura, *Bleak House* exploits the latent opposition between the ways the devices work: using the lyric prospect to break open the gap between the imagination and the world that an omniscient narration, founded on free indirect discourse, seeks to close. The baffling of Esther's visionary intuition, in her nocturnal prospect at Saint Albans, expresses the maimed relation between the structural components of free indirect discourse, severed by Dickens into discrete narratives. Her surge of feeling, as she gazes at the night sky over London, presses toward a revelation that does not arrive. Revelation is claimed instead by the impersonal narrative, flaunting the style of omniscience, which remains closed to her. After a resonance of that narrative's visionary style, fittingly attached to an unseen reflection ("the fancy which the redder light engendered of an unearthly fire, gleaming on all the unseen buildings of the city, and on all the faces of its many thousands of wondering inhabitants"), Esther's intuition fetches up against its impenetrable horizon. What might be revealed to the impersonal narrative in its flights across space and time remains inaccessible save as the purely formal impression of being "different from what I then was." The impersonal narrative resumes in the following chapter, which unveils, with sardonic bravura, the allegorical source of the "unearthly fire": the spontaneous combustion of Krook. That revelation wittily combines typological abstraction—Krook's fate rehearses the apocalyptic fate of the system—with hyperbolic sensuous particularity, as Guppy and Jobling smell, touch, and *taste* the combustion's loathsome material residue. Closed to Esther, the event poses a challenge to readers of *Bleak House*, daring us to reach for other aesthetic criteria, such as the devices of allegory, to make sense of what we are reading.[129]

The Noise of the World

Hundreds of pages later, near the end of the novel, Esther's narrative returns to the scene of her mysterious epiphany, only for the explanation discovered to seem disappointingly literal and inconsequential. Conversing with the detective Mr. Bucket, Esther learns the source of one of those ordinary sounds she had heard:

> As we ascended the hill, he looked about him with a sharp eye—the day was now breaking—and reminded me that I had come down it one night, as I had reason for remembering, with my little servant and poor Jo: whom he called Toughey.
> I wondered how he knew that.
> "When you passed a man upon the road, just yonder, you know," said Mr Bucket.
> Yes, I remembered that too, very well.
> "That was me," said Mr Bucket.
> Seeing my surprise, he went on:
> "I drove down in a gig that afternoon, to look after that boy. You might have heard my wheels when you came out to look after him yourself, for I was aware of you and your little maid going up, when I was walking the horse down." (872–73)

The "wheels coming down the miry hill" were Mr. Bucket's, on his way to "move on" the tough subject Jo—too late, however, to prevent Jo from transmitting his fever to Esther. Twenty-six chapters—ten months—after the scene in question, few if any of the first readers of *Bleak House* (or many readers since) will have noticed Dickens's care in tying up this particular loose thread. Like a fragment of debris of the omniscient narration, stranded on Esther's portion of the pages, it deepens the mystery of a revelation withheld.

The teasing edge of Mr. Bucket's explanation is sharpened by its placement amid the suspenseful sequence of Lady Dedlock's flight and pursuit, directly following a moment—unique in the novel—when it seems as though the two narratives might at last converge. The impersonal narrative has relayed Bucket's search for the fugitive up to the point when he decides he needs Esther's assistance and calls for her at Mr. Jarndyce's London lodgings. The chapter (with the part-issue number) breaks off, as he waits for her to come downstairs. When the novel resumes, a month later, it is with Esther's narrative, narrowly averting a seismic collision of its component planes. (Esther never appears in the impersonal narrative—just as the impersonal narrator never shows up in hers.[130]) Right before the break, at the very end of the installment, while Mr. Bucket is waiting for Esther, the impersonal narrator frames his speculations as to where Lady Dedlock might have gone with an extraordinary conjunction of the formal devices of the visionary prospect and free indirect discourse that, failing to be either, presents a parody of both:

> There, he mounts a high tower in his mind, and looks out, far and wide. Many solitary figures he perceives, creeping through the streets; many solitary figures out on heaths, and roads, and lying under haystacks. But the figure that he seeks, is not among them. Other solitaries he perceives, in nooks of bridges, looking over; and in shadowed places down by the river's level; and a dark, dark, shapeless object drifting with the tide, more solitary than all, clings with a drowning hold on his attention.
>
> Where is she? Living or dead, where is she? If, as he folds the handkerchief and carefully puts it up, it were able, with an enchanted power, to bring before him the place where she found it, and the night landscape near the cottage where it covered the little child, would he descry her there? On the waste, where the brick kilns are burning with a pale blue flare; where the straw-roofs of the wretched huts in which the bricks are made, are being scattered by the wind; where the clay and water are hard frozen, and the mill in which the gaunt blind horse goes round all day, looks like an instrument of human torture;—traversing this deserted blighted spot, there is a lonely figure with the sad world to itself, pelted by the snow and driven by the wind, and cast out, it would seem, from all companionship. It is the figure of a woman, too; but it is miserably dressed, and no such clothes ever came through the hall, and out at the great door, of the Dedlock mansion. (864)

Earlier, the impersonal narrator has facetiously tagged Mr. Bucket as a proxy for omniscience within the story, and as such the narrator's nearest rival: "Time and place cannot bind Mr. Bucket. Like man in the abstract, he is here today and gone tomorrow—but, very unlike man indeed, he is here again the next day" (803). Here again the next day, Mr. Bucket is bound by time and place after all. The baffling of his visionary prospect sets off the narrator's successful tracking of the lonely figure, in the ensuing paragraph, according to the operational logic by which omniscient narratives secure their formal authority at the expense of their characters.[131] The gap between Bucket's detective vision and the narrator's, highlighted by the Dickensian metonymy (like an authorial signature) of the magic handkerchief, rehearses within the impersonal narrative itself the disassembly of free indirect discourse that the structure of *Bleak House* enacts at large.

I turn, lastly, to a third instance of the lyric prospect in *Bleak House*: one anchored, this time, to none of the novel's characters, and so belonging wholly to the impersonal narrative. We, the novel's readers, are sole recipients of the opened vision. Dickens expands and darkens the technique of narration from nobody, or nowhere, that is the hallmark of the impersonal narrative at its most assertive:

> A very quiet night. When the moon shines very brilliantly, a solitude and stillness seem to proceed from her, that influence even crowded places full of life. Not only is it a still night on dusty high roads and on hill-summits, whence a wide expanse of country may be seen in repose, quieter and quieter as it spreads away into a fringe of trees against the sky, with the grey ghost of a bloom upon them; not only is it a still night in gardens and in woods, and on the river where the water-meadows are fresh and green, and the stream sparkles on among pleasant islands, murmuring weirs, and whispering rushes; not only does the stillness attend it as it flows where houses cluster thick, where many bridges are reflected in it, where wharves and shipping make it black and awful, where it winds from these disfigurements through marshes whose grim beacons stand like skeletons washed ashore, where it expands through the bolder region of rising grounds rich in corn-field, windmill and steeple, and where it mingles with the ever-heaving sea; not only is it a still night on the deep, and on the shore where the watcher stands to see the ship with her spread wings cross the path of light that appears to be presented to only him; but even on this stranger's wilderness of London there is some rest. Its steeples and towers and its one great dome grow more ethereal; its smoky house-tops lose their grossness, in

the pale effulgence; the noises that arise from the streets are fewer and are softened, and the footsteps on the pavements pass more tranquilly away. In these fields of Mr. Tulkinghorn's inhabiting, where the shepherds play on Chancery pipes that have no stop, and keep their sheep in the fold by hook and by crook until they have shorn them exceeding close, every noise is merged, this moonlight night, into a distant ringing hum, as if the city were a vast glass, vibrating.

What's that? Who fired a gun or pistol? Where was it? (748–49)

Vision expands across country, beyond any local viewpoint, inserting a human perspective only to underscore its bondage to parallax, to the contingent illusion of centeredness (the watcher on the shore who "stands to see the ship with her spread wings cross the path of light that appears to be presented to only him").[132] A social discourse is invoked but to register its emptiness, in the sardonic mock-pastoral of Lincoln's Inn Fields. And, as in Esther's nocturnal prospect at Saint Albans, vision fades to sound, the medium of abstraction—shattered by the (syncopated) gunshot that fells Mr. Tulkinghorn in his chambers, and the inrushing cries, suddenly localized, of his startled neighbors.

This astonishing sequence ("perhaps the most ambitious piece of sustained descriptive writing in the novel"[133]) strikes the opposite chord to the poem it echoes, Wordsworth's sonnet "Composed upon Westminster Bridge, September 3, 1802":

> This City now doth, like a garment, wear
> The beauty of the morning; silent, bare,
> Ships, towers, domes, theatres, and temples lie
> Open unto the fields, and to the sky;
> All bright and glittering in the smokeless air.
> Never did sun more beautifully steep
> In his first splendour, valley, rock, or hill;
> Ne'er saw I, never felt, a calm so deep!
> The river glideth at his own sweet will:
> Dear God! the very houses seem asleep;
> And all that mighty heart is lying still![134]

At once clothed and laid bare, stilled and silent in the revelation of morning, the city regains its primeval human—or superhuman—form: that of the titan Themis, perhaps, since this is the River Thames. It has a "mighty heart," although one, perhaps ominously, that is "lying still." But Dickens's nighttime epiphany discloses the city as inhuman and inorganic.

Even though the towers and dome may "grow more ethereal" in the moonlight, the figure of the "vast glass" does not invoke transparency—a view into the life of things, as Wordsworth elsewhere puts it. The vast glass would have had a particular association for Dickens's first readers, who flocked in their thousands to the Great Exhibition at the Crystal Palace in Hyde Park in 1851, shortly before he began writing *Bleak House*. Scholars have discussed the novel's rebuke to the Exhibition's triumphalist rendering of the world as a visible, consumable spectacle with London at its center.[135] Now, in a nightmarish totalization, the whole city becomes "a vast glass, vibrating": except that the glass is not an optical but an acoustic medium. Sight, with its promise of clarity, cedes to sound, a merging of all particular sounds into a "distant ringing hum"—not the music of the spheres, but an indiscriminate, insensate, universal noise, a vibration that cannot be decoded into any signal we might recognize.

In splitting apart the organic form of free indirect discourse-based omniscient narration, Dickens's technical division of the narrative splits the figure of the human it stands in for—a human nature that realist fiction renders as whole, rounded, integrated with the world (with its social and historical being, with "life"). Each of the narratives of *Bleak House* invokes the visionary rhetoric of the Wordsworthian romantic prospect; but in each case vision fails, and is replaced with something else—the noise of the world, meaning at once the world's particular acoustic traces (in Esther's prospect) and also (in the impersonal narrator's) a manifestation of the world *as* noise. In each case the story resumes, and so does our reading: bereft of the confidence, however, that we will eventually be able to see everything, even if we continue to feel that just beyond the limit of our senses there *is* an everything, a totality, a system—one that, although of our making, is no longer ours.

"The characters cannot perceive the design, but it is really there," writes George Levine of *Bleak House*.[136] In *Vestiges of the Natural History of Creation*, Robert Chambers reflects upon the ultimate order of the universe encoded in the circular system of classification. Stripped of contingent appearances, nature is revealed as art—no longer a living and changing body, but a "rigid" thing, an effigy:

> The system of representation is therefore to be regarded as *a powerful additional proof of the hypothesis of organic progress by virtue of law*. It establishes the unity of animated nature and the definite character of its entire constitution. It enables us to see how, under the flowing robes of nature, where all looks arbitrary and accidental, there is an

artificiality of the most rigid kind. The natural, we now perceive, sinks into and merges in a Higher Artificial. To adopt a comparison more apt than dignified, we may be said to be placed here as insects are in a garden of the old style. Our first unassisted view is limited, and we perceive only the irregularities of the minute surface, and single shrubs which appear arbitrarily scattered. But our view at length extending and becoming more comprehensive, we begin to see parterres balancing each other, trees, statues, and arbours placed symmetrically, and that the whole is an assemblage of parts mutually reflective. The insects of the garden, supposing them to be invested with reasoning power, and aware how artificial are their own works, might of course very reasonably conclude that, being in its totality an artificial object, the garden was the work of some maker or artificer. And so also must we conclude, when we attain a knowledge of the artificiality which is at the basis of nature, that nature is wholly the production of a Being resembling, but infinitely greater than ourselves. (250–51)

In *Bleak House*, the human perspective—however intelligent or virtuous its bearer—scarcely exceeds the insect's. Thus Esther, in her fever delirium:

Dare I hint at that worse time when, strung together somewhere in great black space, there was a flaming necklace, or ring, or starry circle of some kind, of which *I* was one of the beads! And when my only prayer was to be taken off from the rest and when it was such inexplicable agony and misery to be a part of the dreadful thing? (556)

Her dream (traditionally the medium of allegorical vision) yields a glimpse of entrapment in a total order: the form of the novel, *Bleak House*, and the chain of being, "arranged along a series of close affinities," according to Chambers, "*in a circular form*" (238). It accompanies Esther's (temporary) blindness and the obliteration of her human face. Throughout *Bleak House*, we learn that the attempt to command the world as a system, to see it whole, corrodes one's humanity. Better to keep our eyes fixed on the task at hand, to do our duty without complaint. The critical force of Dickens's great novel registers in the terrible discrepancy between this dour ethical charge and the immense, vivid, overwhelming world of life it opens to our reading.

CHAPTER FIVE

George Eliot's Science Fiction

Are there yet to be species superior to us in organization, purer in feeling, more powerful in device and act, and who shall take a rule over us! There is in this nothing improbable on other grounds.... There may then be occasion for a nobler type of humanity, which shall complete the zoological circle on this planet, and realize some of the dreams of the purest spirits of the present race.

—ROBERT CHAMBERS, *VESTIGES OF THE NATURAL HISTORY OF CREATION* (1844)

Do all humans have to be human beings? It is possible for beings quite other than human to exist in human form.

—NOVALIS, "STUDIES IN THE VISUAL ARTS" (1799)[1]

We Belated Historians

The main business of *Middlemarch*, formulated as the premise of its opening rhetorical question, is with a scientific project, "the history of man":

> Who that cares much to know the history of man, and how the mysterious mixture behaves under the varying experiments of Time, has not dwelt, at least briefly, on the life of Saint Theresa, has not smiled with some gentleness at the thought of the little girl walking forth one morning hand-in-hand with her still smaller brother, to go and seek martyrdom in the country of the Moors?[2]

Whereas the sixteenth-century saint, barred from fields of masculine endeavor, "found her epos in the reform of a religious order," modern

Englishwomen—"later-born Theresas"—have no access to an institutional medium within which they might realize "a national idea" (3). Their aspirations and struggles leave no trace on the historical record. Here the novel makes its claim on scientific knowledge: It is the literary genre that can give an account of those invisible destinies, submerged in private life, confined to "unhistoric acts," in the form of a conjectural case history—the life of "a new Theresa," Dorothea Brooke (838).

George Eliot's appeal to the history of man revises Henry Fielding's pledge at the opening of *Tom Jones*: "The provision which we have here made, is no other than HUMAN NATURE."[3] Fielding's manifesto for the rise of the novel dignified the "new province of writing" by annexing it to the philosophical discourse David Hume had called "the science of man" and "the science of human nature." "In pretending to explain the principles of human nature," Hume wrote, "we in effect propose a compleat system of the sciences built on a foundation almost entirely new, and the only one upon which they can stand with any security."[4] This science of all sciences will furnish a new, secure foundation for the novel, a genre lacking high-cultural credentials, such as a classical genealogy.

Fielding had attained classical stature by the time *Middlemarch* was published, an event viewed then and since as confirming the novel's ascendancy in the modern field of literary genres.[5] George Eliot pays tribute to her forebear in one of the essayistic reflections interspersed throughout *Middlemarch*:

> A great historian, as he insisted on calling himself, who had the happiness to be dead a hundred and twenty years ago, and so to take his place among the colossi whose huge legs our living pettiness is observed to walk under, glories in his copious remarks and digressions as the least imitable part of his work.... But Fielding lived when the days were longer (for time, like money, is measured by our needs), when summer afternoons were spacious, and the clock ticked slowly in the winter evenings. We belated historians must not linger after his example.... I at least have so much to do in unravelling certain human lots, and seeing how they were woven and interwoven, that all the light I can command must be concentrated on this particular web, and not dispersed over that tempting range of relevancies called the universe. (141)

Fielding is a great historian and the author of *Middlemarch* a belated one, where "history" occupies the fertile middle ground between literature and science. Fielding called his work a history (*The History of Tom Jones, A Foundling*), meaning the narrative of an individual life, while avowing the unity and consistency of his multifarious theme, human nature. George Eliot, with stylized anxiety (peering and picking at an illimitable text by candlelight), calls her topic "the history of man"—a history in which man is a "mysterious mixture," no longer homogeneous or stable but subject to "the varying experiments of Time." Time makes the difference: Fielding is greater than us because he had more time. The diminished state of the human present is measured by a felt temporal acceleration (and a concomitant intensification of "our needs"), even as the universe, the potential totality of "human lots," expands beyond our field of view. Fielding could enjoy longer days, more spacious summer afternoons, in a universe less than six thousand years old. The scientific revolution issuing in Lyell's *Principles of Geology* in the early 1830s (the period at which *Middlemarch* is set) stretched out the history of the world to an unimaginable magnitude. In 1869, the year Eliot began writing *Middlemarch*, T. H. Huxley was locking horns with William Thomson in a public debate over the age of the earth: Huxley criticized Thomson's estimate (based on his thermodynamic theory) of a range between 20 and 400 million years, with an outer limit of 100 million years for the evolution of life, as insufficient to accommodate the interaction of heritable variation with natural selection prescribed in Charles Darwin's theory of the transmutation of species.[6]

Darwin and George Eliot occupy the horizon of this book's argument. The relation between them has been extensively studied, usually in terms of a congruence of worldviews and rhetorical strategies; it forms the centerpiece of Gillian Beer's classic monograph *Darwin's Plots*. I will be considering a George Eliot who was in some ways less Darwinian, or more troubled by Darwin's radical insight, than some recent accounts suggest. While her literary career coincided with Darwin's, she did not immediately digest his theory; her fiction activates other developmental forces besides natural selection, and deranges the scientific thought it brings into play. In doing so, it churns up the not-yet-settled, volatile currents of that scientific thought—including Darwin's, who was not always (himself) a pure Darwinist. With that, it deranges its own aesthetic protocols, so often read as an Olympian consummation of Victorian realism. "To a degree that the catchall term 'realism' obscures," writes Lauren Goodlad, "Eliot's oeuvre is generically diverse, bold, and experimental."[7] This chapter seeks to recapture the unsettling force of that experimentalism: to make George Eliot strange again.

George Eliot's first full-length novel, *Adam Bede*, was published nine months before *On the Origin of Species*, in February 1859; the first serial installment of *Middlemarch* appeared in December 1871, ten months after *The Descent of Man*. Eliot was halfway through writing *The Mill on the Floss* when she and George Henry Lewes read the *Origin* together. They were primed for a favorable reception. The book "makes an epoch," Eliot told correspondents, as an affirmation of the "Doctrine of Development" by a "long celebrated naturalist." She compares the *Origin* with the anonymous transformist synthesis *Vestiges of the Natural History of Creation*: although his book is "ill-written," lacking Chambers's popular touch, Darwin's professional authority will secure "a great effect in the scientific world."[8] Not yet appreciating the full force of the theory of natural selection, Eliot assimilated Darwin's argument to the more diffuse "Development Hypothesis" promoted in the 1850s by Lewes and Herbert Spencer, who drew on the embryological research of Karl Ernst von Baer and his British followers, among whom were Chambers, William Carpenter, and Richard Owen.[9] Von Baer's account of embryonic formation as a process of "increasing differentiation and specialization" provided the analogy between organic growth and evolutionary history that would constitute the dominant conception of development, as progressive, purposive, and all-pervasive, throughout the nineteenth century: "Just as the embryo grows inevitably toward its mature form, so the history of life ascends through a fixed hierarchy of stages toward its goal."[10]

Expanded to include "the history of man," in Auguste Comte's sociological synthesis, development was an article of faith in the progressive circle of *The Westminster Review* when Marian Evans joined it in 1851. Lewes's articles "Lyell and Owen on Development," "Von Baer on the Development Hypothesis," and "The Development Hypothesis of [Chambers's] 'Vestiges'" appeared in *The Leader* in 1851 and 1853, Spencer's "The Development Hypothesis" in the same journal in 1852.[11] The analogy with growth encouraged "a developmental, or genetical, rather than a truly historical view of the past," according to Peter Bowler, who argues that Darwin's theory served as a catalyst for the wider cultural acceptance of development, rather than installing a new—radically etiological and materialist— paradigm: that would not achieve general currency until the synthesis of natural selection with Mendelian genetics in the 1920s and 1930s.[12] Meanwhile, Victorian scientists and philosophers supplemented natural selection with other agencies, broadly Lamarckian or transcendental, when they did not sideline it altogether.

Lewes's major assessment of 1868, the four-part essay "Mr. Darwin's Hypotheses," is a case in point. Citing Ernst Haeckel, Lewes looks back across the century-long rise of the Development Hypothesis from "the 'Theoria Generationis' of Wolff (1759), which by the doctrine of Epigenesis laid the foundation-stone of the theory of Development, [to] the 'Origin of Species' (1859), which supplied the coping-stone."[13] Lewes allows Darwin to have provided the most comprehensive explanation to date of a mechanism for the evolution of species. He balks, however, at accepting natural selection as sole causal principle, with its entailments of "community of kinship" and a single origin of terrestrial life, and charges Darwin with having underestimated the contribution of the "laws of organic growth," uniformly operative across diverse lines of development stemming from multiple points of origin.[14] Beer calls Lewes's essay, published in the year before she began *Middlemarch*, "a watershed in George Eliot's understanding of the implications of Darwin's thought."[15] Comments in her correspondence show Eliot sharing Lewes's skepticism about the explanatory scope of Darwin's theory via its insistence on a unitary principle.[16] That principle, natural selection, constituted a radical limit, a horizon—or rather a precipice—of thought, for George Eliot as it did for her contemporaries: as it did, indeed, for Darwin himself, who would cautiously admit other evolutionary forces into his argument. Spencer and Lewes, drawing on Von Baer's laws of embryonic development, advocated a universal "law of Progress" from homogeneity to heterogeneity of structure—articulated as a general evolutionary principle, before Von Baer, by Lamarck: "Nature has produced all the species of animals in succession, beginning with the most imperfect or simplest, and ending her work with the most perfect, so as to create a gradually increasing complexity in their organization."[17] Darwin too shared the general commitment to an "inevitable advance from the more general to the more special form"—explicitly characterized, in the second and subsequent editions of the *Origin of Species*, as an advance from lower to higher states.[18] Eventually natural selection would shake the evolutionary process loose from natural-theological and idealist archetypal and teleological scaffolding for an etiological, directionless, open-ended dynamic of infinite random formal differentiation, shaped by the complex, contingent interactions of living forms within a shared environment: but for many Victorian thinkers, and intermittently even for its author, not yet. All the same, the theory was out there, painstakingly and lucidly expounded for those willing to read it, in the 1859 *Origin*, a boundary marker of nineteenth-century thought.

On the Origin of Species made the case for the new natural history; *The Descent of Man* sank human life in it, inundating the high ground of human exceptionalism. The belated historian of *Middlemarch* writes as the history of man is undergoing drastic, epochal change. Preceded by sublime reaches of geological time, untenanted by human life, it is followed by a no less immense, potentially infinite duration. "Judging from the past, we may safely infer that not one living species will transmit its unaltered likeness to a distant futurity," Darwin writes in the conclusion to *On the Origin of Species*.[19] Just as it has emerged out of some prior biological form, humanity will continue to evolve: into a more perfect, superhuman race, as Alfred Russel Wallace, William Winwood Reade, and others surmised (after Chambers and Spencer) in Darwin's wake.[20] Or else it might mutate into forms scarcely if at all imaginable to us now—or disappear altogether: "Of the species now living very few will transmit progeny of any kind to a far distant futurity," Darwin goes on, "[since] the greater number of species of each genus, and all the species of many genera, have left no descendants, but have become utterly extinct."[21] In Elizabeth Grosz's summary:

> Darwin has effected a new kind of humanity, a new kind of "enlightenment," neither modeled on man's resemblance to the sovereignty of God nor on man's presumed right to the mastery of nature, but a fleeting humanity whose destiny is self-overcoming, a humanity that no longer knows or masters itself, a humanity doomed to undo itself, that does not regulate or order materiality but becomes other in spite of itself, that returns to those animal forces that enable all of life to ceaselessly become.[22]

The dispersal of interwoven human lots across the universe daunts the author of *Middlemarch* as a temporal prospect, "at once alluring and yet artistically and existentially threatening,"[23] more than as a spatial one.

Knowledge and Its Languages

Grosz's account of the philosophical fallout of Darwin's theory clarifies its radical force by abstracting the logic of the theory from Darwin's writing of it. Darwin expanded the role of (Lamarckian) heritable use as a causal agency alongside natural selection, in the absence of experimentally verified laws of heredity, in later editions of *On the Origin of Species* as well as in subsequent works such as *The Variation of Animals and Plants under Domestication* and *The Descent of Man*, along with an increasingly progressive and teleological emphasis. *The Descent of Man* amplifies Darwin's account of sexual selection, an ancillary evolutionary agency sketched in the *Origin of Species*, to grant it quasi-autonomous force—in dialectical interaction with natural selection—and make it the main motor of human racial differentiation and cultural development.[24] Darwin's writings also bear witness to their author's self-conscious struggle with his medium, in a fraught relation between the theory and the language of its articulation. It is here, in the knot between language's transparency to thought and its muddy figurative matter, that the projects of Darwin and Eliot become most intricately entangled.

The "considerable revolution in natural history" Darwin foresees his argument initiating will be, in the first place, a rhetorical revolution:

> We shall have to treat species in the same manner as those naturalists treat genera, who admit that genera are merely artificial combinations made for convenience.... The terms used by naturalists of affinity, relationship, community of type, paternity, morphology, adaptive characters, rudimentary and aborted organs, &c., will cease to be metaphorical, and will have a plain signification.[25]

Darwin invokes the denotative register of scientific language aspired to by natural philosophers since Thomas Sprat praised the Royal Society, two centuries earlier, for its "Resolution, to reject all the amplifications, digressions, and swellings of style: to return back to the primitive purity, and shortness, when men deliver'd so many *things*, almost in an equal number of *words* . . . bringing all things as near the Mathematical plainness, as they can."[26] Darwin also recognizes that a revolution is a two-way transformation, casting some down as it lifts others up. At the same time as he promotes the language of affinity and kinship from metaphorical to literal rank, he downgrades "the term species" to an "artificial [combination] made for convenience."

For now, though, plain signification lies in the future, and Darwin must keep wrangling with his argument's figural substance.[27] He opens *The Variation of Animals and Plants under Domestication* (1868) with an apology:

> The term "natural selection" is in some respects a bad one, as it seems to imply conscious choice; but this will be disregarded after a little familiarity.... For brevity's sake I sometimes speak of natural selection as an intelligent power;—in the same way as astronomers speak of the attraction of gravity as ruling the movements of the planets, or as agriculturists speak of man making domestic races by his power of selection.... I have, also, often personified the word Nature; for I have found it difficult to avoid this ambiguity; but I mean by nature only the aggregate action and product of many natural laws,—and by laws only the ascertained sequence of events.[28]

(The declension of "Nature" via "laws" to "the ascertained sequence of events" is classically Humean.) Darwin was far from alone in pondering the issue. William Whewell, who introduced German hermeneutics to British scientific thought, identifies the refinement of a literal language as essential to the advancement of knowledge in his magisterial reckoning with the early Victorian disciplines, *The Philosophy of the Inductive Sciences*.[29] Himself responsible for such coinages as "anode," "cathode," "ion," and "scientist," Whewell affirms that "almost every step in the progress of science is marked by the formation or appropriation of a technical term." Everyday knowledge-work involves the affections and fancy as well as the intellect; hence "common language" always bears "a tinge of emotion or of imagination":

> But when our knowledge becomes perfectly exact and purely intellectual, we require a language which shall also be exact and intellectual;— which shall exclude alike vagueness and fancy, imperfection and superfluity;—in which each term shall convey a meaning steadily fixed and rigorously limited. Such a language that of science becomes, through the use of Technical Terms.[30]

For Whewell, as for Darwin, scientific language is not found in the world, a spontaneous emanation of data accurately observed. It has to be made: wrought from the "vagueness and fancy, imperfection and superfluity" of vernacular usage, language's natural state. In the original Greek, the terms of geometry, "besides the designation of form, implied some use or application": a sphere was "a hand-ball used in games, a cone . . . a boy's spinning-top, or the crest of a helmet," and so on, "till these words were adopted by the geometers, and made to signify among them pure modifications of space."[31]

Two linguistic stages, according to Whewell, articulate the formation of scientific knowledge. First, a breach in the received knowledge-system, made by a discoverer's inspired act of figuration: a personification, like "natural selection," an abstraction, like "cone," or one of the other kinds Whewell surveys in his concluding "Aphorisms Concerning the Language of Science." Second, the naturalization of the figure, and hence a new order of knowledge, by familiar usage, until its original associations—the aura of its artifice—are worn away, and it can function literally. Technical language is a second-order figuration that conceals its figural origins. Meanwhile, the old terminology loses its denotative edge and dissolves back into the state of metaphor—of "artificial combinations made for convenience."

Shortly before she turned to writing fiction, George Eliot repudiated the scientific ideal of a "perfectly exact and purely intellectual" language:

Suppose ... that the effect which has been again and again made to construct a universal language on a rational basis has at length succeeded, and that you have a language which has no uncertainty, no whims of idiom, no cumbrous forms, no fitful simmer of many-hued significance, no hoary archaisms "familiar with forgotten years"—a patent de-odorized and non resonant language, which effects the purpose of communication as perfectly and rapidly as algebraic signs. Your language may be a perfect medium of expression to science, but will never express *life*, which is a great deal more than science.[32]

Victorian reviewers noted George Eliot's recourse to "the high-scientific style" in her great novels of the 1870s, *Middlemarch* and *Daniel Deronda*, with some applauding her enlargement of fiction's lexical range and others complaining of an encroachment of alien jargon.[33] Recent scholarship has analyzed Eliot's usage of terms and concepts from evolutionary and cell biology, clinical pathology, neurophysiology, psychology, physics, mathematics, astronomy, chemistry, political economy, sociology, comparative mythology, and anthropology.[34] Much of that language, as Beer points out, has long since shed the "freight of controversy" that accompanied its technical appropriation.[35] Some terms have become so naturalized that they no longer register as technical, while others, superseded by developments in their field, have relapsed into a quaint metaphoricity. One aim of the present chapter is to restore the strangeness of scientific language in George Eliot's fiction by attending to its insistent oscillation between literal and figural registers—an oscillation that is intrinsic to the fiction's serious claim on, rather than skeptical disavowal of, scientific truth. Eliot's novels claim as their symbolic field the temporality of the early formation of knowledge, before its fixation into a "patent de-odorized and non resonant language": a temporality in which emergent knowledge may lose form, devolving back into an uncanny figurality, as well as gain it, coalescing into the contours of the real.[36] The novels assume a dissonant, competitive, even combative relation to the great scientific projects of the age. Far from thereby forfeiting scientific status, they reclaim it in their own distinctively novelistic terms. In Darwin's argument, Beer remarks, "the gap between metaphor and actuality was closed up, the fictive became substantive. Fictional insights were confirmed as physical event."[37] But the gap keeps reopening, within as well as between each writer's practice; it charges Eliot's fiction with its imaginative force and, even, occasional violence. We read the emergence of the literal on the far side of figuration and its relapse into a now grotesque figurality.

This chapter views Eliot's realism, the triumphant achievement of the mode in English, as an invention meant to shore up human nature against its prospective dissolution—rather than as an organic or teleological principle in the history of the novel, expressive of that nature. Eliot's high technical prowess, philosophical ambition, and moral seriousness combine to fix the fleeting figure of man, the mysterious mixture, subject to the varying experiments of time. The historicization of nature and humanity and the invocation of a dynamic, transhuman principle of life, radical principles of the new biology, supply the philosophical conditions for the humanism of *Middlemarch* and of Victorian realism more generally. Eliot's fiction extends its imaginative scope all the way down, to the organic stuff of life, as well as upward and outward, to the macroscopic register of the universe—strategically circumscribed, in her novels with English settings, within the fold of "provincial life," until broken open to a vaster world-horizon in *Daniel Deronda*.

George Eliot's later novels exert critical pressure on the developmental analogy between organic formation and historical progress; Devin Griffiths has characterized their experimental twisting and breaking of the conjunction as, instead, "disanalogy," a generation of knowledge through "productive error."[38] The analogy comes oppressively to bear on women, blocked by their gender from the vocational paths of *Bildung* and driven, instead, into the marriage plot—to biological reproduction as the end of life. Organic form asserts itself as a biopolitical imperative over and against the humane ideal of *Bildung* as self-culture. *The Mill on the Floss*, Eliot's revision of the tragic female Bildungsroman pioneered by Mme. de Staël in *Corinne*, worries at the conflation of biological with social formation—masking their analogical, hence figural relation—in the evolutionary argument of Spencer.[39] Eliot treats *Bildung* as the struggle of fully individuated human being to emerge from a primitive organic state. Located in childhood, that organic state also invests the conservative social medium of provincial life, which retards and thwarts the full development of individual personality. *Bildung*, the aspiration to uniquely human species being through individuation, ends up reproducing the struggle internally as a tragic knowledge of self-division. That knowledge's dialectical alternate, a poetic state of suspended consciousness, reiterates in a higher, spiritual key the mindless organicism of provincial society: literalized with the heroine's annihilation by a catastrophic resurgence of "nature" at the novel's close.

With *Middlemarch*, George Eliot realizes in full the project of the novel as a medium of scientific inquiry. Eliot mobilizes the techniques of realism for a total literary form enclosing an array of multiple life stories, a representative sample of human life. Again the imperatives of *Bildung* and marriage collide, bringing to grief the novel's case histories of (this time) masculine scientific inquiry. Meanwhile Dorothea, sentenced to marriage and domesticity, finds recompense in a strenuous ethical commitment to organic life that overcomes tragic resignation to it. The novel awards her, and us, an intuition of the immanence of life itself, which constitutes a vanishing point for the novel's realism and its condition, the human form and scale. Organic life looms as a global horizon or limit for the novel's representational domain, identified in the subtitle of *Middlemarch* as "provincial life," a synecdoche for national life and its tenant, human nature.

The condition of provincial life as a limit—categorically excluding a totality—comes into disconcerting focus in *Daniel Deronda*. In contrast to its predecessors, *Daniel Deronda* confronts the historical present, and assumes the advanced scientific discourses of evolutionary anthropology and biology. Rather than assigning them to fallible characters, the novel distributes these discourses across its narrative, where they swivel between literal and figural registers, forestalling the consolidation of a synthetic or omniscient viewpoint. Organic life now subsumes *Bildung*, for the eponymous hero, in the form of race: Daniel finds his vocation, and a world-historical destiny, in the embodiment of a revealed Jewish ancestry. The (perverse) marking of this Jewish ancestry as paternal, patriarchal, underscores the sacrificial logic whereby women's destinies are once again subdued to the masculine master-plot of *Bildung*. Gwendolen Harleth, the novel's rival protagonist, is relegated to a twilit half-life of penitence; Daniel's bride, Mirah, sinks her life in their marriage; and Daniel's mother, who rebelled (Corinne-like) against the patriarchal sentence of organic destiny, ends in tragic desolation. Weird distortions of the novel's representational norms ensue. Language of organic development, progress, and revival warps into that of survival, atavism, and reversion; effects become causes; metaphors events. Human nature, the experimental subject of the English novel in its classical phase, mutates into strange forms, as Eliot's last novel presses beyond the bounds of realism into a kind of science fiction. In its rearview mirror, *Daniel Deronda* discloses the kinship even of *Middlemarch* with those overtly fantastic works—the Gothic novella "The Lifted Veil," the dystopian fable "Shadows of the Coming Race"—that are usually read as outliers in the George Eliot canon.

Species Consciousness

The Development Hypothesis involved not only pre-Darwinian articulations of epigenetic embryology and Lamarckian transformism but also the application of laws of organic development to social systems.[40] Comte's *Cours de philosophie positive* (1830–42) was popularized in Great Britain by the *Westminster Review* circle, with Harriet Martineau's free translation, *The Positive Philosophy of Auguste Comte*, jostling beside Lewes's critical exposition, *Comte's Philosophy of the Sciences*, in 1853. Comte based his "social physics" on the principle that "the succession of social states exactly corresponds, in a scientific sense, with the gradation of organisms in biology."[41] He derived the schema of a progressive organic scale from Lamarck, whom Comte hailed (although rejecting species transmutation) as having "by far the clearer and profounder conception of the organic hierarchy" than Lamarck's opponent Cuvier.[42] Adopting Comte's conception of the "social organism," freely drawing also on Von Baer's embryology, Herbert Spencer stiffened the hypothesis of a biological determination of social forms into a universal "law of organic progress," consisting of "an advance from homogeneity of structure to heterogeneity of structure":

> Whether it be in the development of the Earth, in the development of Life upon its surface, in the development of Society, of Government, of Manufactures, of Commerce, of Language, Literature, Science, Art, this same evolution of the simple into the complex, through a process of continuous differentiation, holds throughout.[43]

Evolution toward complexity, posited as an immanent principle in nature by Lamarck, entails a "tendency to individuation" accompanied by an increased interdependency among the component parts. "In man we see the highest manifestation of this tendency. By virtue of his complexity of structure, he is furthest removed from the inorganic world in which there is least individuality."[44] In his study of Comte, Lewes stresses a dialectical relation between the individual and its medium, including its "social medium": "so far from organic bodies being independent of external circumstances, they become more and more dependent on them as their organization becomes higher, so that *organism* and *medium* are the two correlative ideas of life."[45] These articulations of the Development Hypothesis express key principles for George Eliot's art. The more evolved an organism is, the more complex it is, with complexity entailing the organism's stronger individuation and, at the same time, its increased interdependency with its medium. "The highest example of Form," Eliot writes in her unpublished essay "Notes on Form in Art," consists in "the relation of multiplex interdependent parts to a whole which is itself in the most varied & therefore the fullest relation to other wholes."[46] Individual and society develop together, braided in reciprocal modification, smoothly and continuously rather than by jagged jumps.

George Eliot appealed to "the Law of Progress," justifying the journal's commitment to a politics of gradual reform, in her prospectus for the relaunch of *The Westminster Review* in January 1852.[47] Four years later, shortly before she began writing fiction, Eliot summarized the Development Hypothesis in her major *Westminster Review* article "The Natural History of German Life":

> The external conditions which society has inherited from the past are but the manifestation of inherited internal conditions in the human beings who compose it; the internal conditions and the external are related to each other as the organism and its medium, and development can take place only by the gradual consentaneous development of both.[48]

"The Natural History of German Life" raises a crucial question for George Eliot's novelistic practice and, more largely, for the nineteenth-century novel's assumption of the history of man as its scientific ground or premise. Eliot poses the question in two complementary ways: that of the relation between the individual and the race or species, and that of the relation between the developmental drives of natural history and human history.[49] Both relations, as we saw in chapter 1, are fundamental to the discourse of a natural history of man, and both bear a potentially disabling tension or contradiction. Eliot's fiction goes on to explore that tension by disarticulating the analogy (most heavily pressed by Spencer[50]) between biological formation (organic development, growth) and historical progress, relegating them to different—perhaps structurally incommensurable—temporal scales: the former cyclical, reiterative, the latter linear and progressive.[51]

George Eliot reflects upon the premodern character of the rural peasantry in the books she is reviewing, volumes 1 and 2 (*Land and People, Civil Society*) of Wilhelm Heinrich Riehl's *Natural History of the German Peoples* (*Die Naturgeschichte des Volkes*). The lives of the peasantry are bound by an organic temporality of the race that is anterior to—and resistant to—individual development:

> In Germany, perhaps more than in any other country, it is among the peasantry that we must look for the historical type of the national

physique. In the towns this type has become so modified to express the personality of the individual, that even "family likeness" is often but faintly marked. But the peasants may still be distinguished into groups by their physical peculiarities. In one part of the country we find a longer-legged, in another a broader-shouldered race, which has inherited these peculiarities for centuries. . . . The cultured man acts more as an individual; the peasant, more as one of a group. Hans drives the plough, lives, and thinks just as Kunz does; and it is this fact, that many thousands of men are as like each other in thoughts and habits as so many sheep or oysters, which constitutes the weight of the peasantry in the social and political scale.[52]

Enlightenment stadial history had characterized the peasantry as a residual mass in the uneven development of nations. Riehl's natural history consigns them to a prehistoric timescale regulated by biological laws of inheritance, the cycle of seasons, and the inertial force of custom. The antithetical temporality of culture, comprising both national-historical progress and individual development, belongs to civil society, to urban life. We glimpse the plan of the nineteenth-century Bildungsroman (Balzac through Hardy), in which protagonists strive to escape rustic stagnation and enter the accelerated time of modernity—the time of the city and its institutions, of human history as change rather than continuity, of individual experience as growth and choice. A paradox, with roots in Rousseau's conjectural anthropology, takes shape. Human nature, species being, resides with the communal, organic life of the people; yet one may become fully human only by quitting that life, struggling into individuation, into historical and ethical identity—to join the vanguard (perhaps) of a new racial formation.

The project of a natural history of human life informs the fictions Eliot wrote after her review of Riehl: *Scenes of Clerical Life* (1857), *Adam Bede* (1859), *The Mill on the Floss* (1860). As Sally Shuttleworth notes, the sequence of tales admits an increasing tension between the imperatives of organic continuity at the level of collective life and the developmental differentiation of the individual.[53] Idyllic set pieces of daily life at the Hall Farm exemplify the social organism in *Adam Bede*, which opens in 1799 (the canonical "sixty years since" of *Waverley*) and closes with a present-tense glimpse of Adam's and Dinah's ongoing domestic felicity, grammatically synchronized with our time of reading. Individual and communal destinies merge once the self-destructive egotists, Hetty Sorrel and Arthur Donnithorne, are expelled from the novel's world.

But the dragging, clogging force of the social organism predominates in *The Mill on the Floss*. Characterizing individual growth as a recapitulation of phylogenetic evolution, Eliot literalizes Karl Morgenstern's formula, "the harmonious formation of the purely human," as a zoological emergence.[54] The Tulliver children are "still very much like young animals," with a "resemblance to two friendly ponies";[55] Maggie is likened to "a small Shetland pony" (13) and "a Skye terrier" (16, 28); Tom is "one of those lads that grow everywhere in England, and at twelve or thirteen years of age look as much alike as goslings" (33); Maggie and her cousin Lucy are like "a rough, dark, overgrown puppy and a white kitten" (61). The narrative tracks the painful growth of Maggie and Tom from this amorphous animal life into the complex sexual, social, and ethical differentiations of adulthood. In Maggie's case, especially, the developmental imperative pushes against the conservative, inertial gravity of the social organism, which thwarts her struggle to become fully human—not because society is "unnatural," but just the opposite.

The provincial setting of *The Mill on the Floss* precludes the appeal to national history, installed in *Waverley* and pervasive in *Adam Bede*, as a mediating term between the disparate scales of individual history and natural history. Eliot's evocation of the setting amplifies, instead, the latter:

> It is one of those old, old towns which impress one as a continuation and outgrowth of nature, as much as the nests of the bower-birds or the winding galleries of the white ants; a town which carries the traces of its long growth and history like a millennial tree, and has sprung up and developed in the same spot between the river and the low hill from the time when the Roman legions turned their backs on it from the camp on the hillside, and the long-haired sea-kings came up the river and looked with fierce, eager eyes at the fatness of the land. It is a town "familiar with forgotten years." (115–16)

The town's population inhabits a collective mentality that is closed to its own past: "The mind of St. Ogg's did not look extensively before or after. It inherited a long past without thinking of it, and had no eyes for the spirits that walk the streets" (118). To live in organic time is to live outside historical consciousness—and outside what Ludwig Feuerbach called "species consciousness," the mental condition of fully achieved human being.

George Eliot (as Marian Evans) had translated Feuerbach's *The Essence of Christianity* into English in 1854, following the translation of David Friedrich Strauss's *Life of Jesus* with which she had first appeared in print in 1846. Both works are landmarks of the other major contemporary intellectual movement that preoccupied Eliot's early career, the German Higher Criticism with its issue, the Young or Left Hegelian critique of religion.[56] Seeking to redeem the ethical spirit of Christianity from dogmatic theology by translating it to the ground of anthropology and myth, Strauss and Feuerbach develop the Herderian, Hegelian idea of a *Bildung der Humanität*, a "formation of humanity" or evolutionary perfection of species being, as substitute for a providential master-plot of human destiny. Designating humanity as the proper object of religious knowledge and feeling, Feuerbach proposes a solution to the vexed question of human exceptionalism. Religion is what sets humans apart from other animals: but not because we are supernaturally endowed with the faculty of veneration, reason, language, or some other divine prosthesis. Religion inheres in a modality of consciousness that is unique to humans, "species consciousness" (*Bewußtsein der Gattungen*)—in Marian Evans's translation, "cognizance of species":

> Consciousness in the strictest sense is present only in a being to whom his species, his essential nature, is an object of thought. The brute is indeed conscious of himself as an individual . . . but not as a species: hence, he is without that consciousness which in its nature, as in its name, is akin to science. . . . Science is the cognizance of species. In practical life we have to do with individuals; in science, with species. But only a being to whom his own species, his own nature, is an object of thought, can make the essential nature of other things or beings an object of thought.[57]

Consciousness of species constitutes human nature as internally double:

> Hence the brute has only a simple, man a twofold [*zweifaches*] life: in the brute, the inner life is one with the outer; man has both an inner and an outer life. The inner life of man is the life that has relation with his species, to his general, as distinguished from his individual, nature. Man thinks—that is to say, he converses with himself. (2)

Feuerbach repurposes the traditional figure of *Homo duplex*, a double human nature. Duplicity consists in the uniquely human "consciousness of the infinite," a consciousness not of some external supernatural presence but "the consciousness which man has of his own—not finite and limited, but infinite nature" (2). The contrast Feuerbach draws between human and animal consciousness derives (via Hegel) from Herder's conception, in his *Treatise on the Origins of Language*, of the sphere or circle (*Sphäre*,

Kreis) that bounds the sensuous horizon of a creature's being. Feuerbach writes,

> The consciousness of the caterpillar, whose life is confined to a particular species of plant, does not extend itself beyond this narrow domain. It does, indeed, discriminate between this plant and other plants, but more it knows not. A consciousness so limited, but on account of that very limitation so infallible, we do not call consciousness, but instinct. (2)

Jakob Uexküll's *Umwelt*, expounded through an evocation of the sensuous-semiotic "bubble" comprising the lifeworld of the tick, will later develop this conception.[58]

Midway through *The Mill on the Floss*, the narrator reflects upon "the mental condition of these emmet-like Dodsons and Tullivers" (little different from Feuerbach's caterpillar, Herder's bee, Uexküll's tick), with their custom-bound, unreflective, "semi-pagan" lives:

> I share with you this sense of oppressive narrowness; but it is necessary that we should feel it, if we care to understand how it acted on the lives of Tom and Maggie,—how it has acted on young natures in many generations, that in the onward tendency of human things have risen above the mental level of the generation before them, to which they have been nevertheless tied by the strongest fibres of their hearts. The suffering, whether of martyr or victim, which belongs to every historical advance of mankind, is represented in this way in every town, and by hundreds of obscure hearths; and we need not shrink from this comparison of small things with great; for does not science tell us that its highest striving is after the ascertainment of a unity which shall bind the smallest things with the greatest? In natural science, I have understood, there is nothing petty to the mind that has a large vision of relations, and to which every single object suggests a vast sum of conditions. It is surely the same with the observation of human life. (272–73)

Darwin makes the aesthetic imperative of formal unity that governs the natural scientist's "large vision of relations" explicit at the close of the *Origin of Species*:

> There is grandeur in this view of life, with its several powers, having been originally breathed into a few forms or into one; and that, whilst this planet has gone cycling on according to the fixed law of gravity, from so simple a beginning endless forms most beautiful and most wonderful have been, and are being, evolved.[59]

But George Eliot restores the tragic ethos that Darwin's appeal to the beautiful grandeur of his vision seeks to deflect. Suffering attends human progress—which occurs through the individual's developmental struggle with local conditions.

Earlier, *The Mill on the Floss* has invoked the Aristotelian, plot-based understanding of tragedy as the downfall of a great man, whose rash action recoils on his community as well as on himself. The lesson of the first half of the novel is that Maggie's father is unworthy of tragic dignity, despite claims on (or feints at) the role. "Mr. Tulliver had a destiny as well as Œdipus, and in this case he might plead, like Œdipus, that his deed was inflicted on him rather than committed by him" (130): the allusion cancels itself, since tragedy now requires, in the modern view, a concomitant quality of consciousness, of understanding and acceptance of moral responsibility for the disastrous event. Mr. Tulliver, the narrator editorializes again, "though nothing more than a superior miller and maltster, was as proud and obstinate as if he had been a very lofty personage, in whom such disposition might be a source of that conspicuous, far-echoing tragedy, which sweeps the stage in regal robes" (197). The gesture of democratization, apparently investing Mr. Tulliver with tragic heroism, again withholds it: "such tragedy," the narrator goes on, "lies in the conflict of young souls, hungry for joy, under a lot made suddenly hard to them"—Mr. Tulliver's children, upon whose lives his folly has brought adversity (197). Redistributing the tragic from action to feeling, suffering, Eliot's narrative elects Maggie as its fit protagonist, "martyr or victim"—while Tom, taking over his father's role, reiterates a hard masculine arrogance that shuts out reflective pathos.

John Kucich analyzes the logic of George Eliot's reformulation of the tragic through Feuerbachian species consciousness. "Feuerbach stands at the head of a liberal humanist tradition that shifts value away from action and toward a kind of reverence for sensibility," he writes, developing Karl Marx's critique in the "Theses on Feuerbach."[60] Feuerbach posited a "human essence," according to Marx (who adapted his conception of species being, *Gattungswesen*, from Feuerbach),[61] by removing human life from "the ensemble of the social relations" that forms it and that constitutes its totality, to produce a dualistic relation between the "abstract—*isolated*—human individual" and the "dumb generality" of species.[62] Species, in other words, is an empty projection of the abstract isolated individual, hypostatized as a human essence. Hence species consciousness, "preoccupied with self-relation, to the exclusion of any kind of dynamic interdependence" within a social medium, entails "a kind of inward mirroring in which man comes to recognize the impersonal forces inherent within his own nature" by internalizing the social contradictions of his existence (misrecognized, however, as a kind of fatality).[63] George Eliot's heroines are virtuosos of this internalization: "they manage to satisfy a displaced need for external confrontation by becoming internally divided"—the condition of species consciousness as tragic feeling.[64]

Maggie Tulliver's fitness for species consciousness accords with Feuerbachian criteria: a sensibility to music,[65] an aptitude for love,[66] a capacity to view her own life as part "of a divinely guided whole" (269), and a (concomitant) susceptibility to "mental conflict" (325)—a painful self-division between competing imperatives of individual desire and duty to others. "The tragedy of our lives is not created entirely from within," the narrator comments (401): it requires the clash between character and social medium. Species consciousness produces itself, moralized, as conscience, the painful submission of our wants to "the reliance others have in us ... those whom the course of our lives has made dependent on us" (475). The dialectic between individuation and interdependency energizes the latter half of the novel in an ongoing debate between Maggie and her lovers, Philip Wakem and Stephen Guest, who both seek to persuade her that her renunciation of desire for duty is "unnatural," an act of violence against her authentic self (329, 449, 475). Maggie's retort is decisive: "Love is natural; but surely pity and faithfulness and memory are natural too" (450).[67] Her reward is "the gift of sorrow—that susceptibility to the bare offices of humanity which raises them into a bond of loving fellowship" (191). The infusion of bare duty with loving fellowship entails the renunciation of individual gratification.

Early in *The Mill on the Floss*, George Eliot awards her childish heroine "that superior power of misery which distinguishes the human being, and places him at a proud distance from the most melancholy chimpanzee" (46–47). Here again we encounter that recurrent figure in the natural history of man, Herder's dejected ape—close enough to human consciousness to intuit his exile from it, like the noble pagans in Dante's Limbo, whose anguish consists in the knowledge of their separation from God. If the phrase seems ironical at Maggie's expense, the irony saddens into truth as she grows older, and takes her place among "the dark unhappy ones," Staël's Corinne and Scott's Rebecca and Flora Mac-Ivor, excluded heroines of nineteenth-century fiction (332). "To suffer is the highest command of Christianity," Feuerbach had exhorted: since God suffered "for men, for others, not for himself," then "he who suffers for others, who lays down his life for them, acts divinely, is a God to men."[68] *She*, Eliot's novel insists. To suffer is to be human: it is the authentic, inward stigma of development.

The sensitive individual internalizes the antagonism between herself and the social organism, producing that double consciousness of the self in relation to an ideal image of human possibility as a relation of frustration and oppression, of foiled ambition and disappointed love. To suffer, above all, is to be a woman, more painfully enmeshed than men are in custom and convention. Blocked from the social paths of *Bildung*, confined to the meager chances of the marriage plot, women nevertheless become the most fully human beings in these novels by virtue of their suffering, their depth of feeling, as they assume the sacrificial role of "martyr or victim" and the aesthetic mantle of tragedy.[69]

"Nature repairs her ravages—but not all" (521): individual tragic consciousness is obliterated, at the close of the novel, in a catastrophic reassertion of organic time, cyclical like Lyell's earth history. Also obliterated, as Tom and Maggie sink into an idyllic depth anterior to their narrated history, is the time of *Bildung* coterminous with our reading: "brother and sister had gone down in an embrace never to be parted: living through again in one supreme moment the days when they had clasped their little hands in love, and roamed the daisied fields together" (521). That primordial union makes absolute the suspension of consciousness, in Wordsworthian spells of reverie and "unknowing," that the novel has represented as the inner, authentic fold of Maggie's mental life. Aligned, in Elisha Cohn's subtle analysis, with the generation of the story itself (in the novel's opening chapter), that inward reverie precedes the self-division of tragic knowledge—the painful enlargement and intensification of cognition that constitutes species consciousness.[70] Maggie's inner life reiterates in a spiritual key the organic resistance to *Bildung* more coarsely engrained in her social medium.

An Intellectual Passion

In *Middlemarch*, the heroine's ethical commitment to a greater organic world (as opposed to her hapless submission to it) provides ambiguous recompense for her banishment from the historical avenues of *Bildung*. George Eliot's narrator expresses the relation between "the natures of women" and their "social lot" in the Darwinian language (as Beer notes) of a drive to variation, the motor of development,[71] which is dynamically at odds with the retentive force of custom and convention (troped as fashion, a formal rather than structural change that is no change at all):

> The limits of variation are really much wider than anyone would imagine from the sameness of women's coiffure and the favourite love-stories

in prose and verse. Here and there a cygnet is reared among the ducklings in the brown pond, and never finds the living stream in fellowship with its own oary-footed kind. (4)

The novel poses its question: does Dorothea's fineness of character, setting her apart within her family and social circle, make her a freak, a sport (as her nickname, "Dodo," suggests)—or the prototype of a more evolved human being?

Her election as protagonist, exceptional by virtue of the role, implies the answer. Dorothea is the character in the novel best equipped for that enhanced consciousness of self that opens onto consciousness of the infinite and hence constitutes, in Feuerbach's argument, "consciousness of species." Like Maggie Tulliver, but on a scale befitting her rank, Dorothea has a passionate sense of duty to her fellow beings, energized by a capacity for "vivid sympathetic experience" (788). At the climax of her story, following a nightlong spiritual ordeal, she casts off the tragic burden of her suffering by a powerful exertion of that sympathetic faculty. Her sympathetic exertion affords a sublime expansion of vision and feeling:

> She opened her curtains, and looked out towards the bit of road that lay in view, with fields beyond outside the entrance-gates. On the road there was a man with a bundle on his back and a woman carrying her baby; in the field she could see figures moving—perhaps the shepherd with his dog. Far off in the bending sky was the pearly light; and she felt the largeness of the world and the manifold wakings of men to labor and endurance. She was a part of that involuntary, palpitating life, and could neither look out on it from her luxurious shelter as a mere spectator, nor hide her eyes in selfish complaining. (788)

The accession of consciousness swells beyond Dorothea's social medium (Middlemarch: provincial life) and beyond human species being ("the manifold wakings of men to labor and endurance") to life itself, life as such, "involuntary, palpitating life." The phrase, with its arresting insistence on life as dynamic material process, discloses the far horizon of the topic named in the novel's opening sentence: "the history of the man, and how the mysterious mixture behaves under the varying experiments of Time." That far horizon is a boundary, where consciousness fails, as the history of man warps into some other, unthinkable formation. What *Middlemarch* presents as the culmination of the heroine's *Bildung*—a powerfully moralized accession of species consciousness—opens onto the end of its project, the dissolution of humanity into a radical, unknowable difference. "Every limit is a beginning as well as an ending," the narrator reminds us in the novel's "Finale" (832).

While *Middlemarch* maintains Dorothea's status as protagonist, hers is just one among several plots of troubled vocation. Two are case histories of scientific inquiry, Casaubon's quest for the "Key to all Mythologies" (279) and Lydgate's for "the primitive tissue" (148). Reckoning both projects against the modern revolutions in their fields, consolidated during George Eliot's career, the novel forges its own claim on the production of scientific knowledge. Casaubon toils in the theological old regime of allegorical mythography, ignorant of contemporary German innovations in philology, biblical criticism, and religious anthropology that will issue in F. Max Müller's "science of language" and Edward B. Tylor's "science of culture" in the decade preceding *Middlemarch*.[72] Lydgate seeks to extend Xavier Bichat's tissue theory on the historical cusp of its transformation in the 1830s by the pioneers of cell biology, Matthias Schleiden and Theodor Schwann, whose work will be expanded by Rudolf Virchow in the 1860s.[73] *Middlemarch* presents itself as a superior representation of spiritual and organic life because it is up to date, written from the perspective of those scientific revolutions, and also, more crucially, because it frames them within the greater horizon of knowledge formed by its literary "enterprise of totalization," remarked by so many critics.[74] The novel's formal design, a scientific as well as poetic synthesis constituted by the techniques of Victorian realism (omniscient narration, free indirect discourse, polyphonic plotting, analogical patterning), affords a more comprehensive measure of the characters' failures than the record of their intellectual errors.

Casaubon's adherence to an obsolete paradigm does not disqualify comparative mythography and the history of religion from scientific status. "The subject Mr. Casaubon has chosen is as changing as chemistry," Ladislaw tells Dorothea: "new discoveries are constantly making new points of view" (221–22). It was the field through which George Eliot entered print, in her translations of Strauss and Feuerbach and her first article for the *Westminster Review*, on Robert William Mackay's *The Progress of the Intellect, as Exemplified in the Religious Development of the Greeks and Hebrews* (1850). Mackay's book, "the nearest approach in our language to a satisfactory natural history of religion,"[75] marks a midpoint in the Victorian uncoupling of mythology from doctrinal allegory. Attending to Friedrich Creuzer's epochal *Symbolism and Mythology of Ancient Peoples* (*Symbolik und Mythologie der alten Völker*, third edition, 1837–42), Mackay turns away from Jacob Bryant's *A New System; Or, an Analysis of Antient Mythology* (three editions, 1774–76 to 1807), identified as the

British monument of the prescientific era by Eliot in her review (and by Ladislaw in *Middlemarch*, 222), in anticipation of the progressively naturalistic approaches of Müller and Tylor.[76] Creuzer and Mackay find the origins of myth not, as Bryant does, in a savage state of mental confusion—the symptom of heathen peoples' loss of access to revealed truth—but in a primal, authentic engagement of the human imagination with external reality. Placed at the center of creation, man "sees all nature in his own nature, and thus cannot see otherwise except according to the laws of his own being," writes Creuzer: "What the abstract mind calls an active force is only a person to the primitive native way of seeing.... Thus it follows that what we call the figurative is nothing else than the impress of a form of our own thinking" upon the natural world.[77] "The inspiration of antiquity, and the poetical imagery which flowed from it were not understood figuratively, but felt literally," writes Mackay.[78] Mackay follows Creuzer's distinction between an original symbolic apprehension of the world and its fixation into mythological systems, which occupy a transitional stage between the "use of a metaphorical symbolism, and the formation of an abstract theology."[79] Intellectual degeneration ensues: "Symbols thus came to usurp an independent character as truths and persons.... [M]any, mistaking a sign for the thing signified, fell into a ridiculous superstition."[80] Mackay anticipates Müller's and Tylor's contention that myth is the product of a universal developmental stage of human history, a "*Mythological* or *Mythopoeic Age*," where it constitutes an attempt to understand natural phenomena through the invention of "poetical names."[81] "The sunrise was the revelation of nature," in Müller's naturalization of Bryant's solar hypothesis: "it was the simple story of nature which inspired the early poet, and held before his mind that deep mirror in which he might see reflected the passions of his own soul."[82] Myth arises as a primitive (animistic) interpretation of natural phenomena through anthropomorphic techniques of personification and analogy.

Consonant with these accounts, Eliot's review of *The Progress of the Intellect* leans on a Romantic distinction (Schiller's, Coleridge's) between the naïve or organic "conception of a symbol" and "the allegorizing mania" of belated, alienated interpreters (among whom Eliot includes Mackay):

> Owing to the manysidedness of all symbols, there is a peculiarly seductive influence in allegorical interpretation; and we observe that all writers who adopt it, though they set out with the largest admissions as to the spontaneous and unconscious character of mythical allegory,... acquire a sort of fanatical faith in their rule of interpretation, and

fall into the mistake of supposing that the conscious allegorizing of a modern can be a correct reproduction of what they acknowledge to be unconscious allegorizing in the ancients. We do not see what unconscious allegory can mean, unless it be personification accompanied with belief, and with the spontaneous, vivid conception of a symbol, as opposed to the premeditated use of a poetical figure.[83]

The criticism looks forward to Casaubon: temperamentally incapable of the spontaneous, vivid conception of a symbol, hunched in allegorical exegesis, he himself has become "a quasi-allegorical figure, the personification of the dead letter, the written word."[84] The novel diagnoses Casaubon's scientific failure, in short, as an imaginative failure to recover a vital relation to meaning: to interact sympathetically with signs, to inhabit representation as a living system, as it was for its first interpreters.

It is less easy to discern Lydgate's error. While Eliot treats his case with greater nuance and sympathy, the design of *Middlemarch* prompts readers to detect an analogy between Lydgate's objective and the key to all mythologies. Lydgate has posed his research question, "what was the primitive tissue? . . . not quite in the way required by the awaiting answer," the narrator remarks, without explaining how he has gone astray or by how far (148). Two explanations have prevailed: that Lydgate misses the advent of Schleiden and Schwann's cell theory (because he is too early for it, or because his commitment to Bichat's tissue anatomy closes his mind to the new paradigm); or, that the idea of a unitary foundation or origin is categorically mistaken.[85] L. S. Jacyna has shown that "the quest for a primitive type that would unite all living beings," "the primal form that could be considered as the foundation of both the embryonic and the biological series," dominated the research agenda of the emergent biomedical sciences in the 1830s.[86] Lydgate is presumably one of those medical researchers who "preferred the notion that fibres, rather than globules, were the primitive tissue element," while the globular hypothesis—anticipating the new cell theory—was favored by philosophical anatomists.[87] Schwann himself, however, acknowledged Bichat's tissue theory as laying the foundation for his identification of the nucleated cell as the fundamental unit of life—hence, a commitment to it would not necessarily disqualify Lydgate's project. Although British scientists were quick to embrace their work, T. H. Huxley, after initial enthusiasm, criticized Schleiden and Schwann in his influential (if controversial) review of the field, "The Cell-Theory" (1853), which George Eliot cites in her *Quarry for* Middlemarch.[88]

Huxley's essay complicates the diagnosis of Lydgate's error as a failure to anticipate the nascent cell theory. Huxley refutes the Schleiden-Schwann conception of the nucleated cell as an autonomous, generative unit, a "centre of force," on the grounds that it violates the more rigorously developmental, epigenetic understanding of organic systems theorized in 1759 by Wolff—which Lewes, as we saw, would cite as the scientific foundation of the "Development Hypothesis" culminating in Darwin's work a hundred years hence.[89] "For Huxley, cells were not separate and independent entities—they were interconnected elements within the integrated organism, seamlessly interfacing with specialized body parts, such as tissues and organs . . . the cell's periphery and the interstitial region between cells, not the nucleus, were the sites of future organismic development."[90] Lydgate's desire "to demonstrate the more intimate relations of living structure" (148) harmonizes with the larger narrative emphasis, throughout *Middlemarch*, on the dynamic interaction among distributed parts of social life within a complex totality. Eliot's novel unfolds a "network" rather than a "membrane" model of the social organism, as Laura Otis has argued.[91]

In short, it is by no means clear that an intrinsic error disables Lydgate's project.[92] Michael York Mason characterizes him as "a Whewellian scientist with the Whewellian characteristics of fertility of imagination and discipline of mind."[93] Lydgate's commitment to "that arduous invention which is the very eye of research, provisionally framing its object and correcting it to more and more exactness of relation" (165), keeps faith with Whewell's emphasis, in *The Philosophy of the Inductive Sciences*, on the primacy of "a constant invention and activity, a perpetual creating and selecting power," in scientific discovery, reiterated in George Eliot's generation by Huxley, Lewes, and Tyndall.[94] Eliot's narrative of "the growth of an intellectual passion" in chapter 15 of *Middlemarch* suggests not only Lydgate's capacity for original thought but also a hermeneutic procedure. The scene of origins is a scene of reading:

> The page he opened on was under the heading of Anatomy, and the first passage that drew his eyes was on the valves of the heart. He was not much acquainted with valves of any sort, but he knew that *valvae* were folding doors, and through this crevice came a sudden light startling him with his first vivid notion of finely adjusted mechanism in the human frame. (144)

Revelation comes through the momentary defamiliarization of a technical term. In a complex event of translation, Lydgate's grasp of the literal, etymological meaning of *valvae* opens his mind to the original, figurative force of its scientific usage—opening, in Eliot's metaphoric amplification, the doors to his calling.[95] This ability to glimpse the metaphoric strangeness of a scientific concept, before its resolution into fact, is the precondition for its reinterpretation. Scientific discovery recapitulates "the spontaneous, vivid conception of a symbol"—the organic fusion of feeling and thinking, of figural and literal meaning—that the comparative mythologists ascribed to the first "poet-priests of nature."[96]

The problem, familiar to readers of *Middlemarch*, is that Lydgate's "[failure] to deploy his scientific imagination in his personal life" impedes not only his professional ambitions but also, beyond that, the full and harmonious formation of humanity, which is the goal of *Bildung*.[97] His predicament falls under the larger methodological principle in the formation of scientific knowledge that Whewell called the "consilience of inductions." The final verification of a hypothesis requires not just that it be able to predict unobserved cases, according to Whewell, but that it also "explain and determine cases of a *kind different* from those which were contemplated in the formation of our hypothesis."[98] In other words, its determinative authority should reach across other domains. Lydgate, his scientific project hobbled not by any theoretical flaw but by his failure to apply his intelligence across the range of cases (social, erotic) that comprise his lifeworld, exemplifies a novelistic extension of Whewell's principle beyond the genres of the natural sciences.

The "consilience of inductions" is what the author of *Middlemarch*, in contrast to her character, is able to accomplish. The novel teaches its reader that this capacity to "explain and determine cases" across different domains is its distinctive feature. "The highest Form," as Eliot wrote in "Notes on Form in Art," "is the highest organism, that is to say, the most varied group of relations bound together in a wholeness which again has the most varied relations with all other phenomena": individuation enhanced by interdependency.[99] Eliot's essay theorizes her endeavor to realize that "highest Form" in *Middlemarch*, with its complex interrelation among local plots of *Bildung* and other patterns and systems.[100] Whewell describes the epistemological formation of this "wholeness": "The Consiliences of our Inductions give rise to a constant Convergence of our Theory towards Simplicity and Unity," in "a succession of steps by which we gradually ascend in our speculative views to a higher and higher point of generality."[101] With this synthetic principle, *Middlemarch* makes its claim on a knowledge that exceeds the particular objectives, Lydgate's as well as Casaubon's, that it encompasses. Their failures are constitutional and characterological. It is not just that Casaubon is mistaken: his mistake is symptomatic, psychosomatic, conditioned by the "melancholy absence of passion" that forestalls his being able to take an intuitive leap beyond the dead letter of exegesis (417), even as Lydgate's imagination, energized by an intellectual passion within his professional life, is blighted by "spots of commonness" outside it (150). Character is the novel's subject, the resource that equips it better than other kinds of writing "to know the history of man, and how the mysterious mixture behaves under the varying experiments of Time." The novel can run conjectural simulations across its range of case studies of all those domains—erotic, domestic, social, political, artistic, philosophical—in which "character too is a process and an unfolding" (149). "Within the novel's system, the novel itself stands as that perfect form against which the failures of the characters that inhabit it must be measured," writes Kent Puckett, describing a formal logic that also governs the novel's constituent discourses.[102] The "perfect form" the novel lays claim to is not a resolved, determinate scheme, but form as process, dynamic and evolving—a symbolic system that is imaginatively grasped as figural rather than as literal, or rather, a knowledge that inheres in the vibration between literal and figural senses, not yet hardened into habit and fact.

Involuntary, Palpitating Life

Vibration between literal and figural sustains Dorothea's epiphany:

> Far off in the bending sky was the pearly light; and she felt the largeness of the world and the manifold wakings of men to labor and endurance. She was a part of that involuntary, palpitating life, and could neither look out on it from her luxurious shelter as a mere spectator, nor hide her eyes in selfish complaining. (788)

Her surge of feeling exceeds the human realm, economic and biological, of work and the family made visible in the scene: the man with the bundle on his back, the woman with her baby, the shepherd and his dog. These figures encode that human realm as pastoral, rather than epic, hence as miniature, limited, *less* than the world's intuited largeness. "She was a part of that involuntary, palpitating life": the phrase designates the entirety of terrestrial living forms and their history and futurity, a phenomenon that cannot be seen or apprehended through the senses, hence, an abstraction, a figure, synecdoche. At the same time, "involuntary, palpitating life" consists physically in a subcellular order of organic matter, discernible through the modern achromatic microscopes coming into use in the late 1820s.[103] It is the "primitive tissue" of Lydgate's quest: no tissue, but "a microscopic lump of jelly-like substance, or protoplasm," in Lewes's paraphrase, evincing "the cardinal phenomena of life: Nutrition, Reproduction, and Contractility."[104] In his 1868 lecture "On the Physical Basis of Life," Huxley defined the protoplasm as the "one kind of matter which is common to all living beings," through which "their endless diversities are bound together by a physical, as well as an ideal, unity," Dorothea as well as dogs and daisies.[105] "Involuntary, palpitating life" is George Eliot's translation of the neologism "protoplasm," no less (and no more) figurative than Huxley's "physical basis of life," or Haeckel's *Urschleim* (primordial slime), "the evolutionary basis of all subsequent life": a specimen of which, dredged from the depths of the North Atlantic, Huxley described and named in Haeckel's honor.[106]

Involuntary, palpitating life, microscopic and macroscopic, internal to each individual body and enfolding all of them, designates the formal horizon of totality the novel lays claim to. It designates a primal matter and also that matter's conceptual, value-laden abstraction: "THERE IS NO WEALTH BUT LIFE," John Ruskin exhorted in 1860.[107] Translating a technical term (protoplasm) into its figural and literal senses, George Eliot's phrase invites us to contemplate the symbolic logic of organic form. Where "an allegory is but a translation of abstract notions into a picture-language which is itself nothing but an abstraction from objects of the senses," in Coleridge's famous definition, a symbol "is characterized by a translucence of the Special in the Individual or of the General in the Especial or of the Universal in the General.... It always partakes of the Reality it renders intelligible; and while it enunciates the whole, abides itself as a living part in that Unity, of which it is the representative."[108] The Romantic symbol, according to the comparative mythologists, rehearses the imagination's original act of apprehension of the world; it rehearses, also, the imaginative act that generates scientific knowledge, according to Whewell and other theorists of scientific method. Casaubon fails to reclaim that original imaginative act; Lydgate succeeds, but fails to translate his intuition across other domains of experience to achieve the "consilience of inductions," the condition of a formal totality. George Eliot grants the reader of *Middlemarch* access to Dorothea's spontaneous—organic—intuition of such a totality: seeing through her intuition, or across it, since (in the logic of free indirect discourse discussed in chapter 4) the intuition arrives via what Ann Banfield calls an "unspeakable sentence," an insight exceeding the character's possible articulation.[109] (We cannot imagine

Dorothea saying, even to herself, "I am a part of that involuntary, palpitating life.") Our reading inhabits the phrase's shimmer between registers, literal and figurative, material and formal, in its simultaneous denotation of an abstract whole (all living beings) and a universal substance (protoplasm).[110]

The scene of provincial life (the merely pastoral social medium) opens onto a sublime destiny that awaits the heroine's and our attention. The vision of organic life charges Dorothea's re-engagement with the world with heroic grandeur. ("There is grandeur in this view of life."[111]) Yet the phrase "involuntary, palpitating life" bears a hint at least faintly monstrous, even if we do not read it in light of Lewes's conjecture that "the earth at the dawn of life was like a vast germinal membrane, every slightly diversified point providing its own vital form," or of Haeckel's evocation of "huge masses of naked, living protoplasm [covering] the greater ocean depths."[112] The figure reverses the tropism of development, toward specialization and individuation, for the intuition of an elemental state of undifferentiation. Manifest over time as a force that (in Darwinian logic) overcomes the human form and scale, involuntary, palpitating life dissolves as it absorbs what realist fiction has been teaching us to value: any particular, individual life, with its sustaining sympathetic properties of consciousness and agency. And it reiterates, with the coercive weight of universal truth, biology over history as the medium of women's destiny, "labor and endurance" not as work but as childbearing.

The closing paragraph of *Middlemarch* resolves Dorothea's *Bildung* into the "unhistoric" forum of organic growth and private influence:

> Her finely-touched spirit had still its fine issues, though they were not widely visible. Her full nature, like that river of which Cyrus broke the strength, spent itself in channels which had no great name on the earth. But the effect of her being on those around her was incalculably diffusive: for the growing good of the world is partly dependent on unhistoric acts. (838)

The simile of the river, broken for military conquest[113] but thence useful for agricultural irrigation, quietly rewrites the great flood at the end of *The Mill on the Floss*. It converts tragic catastrophe into a tonally ambivalent—comic yet melancholic—evocation of ongoing domestic life. Event, the unit of historical reckoning, is subsumed into metaphor, with the "diffusive" maze of fertile channels reiterating, for the last time, the novel's master figure for organic form, the web—or rather, and with a keener irony, its variant in contemporary development theory, the plan of branching lines of descent.[114]

The microscopic view of pulsing protoplasm, more than the ennobling abstraction "life," trains us to a literal as well as figurative reading of incipient events of organic transformation in *Middlemarch*:

> She might have compared her experience at that moment to the vague, alarmed consciousness that her life was taking on a new form, that she was undergoing a metamorphosis in which memory would not adjust itself to the stirring of new organs. Everything was changing its aspect: her husband's conduct, her own duteous feeling towards him, every struggle between them—and yet more, her whole relation to Will Ladislaw. Her world was in a state of convulsive change. (490)

"Dorothea, it seems, experiences not just a reorganization of her consciousness but its annexation of a desiring body," writes Catherine Gallagher: the end of *Bildung* (affectively commanding readers) is the character's emergence into biological life.[115] Dorothea has caught this metamorphic potential from Ladislaw, the novel's spokesman for an organic-form aesthetic. Ladislaw claims for the poet that primal bardic intuition posited by the mythographers:

> To be a poet is to have a soul so quick to discern that no shade of quality escapes it, and so quick to feel, that discernment is but a hand playing with finely ordered variety on the chords of emotion—a soul in which knowledge passes instantaneously into feeling, and feeling flashes back as a new organ of knowledge. One may have that condition by fits only. (223)

A powerful sympathetic circuit between knowledge and feeling triggers organic change in the sensitive individual, in a quasi-Lamarckian process of modification.

Embodying the poetic type he celebrates, Ladislaw instantiates a fusion of mythological and biological registers:

> The first impression on seeing Will was one of sunny brightness, which added to the uncertainty of his changing expression. Surely, his very features changed their form; his jaw looked sometimes large and sometimes small; and the little ripple in his nose was a preparation for metamorphosis. (209)

The description adorns him with symbolic highlights from Max Müller's solar mythos.[116] As well as the mythographic allusion (pointing back to Ovid), the language evokes evolutionist natural history and cell biology, which appropriated "metamorphosis" as a technical term in the mid-1830s and 1840s.[117] We glimpse Ladislaw quivering on the brink of mutation into some new organic form. The concord of discourses typifies Eliot's synthetic project in *Middlemarch*, while leaving vague what future human type Ladislaw's physiognomy might be preparing.

Perhaps we would not notice this if it were not for another interloper in Middlemarch, Ladislaw's analogical opposite, another son at odds with legitimacy and inheritance—and the secret sharer of a genetic link in the murky back-story of Bulstrode's past.[118] The "alien and unaccountable" Joshua Rigg (427) arrives at Peter Featherstone's funeral "as if from the moon":

> This was the stranger described by Mrs Cadwallader as frog-faced: a man perhaps about two or three and thirty, whose prominent eyes, thin-lipped, downward-curved mouth, and hair sleekly brushed away from a forehead that sank suddenly above the ridge of the eyebrows, certainly gave his face a batrachian unchangeableness of expression. (332)

If Ladislaw embodies the Apollonian promise of a future human type, Rigg exhibits the "half ichthyic and half batrachian" aspect that H. P. Lovecraft, pulp-modernist master of weird fiction, will call the "Innsmouth look":

> He had a narrow head, bulging, watery-blue eyes that seemed never to wink, a flat nose, a receding forehead and chin, and singularly undeveloped ears. . . . His hands were large and heavily veined, and had a very unusual greyish-blue tinge. The fingers were strikingly short in proportion to the rest of the structure, and seemed to have a tendency to curl closely into the huge palm.[119]

Once again, Eliot mixes mythographic and biological registers. Rigg appears to be an avatar of the "Near Eastern mermen-deities, the Assyrian Oannes and the Phoenician god Dagon," which obsessed nineteenth-century mythographers,[120] among them Mr. Casaubon, with his "entirely new view of the Philistine god Dagon and other fish deities" (484). (Readers of Lovecraft will recall that the Innsmouth churches are dedicated to "the Esoteric Order of Dagon."[121]) But perhaps we need not infer that the "frog-feature[d]" woman with whom Peter Featherstone had congress at an unnamed seaport (413) was one of Lovecraft's Deep Ones. Rigg's physiognomy also conforms to the scientific account of monstrosity as a heritable retardation of fetal development, established in the 1820s by Geoffroy and Isidore Saint-Hilaire, adopted by Chambers and Owen: arrested at the batrachian stage, in this case, in one of Rigg's ancestors.

An Inherited Yearning

Deformation of humanity shadows the radical outsiders in *Middlemarch*. These hints of mutation into some other biological form—more highly evolved in Ladislaw's case, degenerate in Rigg's—do not precipitate into a literal register, one of material embodiment, but remain symbolic or metaphoric. Such a metamorphosis—a translation from figure to event— would transform the novel's experimental subject, human nature, with catastrophic consequences for its signature literary realism. *Middlemarch* upholds a realist account of human nature and its habitat, specified in the novel's subtitle, provincial life. Eventually the outsiders leave Middlemarch, as though expelled from its regime of representation: Rigg back to Innsmouth, Ladislaw to London, the milieu of progress.[122] The conservative force field of provincial life may restrict and stultify human nature, but it also stabilizes it, sustains it. We begin to see why the great achievements of nineteenth-century realism in English, from Jane Austen to George Eliot, require a provincial setting, while novels of metropolitan life (Dickens's) break with realist protocol. Like a Darwinian island or mountaintop, provincial life affords a gradualist preserve of human nature, a sanctuary of slow time where the history of man is all but suspended and, notwithstanding local quirks of eccentricity or inbreeding, the range of morphological possibilities is reassuringly narrow. Realism starts to lose its grip as we move away from "the Midland counties of England,"[123] and human nature writhes into exotic types and shapes, no longer amenable to the normative techniques of mixed characterization, free indirect discourse, and so on.

Figure becomes event, with other disruptions of realist form, in *Daniel Deronda*. George Eliot's last novel unmoors its characters from the stabilizing bonds of provincial life (synecdochic for national life) and its temporal corollary, the intimate historical remove of a generation. We confront, instead, an onrushing cosmopolitan present, set against the deep time of a racial—Jewish—ancestry receding beyond English history.[124] Eliot names the provincial setting of her early books "Wessex," kingdom of a vanished past. An early descriptive passage sets the elegiac key:

> Pity that Offendene was not the home of Miss Harleth's childhood, or endeared to her by family memories! A human life, I think, should be well rooted in some spot of a native land, where it may get the love of tender kinship for the face of earth, for the labors men go forth to, for the sounds and accents that haunt it, for whatever will give that early

home a familiar unmistakable difference amid the future widening of knowledge. (22)

The future widening of knowledge, which must include (the novel's networks of allusion tell us) Darwinian natural history, abrades the ontological security of that "familiar unmistakable difference."[125] Pity indeed. Unprotected by familiar associations, the heroine suffers "fits of spiritual dread" at the sudden opening of her horizon: "Solitude in any wide scene impressed her with an undefined feeling of immeasurable existence aloof from her" (63). It is as though the cosmic wilderness of organic life intuited by Dorothea in *Middlemarch* looms closer, larger, filling or rather emptying the sky. Beyond the ethical overthrow of her egoism, Gwendolen's agoraphobia forecasts the dissolution of the forms of an English national narrative—provincial life, the country house, domestic fiction, the marriage plot—for an inchoate world-historical destiny and, beyond it, sublime reaches of evolutionary drift in which human life itself will fray away.

Daniel Deronda opens a retrospect in which we view the contingency rather than inner necessity of the realism of *Middlemarch*. The novel is (famously) preoccupied with inheritance and descent: with the terms "affinity, relationship, community of type, paternity, morphology, adaptive characters, rudimentary and aborted organs, &c.," which Darwin had wished to reduce to "plain signification."[126] In *On the Origin of Species* (following Owen), Darwin insists on "the very important distinction between real affinities," marked by homological structures, "and analogical or adaptive resemblances."[127] Homology, expressing genetic descent, is the vehicle of plain signification, while analogy works like metaphor, by morphological substitution—the conversion of limbs to fins, for instance, yielding the adaptive resemblance between whales and fishes. The relation between genealogy and resemblance is everywhere at stake in *Daniel Deronda*, which renders descent, marked by a recurrence of ancestral traits, as at once a biological and a cultural phenomenon. George Eliot now draws on the advanced discourses of Darwinian biology and Tylorian anthropology, which she had been reading since the completion of *Middlemarch*.[128] Her description of the Cohen children, "looking more Semitic than their parents, as the puppy lions show the spots of far-off progenitors," alludes to Darwin's observation on the "*tendency* in the young of each successive generation to produce the long-lost character," as in "the stripes on the whelp of a lion, or the spots on the young blackbird."[129] The Cohens' distribution of "the thin tails of the fried fish" to their poor relation Mordecai may be "a 'survival' of prehistoric practice, not yet generally admitted to be superstitious."[130] "Survival" is a key term in Tylor's *Primitive Culture*, where it designates "primaeval monuments of barbaric thought and life," persisting into modernity as unconscious habits or rituals.[131]

Elsewhere, George Eliot's narrative requires us to interpret not just particular terms but the system or register that generates them—bringing into view tensions or contradictions among them.[132] As with Ladislaw and Rigg in *Middlemarch*, taxonomic heterogeneity marks limit cases in the novel's character system. Endowing its glamorous couple, Gwendolen and Grandcourt, with reptilian traits, *Daniel Deronda* assigns the lexicon of anthropology or comparative mythology to one and natural history or biology to the other, expressive of divergent modalities of fatality and perversity. Gwendolen, "[got] up as a sort of serpent," exhibits "a sort of Lamia beauty" (12), "attractive to all eyes except those which discerned in them too close a resemblance to the serpent, and objected to the revival of serpent-worship" (19). Grandcourt strikes his bride-to-be as "a handsome lizard of a hitherto unknown species" (137), looks "as neutral as an alligator" (157), and (once they are married) exhibits "a will like that of a crab or a boa-constrictor, which goes on pinching or crushing without alarm at thunder" (423). Far from modeling a synthesis, the different registers disarray the analogical pattern they ostensibly serve.[133] The zoological language, with its Darwinian cladding, encourages a reading of Grandcourt's case as one of reversion to a physiologically latent, prehuman, indeed premammalian ancestry: developmentally arrested at the saurian stage, his coldblooded will to power unchecked by the gentle faculties of reason or sympathy.[134]

As for Gwendolen, serpent-worship is a recurrent topic of comparative mythology, from Bryant through Mackay to Tylor.[135] The allusive context makes her reptilian likeness biologically less fixed than her husband's. Analogical or allegorical, rather than homological and literal, it is (hence) open to redemptive humanization and the reader's sympathy. Tylor defines the "magical arts" as those "in which the connexion is that of mere analogy or symbolism," and his account of animism as the typical mode of primitive thought fits Gwendolen's case.[136] The narrator, commenting on "the undefinable stinging quality—as it were a trace of demon ancestry—which made some beholders hesitate in their admiration of Gwendolen" (68), adumbrates her "fits of spiritual dread" (63), "streak of superstition" (276), affinity for second sight, and other regressive mental habits as survivals of an ancient Celtic heritage (implicit also in the setting for Gwendolen's tryst with Lydia Glasher, the Whispering Stones).[137] The prehistory that lingers in Gwendolen, in short, is a human one, pagan and mythic. Her progress in the novel can follow the program of Christian conversion, appropriately (after Strauss and Feuerbach) humanized, that is, stripped of dogmatic trappings.

The novel's eponymous hero, in contrast to these primitives, embodies a future human type around whom the withering formations of national destiny may be regenerated. Endowed with a "calm intensity of life and richness of tint in his face" (162), Daniel Deronda is a warm-blooded, hypermammalian development of *Middlemarch*'s Ladislaw, a New Man whose metamorphic potential expresses itself in an acute sympathetic porousness to the lives of others. With their assumption of feminine qualities of sensitivity and receptiveness, both are evolved types of the Bildungsroman protagonist: Ladislaw is "a creature who entered into every one's feelings, and could take the pressure of their thought instead of urging his own" (*Middlemarch*, 496); Deronda possesses a "many-sided" and "plenteous, flexible sympathy" (364), "an activity of imagination on behalf of others" (178), which expresses itself in "a half-speculative, half-involuntary identification of himself with the objects he was looking at" (189). In *The Descent of Man*, Darwin identifies sympathy as one of the prehuman "social instincts." By no means a uniquely human faculty, its cultivation and expansion nevertheless mark a higher stage of moral development through which (in language echoing the Romantic *Bildung der Humanität*) human beings may become more completely themselves: "As man advances in civilization," his sympathetic capacity reaches not just beyond tribe and nation but "beyond the confines of man," to realize "the very idea of humanity."[138] Sympathy, the quintessentially human attribute of Scottish Enlightenment moral philosophy, becomes fully human when it crosses species boundaries: "that disinterested love for all living creatures, the most noble attribute of man . . . seems to arise incidentally from our sympathies becoming more tender and more widely diffused, until they are extended to all sentient beings."[139] Darwin affirms the totality of (sentient) life as an ethical telos.

Daniel Deronda's case is one of hypertrophied sympathy. "Too reflective and diffusive," it blocks him from being "an organic part of social life," so that he "[roams] in it like a yearning disembodied spirit" (364–65). Daniel reclaims an "organic centre" in the discovery of his Jewishness, which supplies him—and the novel—with the future-oriented national destiny that a provincial Englishness no longer affords. *Daniel Deronda*, writes John Plotz, is "utterly original in asking what happens when culture is internalized so successfully that that it enters into the body of the consciously nationalized subject."[140] The language of organic life, clustering around the Jewish theme of the second half of the novel, comes to fruition in Daniel's willing embrace of his heritage, in which (again) cultural and biological determinants are mixed, this time to regenerative rather than monstrous effect.

The ethical trajectory of Deronda's *Bildung* has been much discussed. Amanda Anderson argues that his "reflective dialogism," the rational and critical analogue to his sympathetic faculty, counterbalances the blind organicism preached by his mentor Mordecai, to produce a liberal redemption of an ethnically based religious tradition.[141] Strange formal mutations dishevel this Arnoldian dialectic. The novel's interest in the recurrence of ancestral identities as a biological matrix for future development yields startling rhetorical symptoms, notably the full-scale activation of the trope of metalepsis, a transference of effects into causes and metaphors into events, analyzed in Cynthia Chase's virtuosic deconstructive reading of *Daniel Deronda*, and elsewhere characterized as a modulation from realism to romance or allegory.[142] "Superstitions carry consequences which often verify their hope or their foreboding" (330): what were Gothic shadows and omens in Gwendolen's story, associated with her susceptibility to second sight, migrate into the literal register of plotting in Daniel's. Nor does the metalepsis stop there. Daniel's accession of spiritual authority over Gwendolen coincides with a transference of portent into event in her story too. In one of her therapeutic encounters with Daniel, Gwendolen explains the traumatic fulfillment of her premonitions of the "dead face" (Grandcourt's, as it has turned out) by a recourse to magical thinking: "I did, I did kill him in my thoughts" (695). Committing murder in one's thoughts, like committing adultery in one's heart, follows a recognizably Christian logic of internalization (Matthew 5: 21–30), even if in this case the thoughts are followed by an actual result, as in witchcraft. (The narrator twice calls Gwendolen, early on, a "young witch," 77, 95.) Accordingly, Gwendolen's remorse forms the hinge of her conversion, "the culmination of that self-disapproval which had been the awakening of a new life within her" (697).[143]

In a further paradox, Gwendolen's conversion shadows Deronda's metamorphosis into a Jew. The latter process should entail, Chase insists, a physiological as well as moral transformation, from "disembodied spirit" to organic body, even as its referent (Deronda's phallus) is occluded.[144] Instead, a different kind of scandal articulates Daniel's metamorphosis, in his Messianic fulfillment of Mordecai's vision of "the prefigured friend." Mordecai is himself a survival, one of those aboriginal "poet-priests" conjectured by the mythographers. His "mind wrought so constantly in images, that his coherent trains of thought often resembled the significant dreams attributed to sleepers by waking persons in their most inventive moments; nay, they often resembled genuine dreams in their way of breaking off the passage from the known to the unknown," the narrator explains (473): echoing Comte's definition (in Lewes's paraphrase) of the "theological stage" of the human understanding of reality, "the primitive spontaneous exercise of the speculative faculty, proceeding from the known (i.e. consciousness) to the unknown."[145] The medium of cognition is an exquisitely developed aesthetic—indeed synaesthetic—sensibility:

> He was keenly alive to some poetic aspects of London. . . . Leaning on the parapet of Blackfriar's Bridge, and gazing meditatively, the breadth and calm of the river, with its long vista half hazy, half luminous, the grand dim masses of tall forms of buildings which were the signs of world-commerce, the oncoming of boats and barges from the still distance into sound and color, entered into his mood and blent themselves indistinguishably with his thinking, as a fine symphony to which we can hardly be said to listen, makes a medium that bears up our spiritual wings. (474)

This Whistlerian reverie is as up to date as it is primitive. The narrator goes on to compare Mordecai's imagination with that of the experimental scientist, invoking (like the narrator of *Middlemarch* in Lydgate's case) the paradigm of Whewellian hypothesis:

> His exultation was not widely different from that of the experimenter, bending over the first stirrings of change that correspond to what in the fervor of concentrated prevision his thought has foreshadowed. The prefigured friend had come from the golden background, and had signaled to him: this actually was: the rest was to be. (493)

It is as though a three-dimensional person, herald of a realist mimesis, should lean forth from the typological pattern of a Byzantine mosaic or Sienese altarpiece—even as the reverse process is taking hold, in the claim of a prophetic authority over realist representation.

Already hyperbolic, the comparison escalates:

> Perhaps his might be one of the natures where a wise estimate of consequences is fused in the fires of that passionate belief which determines the consequences it believes in. The inspirations of the world have come in that way too: even strictly-measuring science could hardly have got on without that forecasting ardor which feels the agitations

of discovery beforehand, and has a faith in its preconception that surmounts many failures of experiment. (513)

In a shamanistic blaze of enthusiasm, hypotheses become self-verifying. This extraordinary succession—from "yearning" through "hope" to an "expectant faith" (472) so passionate and at the same time so disciplined it "determines the consequences it believes in"—effects an alchemical transmutation of Comtean evolutionary stages, in which vatic intuition generates scientific knowledge. "The distinguishing characteristic of science is, that it sees and *foresees*. Science is *prevision*."[146]

Daniel's own critical doubts about Mordecai's pronouncements, and the narrator's equivocations, scarcely soften the force with which "that wish-begotten belief in his Jewish birth, and that extravagant demand of discipleship," do actually turn out "to be the foreshadowing of an actual discovery and a genuine spiritual result" (512)—since they are endorsed by the novel's plot:

> The more exquisite quality of Deronda's nature—that keenly perceptive sympathetic emotiveness which ran along with his speculative tendency—was never more thoroughly tested. He felt nothing that could be called belief in the validity of Mordecai's impressions concerning him or in the probability of any greatly effective issue: what he felt was a profound sensibility to a cry from the depths of another and accompanying that, the summons to be receptive instead of superciliously prejudging. (496)

He does not believe, but he feels—he sympathizes: and sympathy can be more powerful than belief, since the narrative does his believing for him, realizing literally what Deronda is allowed to take figuratively.

The narrator fuses the language of artistic imagination, scientific hypothesis, and prophetic revelation:

> And since the unemotional intellect may carry us into a mathematical dreamland where nothing is but what is not, perhaps an emotional intellect may have absorbed into its passionate vision of possibilities some truth of what will be—the more comprehensive massive life feeding theory with new material, as the sensibility of the artist seizes combinations which science explains and justifies. (514)

Daniel's sympathetic receptivity marks his evolutionary fitness. Inverting, now, the Christian consignment of Judaism to its superseded past, the typological scheme emergent in the second half of the novel casts Deronda as the new man who overcomes a vitiated Englishness by embracing and embodying ancestral Jewishness. Undertaking a Jewish conversion of Christian culture, as Cynthia Scheinberg argues, the new Messiah redeems the rhetorical trajectory we have been tracing—from figural to literal and carnal—as *the word made flesh*.[147] The conversion entails the novel's removal of its hero, in a seeming paradox, from the effective zone of the reader's sympathy, "*withdrawing* its gaze from Daniel's interior, once his Jewishness is revealed," Plotz remarks: "Eliot proposes that acquiring (or coming into) a national culture can transform a character so completely that he will not only become opaque to other characters, but actually disappear from the narrator's own line of sight."[148] Accession to racial being erases the realist novel's signature of individual character. Daniel is now something else, a type, invested with anagogic purpose. The recurrence of an ancestral, primitive form—allegorical and prophetic rather than mimetic and realist—brings forth the story of the future.

The science that bears upon these turns appears to be the "Pangenesis hypothesis," the controversial theory of the mechanism of inheritance mooted by Darwin in *The Variation of Animals and Plants under Domestication*. Lewes devoted the conclusion of his *Fortnightly Review* essay, "Mr. Darwin's Hypotheses," to a cautious defense of Pangenesis:

> The need for some such hypothesis has long been felt. Every one, as he remarks, must wish to explain, even if imperfectly, how it is possible for a bodily or mental characteristic which distinguished a parent to reappear in the offspring—how the peculiarity of an ancestor suddenly reappears in a descendant after lying dormant through generations; and this peculiarity may be a feature, a predisposition to disease, a monstrous deviation from the typical form, a perverted instinct or a glorious gift.[149]

Or, as Darwin himself summarized the key question: "How can we make intelligible, and connect with other facts, this wonderful and common capacity of reversion,—this power of calling back to life long-lost characters?"[150] It was, we saw in chapter 3, the key question addressed in Geoffroy and Isidore Saint-Hilaire's embryological research; references to the latter's "teratology treatise" (*Histoire générale et particulière des anomalies de l'organisation chez l'homme et les animaux*) are pervasive throughout *The Variation of Animals and Plants under Domestication* (as well as Darwin's other works).

According to Darwin's hypothesis, "the whole organisation, in the sense of every separate atom or unit, reproduces itself. Hence ovules and pollen-grains,—the fertilised seed or egg, as well as buds,—include and consist of a multitude of germs thrown off from each separate atom of the organism" (2: 358). Expressive of each state of the organism's development, including modifications acquired in its lifetime, the submicroscopic germs or "gemmules" are circulated to the reproductive organs, where they become hereditary. "These multiply and aggregate themselves into buds and the sexual elements; their development depends on their union with other nascent cells or units; and they are capable of transmission in a dormant state to successive generations" (2: 402). Transmission need not entail development, since the gemmules, "though generally acting in conjunction, are distinct powers"—as the phenomenon of reversion shows. Darwin reimagines the entire organism as a sort of living archive, which stores and transmits the successive developmental states of itself and its ancestors for potential regeneration. His vision radically disassembles the concept of a unitary, historically singular self or subject:

> The child, strictly speaking, does not grow into the man, but includes germs which slowly and successively become developed and form the man. In the child, as well as in the adult, each part generates the same part for the next generation. Inheritance must be looked at as merely a form of growth, like the self-division of a lowly-organised unicellular plant. Reversion depends on the transmission from the forefather to his descendants of dormant gemmules, which occasionally become developed under certain known or unknown conditions. Each animal and plant may be compared to a bed of mould full of seeds, most of which soon germinate, some lie for a period dormant, whilst others perish. . . . Each living creature must be looked at as a microcosm—a little universe, formed of a host of self-propagating organisms, inconceivably minute and as numerous as the stars in heaven. (2: 403–4)

In this uncanny recapitulation-with-a-difference of an earlier phase of scientific history—preformationist embryology—the body contains past and future multitudes, its ancestors and progeny, as its life (involuntary, palpitating) swarms through deep time.

Gwendolen's fits of spiritual dread come closest, perhaps, to this mind-bending thought: "The little astronomy taught her at school used sometimes to set her imagination at work in a way that made her tremble: but always when some one joined her she recovered her indifference to the vastness in which she seemed an exile" (64). Astronomy remains astronomy, and the metaphoric leap is spared her. For Deronda, the recurrence of an ancestral developmental stage promises the organic embodiment he has been seeking. Pangenesis—"a latent obstinacy of race in him"

(635)—authorizes his rebuke to his mother, checking her defiant assertion of individual autonomy: "the effects prepared by generations are likely to triumph over a contrivance which would bend them all to the satisfaction of self" (663). It authorizes his acknowledgment of Mordecai's visionary persuasion:

> It is you who have given shape to what, I believe, was an inherited yearning—the effect of brooding, passionate thoughts in many ancestors—thoughts that seem to have been intensely present in my grandfather. Suppose the stolen offspring of some mountain tribe brought up in a city of the plain, or one with an inherited genius for painting, and born blind—the ancestral life would lie within them as a dim longing for unknown objects and sensations, and the spell-bound habit of their inherited frames would be like a cunningly-wrought musical instrument, never played on, but quivering throughout in uneasy mysterious moanings of its intricate structure that, under the right touch, gives music. (750)

In this audacious revision of a familiar Romantic symbol, the body is an Aeolian harp,[151] vibrant with affective currents of organic life—expressive of an ontological nostalgia—gusting from past generations. Pangenesis authorizes, too, the activation of a reptilian developmental stage in that "remnant of a human being," Grandcourt (404): since the recurrent condition may be "a monstrous deviation from the typical form, a perverted instinct," as well as "a glorious gift," in Lewes's words. George Eliot asks us to imagine a biologically actual return of ancestral forms as figures of future possibility, monstrous and messianic.

Pangenesis, of course, did not settle into scientific knowledge. Ousted after Darwin's death by August Weismann's restriction of heritable cellular material to the germ-plasm and then by the belated diffusion of Gregor Mendel's genetic research, it retains, for readers who go back to it, its weird figurality. Darwin himself felt it:

> I am aware that my view is merely a provisional hypothesis or speculation; but until a better one be advanced, it may be serviceable by bringing together a multitude of facts which are at present left disconnected by any efficient cause. As Whewell, the historian of the inductive sciences, remarks:—"Hypotheses may often be of service to science, when they involve a certain portion of incompleteness, and even of error." (2: 357)

Lewes concurred, drawing a contrast between Natural Selection and Pangenesis: where the former exhibits "the introduction of one inference in

[a] series of facts, . . . all [of the latter's] elements are inferences; not one of them can be admitted as proven."[152] As Kant had complained with reference to Herder's philosophical anthropology, to found a history "simply and solely [upon] conjectures does not seem much better than to make a draft from a novel."[153] Still in its molten, metaphoric state, not yet cooled and tempered by experimental verification, the hypothesis may revert to its origin as a stroke of the imagination, a wild invention, a figure or fiction—much like George Eliot's realism in *Daniel Deronda*.

Shadows of the Coming Race

In *Daniel Deronda*, George Eliot forged beyond realism into a kind of science fiction. She would venture further, reneging on the promise of evolutionary perfection embodied in her novel's hero. "Believing as I do that man in the distant future will be a far more perfect creature than he now is, it is an intolerable thought that he & all the other sentient beings are doomed to complete annihilation after such long-continued slow progress," Darwin brooded in his late autobiographical sketch, the same year that *Daniel Deronda* completed serialization.[154] Three years later, in "Shadows of the Coming Race" (originally intended as the last chapter of *Impressions of Theophrastus Such*), George Eliot imagines a posthuman, indeed postorganic evolutionary pathway. (The title alludes to the subterranean superhuman civilization, destined to replace our own, in Edward Bulwer Lytton's 1871 novel *The Coming Race*.) It is one in which the forces of natural selection have absolute sway—issuing, disconcertingly, in the extinction not only of humanity but of life as we know it. Eliot's sketch imagines a future in which machines designed to save human labor themselves develop, "by a further evolution of internal molecular movements," into self-reproducing entities: "This last stage having been reached, either by man's contrivance or as an unforeseen result, one sees that the process of natural selection must drive men altogether out of the field."[155]

Eliot draws out the hypothesis, proposed nearly a century ago by Herder and given recent scientific currency by Owen and Tyndall, that biological life is continuous at a base molecular level with processes of mineral formation. However, she capsizes Tyndall's bid to subsume inorganic matter into a transcendental organic principle infusing all of nature.[156] In this mordant gloss on the Development Hypothesis, life's evolutionary advance to greater complexity dispenses with the organic state altogether. We glimpse what may lie beyond the veil of involuntary, palpitating life, to be scrutinized (in Elizabeth Grosz's account) by "Darwin's first philosophical heir," Henri Bergson:

> If Darwin demonstrates man's immersion in and emergence from animal (and ultimately plant) life (or even life before plants and animals separated), it is Bergson . . . who demonstrates man's immersion in and emergence from the inhuman, the inorganic, or the nonliving. . . . The common impetus life carries within it is that of *materiality itself*, the capacity to make materiality extend itself into the new and the unforeseeable.[157]

Eliot's closing evocation assumes an eerie, litotic lyricism, resonant with the rhetorical repertoire of a superseded organicism:

> Thus this planet may be filled with beings who will be blind and deaf as the inmost rock, yet will execute changes as delicate and complicated as those of human language and all the intricate web of what we call its effects, without sensitive impression, without sensitive impulse: there may be, let us say, mute orations, mute rhapsodies, mute discussions, and no consciousness there even to enjoy the silence.[158]

Eliot fashions a cold rejoinder to Darwin's rhapsodic vision of the plenitude of terrestrial life: "from so simple a beginning endless forms most beautiful and most wonderful have been, and are being, evolved." The evolutionary process culminates in a mechanical-mineral culture sans consciousness, sans life, aesthetically no less "delicate and complicated"—no less beautiful and wonderful—than ours.

NOTES

Introduction: The Human Age

1. Some convenient chronological bookends: Henry Fielding's critical reflections on the novel in *Tom Jones* coincide with the first volume of the Comte de Buffon's *Natural History*, 1749; George Eliot's *Middlemarch* begins serialization a few months after the publication of Charles Darwin's *The Descent of Man*, 1871.

2. Foucault, *The Order of Things*, 336, 340, 372–73.

3. Herder, "How Philosophy Can Become More Universal," in *Philosophical Writings*, 29; Kant, letter to Marcus Hertz (1773), in *Anthropology, History, and Education*, 227. On the era of "humanism or anthropologism ... the dominating and spellbinding extension of the 'human sciences' in the philosophical field" (Hegel to Heidegger), see Derrida, "The Ends of Man," in *Margins of Philosophy*, 116–17; on anthropology as "perhaps the fundamental arrangement that has governed and controlled the path of philosophical thought from Kant until our own day," see Foucault, *The Order of Things*, 336–73 (373).

4. Darwin, *The Descent of Man*, 681.

5. Derrida, "The Ends of Man," in *Margins of Philosophy*, 116–18; Foucault, *The Order of Things*, 379.

6. Walter Scott's phrase, in *Ivanhoe* (18), which goes on (we shall see in chapter 3) to unsettle the bounds of "our common nature."

7. Franco Moretti, *Distant Reading*, 19–21; *The Way of the World*, 15.

8. Lukács, *The Historical Novel*, 35.

9. See Johannes Fabian's classic account of anthropology's sequestration of "primitive" peoples in an ahistorical temporality, *Time and the Other*. Thanks to Ayşe Agiş for pressing me on this.

10. Eliot, *Middlemarch*, 3.

11. Ibid., 788. For "species consciousness" (*Bewußtsein der Gattungen*, "cognizance of species," in Eliot's translation), see Feuerbach, *The Essence of Christianity*, 1–2.

12. Thanks to Nancy Armstrong for the formulation here.

13. Lukács, *Studies in European Realism* 3, 8.

14. Greif, *The Age of the Crisis of Man*, 4–9.

15. Derrida, "The Ends of Man," in *Margins of Philosophy*, 117.

16. Kant, "Idea for a Universal History with a Cosmopolitan Aim," in *Anthropology, History, and Education*, 107–20. On the Bildungsroman and modern human rights discourse, see Slaughter, *Human Rights Inc.*, 88–120.

17. Foucault, *The Order of Things*, xxv, 379, 373.

18. Derrida, "The Ends of Man," in *Margins of Philosophy*, 121.

19. See Cleary, "Realism after Modernism and the Literary World System"; Esty and Lye, "Peripheral Realisms Now."

20. *The Human Age* is the title both of a dystopian satire by Wyndham Lewis (*Monstre Gai* and *Malign Fiesta*, 1955; sequels to *The Childermass*, 1928) and of a

recent essay in humanist salvage by Diane Ackerman. On the Anthropocene, see Chakrabarty, "The Climate of History"; Jeremy Davies, *The Birth of the Anthropocene*; Taylor, "Anthropocene."

21. Ghosh, *The Great Derangement*, 7–8. McKibben's *The End of Nature* (1989) was the first popular book-length treatment of anthropogenic climate change.

22. Greif, *The Age of the Crisis of Man*, 11.

23. From Beer's *Darwin's Plots* (first published in 1983) and George Levine's *Darwin and the Novelists* (1988) to, most recently, Griffiths's *The Age of Analogy* (2016).

24. On "the physical basis of mind" (Lewes's or Eliot's phrase), see, e.g., Dames, *The Physiology of the Novel*; Ryan, *Thinking without Thinking in the Victorian Novel*; Benjamin Morgan, *The Outward Mind*; Cohn, *Still Life*; on anthropological themes, Marcus, *Between Women*; Psomiades, "The Marriage Plot in Theory"; Schaffer, *Romance's Rival* (marriage); Lecourt, *Cultivating Belief* (religion and race); Rajan, *A Tale of Two Capitalisms* (religion and economics); Buzard, *Disorienting Fiction* (ethnography and culture); on Victorian biopolitics (and bioeconomics), Gallagher, *The Body Economic*; Steinlight, *Populating the Novel*.

25. See, e.g. (1995–2010), Fox, Porter, and Wokler, eds., *Inventing Human Science*; Nash, *Wild Enlightenment*; Douthwaite, *The Wild Girl, Natural Man, and the Monster*; Davidson, *Breeding*; Laura Brown, *Homeless Dogs and Melancholy Apes*. On the life sciences and (mainly) poetry in the Romantic period, see McLane, *Romanticism and the Human Sciences*; Gigante, *Life*; Mitchell, *Experimental Life*; Goldstein, *Sweet Science*.

26. "I have, therefore, like a maiden knight with his white shield, assumed for my hero . . . an uncontaminated name, bearing with its sound little of good or evil" (Scott, *Waverley*, 3).

27. Chambers, *Vestiges of the Natural History of Creation*, 272: echoing Lamarck's proposition, "That [the human] race having obtained the mastery over others through the higher perfection of its faculties will take possession of all parts of the earth's surface, that are suitable to it" (*Zoological Philosophy*, 170). See Siobhan Carroll's vivid account of the new sense of human planetary occupation and its limits in this period, *An Empire of Air and Water*.

28. Marx, *Capital*, in *Karl Marx*, 76.

29. Shelley, *The Last Man*, 27.

30. Buffon, *Natural History*, 3: 278.

31. Here and later in this introduction I am paraphrasing McLane's powerful reading of *Frankenstein* in *Romanticism and the Human Sciences*, 13–16, 84–108.

32. Fielding, *Tom Jones*, 30.

33. "When the first mean selfish Creature appeared on the human Stage, who made Self the Centre of the whole Creation; would give himself no Pain, incur no Danger, advance no Money, to assist, or preserve his Fellow-Creatures; then was our Lawyer born" (Fielding, *Joseph Andrews*, 164).

34. Fielding, *Tom Jones*, 68.

35. Hume, *A Treatise of Human Nature*, 43.

36. Eliot, *Middlemarch*, 3.

37. On the "Buffonian Revolution," see Sloan, "The Gaze of Natural History," 123, 126–27. I accept Sloan's account over Foucault's identification of Cuvier (overlooking

Buffon) as the author of an epistemic shift from classical natural history to modern biology. See also Reill, *Vitalizing Nature*, 252–53. On the terminology of "natural history," see introduction, note 45, below.

38. Chico, *The Experimental Imagination*, 1; see 17–48. On scientific empiricism and the early novel, see Bender, *Ends of Enlightenment*, 22–28.

39. Bender, ibid., 38–55 (41). On "regulative analogy," with its verbal trigger "as if," as a shared technique of nineteenth-century fiction and scientific writing, see Farina, *Everyday Words*, 95–136. On *Middlemarch* as "a work of experimental science," see Shuttleworth, *George Eliot and Nineteenth-Century Science*, 143; and Beer, *Darwin's Plots*, 148–52.

40. On Hume, fictionality, and the novel, see Duncan, *Scott's Shadow*, 119–30. In a landmark essay, Gallagher traces the emergence of a modern rhetoric of fiction through management of the relation between individual and species or type: "The Rise of Fictionality," 341–45.

41. Whewell, *The Philosophy of the Inductive Sciences*, 2: 42. On the hermeneutic—as opposed to strictly Baconian—tradition of scientific induction, introduced into early Victorian Britain by Whewell, see Smith, *Fact and Feeling*, 16–22, 37–41; Anger, *Victorian Interpretation*, 87–94.

42. Darwin, *On the Origin of Species*, 401, 423.

43. Rousseau, *The Discourses*, 132.

44. By Dugald Stewart: "Account of the Life and Writings of Adam Smith, LL.D.," in *Works*, 7: 31.

45. Dugald Stewart, "A Dissertation Exhibiting the Progress of Metaphysical, Ethical and Political Philosophy," in *Works*, 6: 65. With the term "natural history of man," here and throughout *Human Forms*, I follow the usage established by Buffon and Stewart. In making nature historical, Buffon broke down the taxonomic distinction between "natural history" (description and classification) and "natural philosophy" (causal and mathematical relations) in D'Alembert's *Preliminary Discourse to the Encyclopedia* (1751; see Rudwick, *Bursting the Limits of Time*, 52). Stewart's identification of the "theoretical or conjectural history" of man with "natural history" (alluding to Hume's *Natural History of Religion*) similarly cuts across Gaukroger's distinction between philosophical anthropology and natural history, "the forebears of what came to be known as social anthropology and physical anthropology respectively," a distinction that opens with the formation of nineteenth-century disciplines, as Gaukroger admits (*The Natural and the Human*, 214). Where Gaukroger is attentive to Buffon's key role in the historical science of man, Frank Palmeri elides natural history from his account of conjectural history before Darwin's *Descent of Man* (*State of Nature, Stages of Society*, chapter 5). On the cross-pollination of nineteenth-century literary and scientific discourses with analogical reasoning and comparative historicism, see Griffiths, *The Age of Analogy*, 7–21.

46. Foucault, *Abnormal*, 63.

47. Agamben, *The Open*, 15.

48. Buffon, *Natural History*, 2: 353–68.

49. On attempts to define human uniqueness, see Thomas, *Man and the Natural World*, 30–41.

50. Morgenstern, "On the Nature of the Bildungsroman," 655.

51. Blumenbach, *Anthropological Treatises*, 57.
52. See Stocking, *Victorian Anthropology*, 64–69.
53. See Thomas Richards, *The Imperial Archive*, 45–48. Georges Canguilhem argues that evolutionary adaptation makes monstrosity "the rule and originality a temporary banality" (*Knowledge of Life*, 105–6).
54. Darwin, *On the Origin of Species*, 49–50, 55–56. On Geoffroy's system, see Chai, "Life and Death in Paris," 715.
55. Agamben, *The Open*, 16.
56. Burke, *Selected Letters*, 102.
57. Muthu, *Enlightenment against Empire*, 31–65 (Diderot versus Rousseau); 122–209 (Kant); 210–58 (Herder). On Diderot, see Curran, "Logics of the Human in Diderot's Supplement"; for Kant's argument with Forster, "On the Use of Teleological Principles in Philosophy," in *Anthropology, History, and Education*, 192–218. See also Fulford, Lee, and Kitson, *Literature, Science and Exploration*, 127–38.
58. Festa, *Sentimental Figures of Empire*, 153.
59. Ibid., 132–204.
60. Long, *History of Jamaica*, 356, 365.
61. Thanks, again, to Ayşe Agiş for this.
62. Hume, *Enquiry Concerning Human Understanding*, 75–76.
63. "Those who travel in order to acquaint themselves with the different manners of men might spare themselves much pains by going to a carnival at Venice; for there they will see at once all which they can discover in the several courts of Europe. The same hypocrisy, the same fraud; in short, the same follies and vices, dressed in different habits. . . . [H]uman nature is everywhere the same, everywhere the object of detestation and scorn" (Fielding, *Tom Jones*, 417).
64. Scott, *Waverley*, 5.
65. Austen, *Northanger Abbey*, 147.
66. Hume, "Of National Characters," in *Essays*, 207.
67. See Brantlinger, "Imperial Gothic: Atavism and the Occult in the British Adventure Novel, 1880–1914," in *Rule of Darkness*, 227–53. Compare analyses of the absence of a realist aesthetic in nineteenth-century Irish fiction, symptomatic of the lack of hospitable historical conditions (civil society and a national state), in Lloyd, *Anomalous States*, 129–31; Deane, *Strange Country*, 18–20; Eagleton, *Heathcliff and the Great Hunger*, 181–88; and Priya Joshi's argument that nineteenth-century Anglophone Indian readers' predilection for sensational and melodramatic over realist fiction reflected life "under the shadow of a colonial state in which Indian readers were only ever subjects, never citizens" (*In Another Country*, 80–87 [82]).
68. McLane, *Romanticism and the Human Sciences*, 11–12.
69. Ibid., 32, 19. For the "literary absolute," a term developed by Lacoue-Labarthe and Nancy from Friedrich Schlegel's poetic theory, see chapter 2 of *Human Forms*.
70. Shelley, *Frankenstein*, 121.
71. McLane, *Romanticism and the Human Sciences*, 10–13, 84–108 (15); see also, on *Frankenstein* and human rights, Reese, "A Troubled Legacy"; and species discourse, Douthwaite, *The Wild Girl, Natural Man, and the Monster*, 214–22; and the aesthetics of monstrousness, Gigante, "Facing the Ugly."
72. McLane, ibid., 18.

73. See Moretti, *The Way of the World*, 5–6, 15; *Distant Reading*, 19–21.
74. Moretti, *The Way of the World*, 63–72 (71, 65).
75. Corbett's term, in *Allegories of Union*.
76. James Hutton, "Theory of the Earth," 304. The infinite earth history Hutton posited was cyclical and homeostatic, rather than linear and open-ended: see Rudwick, *Bursting the Limits of Time*, 158–72.
77. Shelley, *Frankenstein*, 90, 141.
78. Bataille, "Formless," in *Visions of Excess*, 31.
79. Murphy, introduction to *The Works of Henry Fielding, Esq., with the Life of the Author* (1762); in Williams, ed., *Novel and Romance 1700–1800*, 261. "Still observing the grand essential rule of unity in the design . . . no author has introduced a greater diversity of characters, or displayed them more fully, or in more various attitudes" (257).
80. Hilles, "Art and Artifice in *Tom Jones*," 919. (Thanks to Jill Campbell for this reference.) See also Douglas Brooks's numerological exegesis, *Number and Pattern in the Eighteenth-Century Novel*, 98–109.
81. Howard Colvin points out that Wood's design does not correspond with the symmetrical scheme reproduced by Hilles, and was in any case imperfectly executed: review of Simon Varey, 262.
82. See Gallagher, "Formalism and Time." I am grateful to Ella Mershon for modeling the temporal dimensions of narrative and narration in *Tom Jones*.
83. British landmarks include William Hogarth's *Analysis of Beauty* (1753), Hume's "Of the Standard of Taste" (1757), and Burke's *Philosophical Enquiry into the Origin of Our Ideas of the Sublime and Beautiful* (1757). Before these, in *Observations on Man, his Frame, his Duty, and his Expectations* (1749), David Hartley's "doctrine of vibrations" offered a nonreductive neurophysiological basis for psychology and metaphysics: see Allen, *David Hartley on Human Nature*.
84. Burke, *Philosophical Inquiry*, 102.
85. Dames, *The Physiology of the Novel*, 13, 21, 44; for the general discussion, see 47–70.
86. Lewes, "Dickens in Relation to Criticism," 576.
87. See Lukács, *Theory of the Novel*, 81.
88. James, preface to vol. 7 of the New York edition (1907), in *Literary Criticism*, 1107.
89. Maxwell, *The Mysteries of Paris and London*.
90. Leavis, *The Great Tradition*, 2.
91. See, e.g., Feltes, *Modes of Production of Victorian Novels*, 49.
92. Goodlad, *The Victorian Geopolitical Aesthetic*, 175–84 (178, 177): citing Lukács, *The Historical Novel*, 350.
93. Feuerbach, *The Essence of Christianity*, 2.
94. Grosz, *Becoming Undone*, 5.
95. See Chai, "Life and Death in Paris," 721–26.
96. Spencer, "The Social Organism," 99.
97. Gallagher, *The Body Economic*, 36–50 (37).
98. See Bewell, "Jefferson's Thermometer," 122.
99. James, "The Art of Fiction," in *The Art of Criticism*, 174, 172.

100. See Fredrickson, *The Ploy of Instinct*.

101. Grosz, *Becoming Undone*, 27. The continuity between inorganic and organic matter—between life and nonlife—is a commonplace of evolutionist thinking, from Diderot and Herder through Huxley and Tyndall: see chapter 5 of *Human Forms*, and (for the Epicurean tradition upon which it draws) Goldstein, *Sweet Science*, 19–24.

102. Grosz, ibid., 21, 25.

103. Canguilhem, *Knowledge of Life*, 119.

104. Meillassoux, *After Finitude*, 5. See also Brassier on "the realist conviction that there is a mind-independent reality, which, despite the presumptions of human narcissism, is indifferent to our existence and oblivious to the 'values' and 'meanings' which we would drape over it in order to make it more hospitable" (*Nihil Unbound*, xi). I am grateful to Julián Heffernan for directing me to Meillassoux and Brassier.

105. Meillassoux, ibid., 11, 10.

106. Brassier, *Nihil Unbound*, 223–30 (224, 223).

107. On Buffon's *Epochs*, see Heringman, "Deep Time at the Dawn of the Anthropocene."

108. Heffernan, "The Stamp of Rarity," 98–113.

109. Marx, "The German Ideology," in *Karl Marx*, 26.

110. See Taylor, "Anthropocene": "If the Anthropocene is 'the very negation of universal species being' then the long nineteenth century is precisely the geohistorical space within which that negation took place. It is also the period in which many of the conceptual tools and rubrics—from the stratigraphic method, to the greenhouse effect, natural selection to anthropogenic extinction—were articulated" (575); and Hensley and James, "Soot Moth." Arguably the global scale of the European remaking of natural as well as human environments had become evident to other populations across the planet long before then.

111. Brassier, *Nihil Unbound*, 223; citing Lyotard, *The Inhuman*, 9.

112. Ruskin, *Fiction—Fair and Foul*, 8–9.

Chapter One: The Form of Man

1. Herder, "On the Change of Taste," in *Philosophical Writings*, 255. © Cambridge University Press 2002, reproduced with permission.

2. Hume, *A Treatise of Human Nature*, 43.

3. Stephen Gaukroger distinguishes "philosophical anthropology" from the "natural history of man" on the grounds of disparate methods and agendas, realized in their subsequent disciplinary trajectories in (respectively) nineteenth-century social and physical anthropology (*The Natural and the Human*, 12–13, 230–31). As he admits, the distinction was not recognized in the late eighteenth century: Scottish conjectural history (Smith, Ferguson, Millar, Robertson, Kames, Monboddo), discussed by Gaukroger under natural history, belongs just as well to philosophical anthropology. The quarrel between Kant and Herder (the main topic of this chapter) can be characterized, in Gaukroger's terms (214), as turning on Kant's resistance to Herder's attempt to integrate natural history with philosophical anthropology. On the Enlightenment project of "Master Systems," see Siskin, *System*, 115–20.

4. Bender, *Ends of Enlightenment*, 54; Reill, *Vitalizing Nature*, 69.

5. Sloan, "The Gaze of Natural History," 123, 126–27; see also Larson, *Interpreting Nature*, 12–15; Reill, *Vitalizing Nature*, 35–42; Mensch, *Kant's Organicism*, 47–50. On the taxonomic grid (*mathesis*), see Foucault, *The Order of Things*, 71–76, 128–57. As Reill implies (252–53), Foucault's account of the epistemic shift from what he calls "natural history" to "biology" elides Buffon for the intellectually less consequential figure of Cuvier (a better fit with Foucault's c. 1800 periodization).

6. Dugald Stewart, "A Dissertation Exhibiting the Progress of Metaphysical, Ethical and Political Philosophy," in *Works*, 6: 65.

7. On this redistribution, see Klancher, *Transfiguring the Arts and Sciences*, 16–19, 154–63.

8. See Kant, "Idea for a Universal History with a Cosmopolitan Aim," in *Anthropology, History, and Education*, 107–75; Zammito, *The Genesis of Kant's* Critique of Judgment, 178–88, 203–8; Zammito, *Kant, Herder, and the Birth of Anthropology*, 189–213; Beiser, *The Fate of Reason*, 153–58, 203–8; Van den Berg, *Kant on Proper Science*, 206–9; Gaukroger, *The Natural and the Human*, 200–216. For the prehistory of the debate, in Kant's early (and abiding) objection to the scientific status of theories of organic generation, see Mensch, *Kant's Organicism*, 62–64.

9. Herder, "How Philosophy Can Become More Universal," in *Philosophical Writings*, 29.

10. Quotation given in Zammito, in *Kant, Herder, and the Birth of Anthropology*, 227.

11. Esterhammer, "Continental Literature, Translation, and the Johnson Circle," 103. *Outlines* was translated by Thomas Churchill with the assistance of Henry Fuseli, who may have prompted Johnson to commission the translation.

12. Kant, letter to Marcus Hertz, cited in *Anthropology, History and Education*, 227.

13. Kant, "Idea for a Universal History with a Cosmopolitan Aim," ibid., 109; future references to this volume will be cited in the text.

14. In *Anthropology from a Pragmatic Point of View*, however, Kant allows that, "while not exactly sources for anthropology, these are nevertheless aids: world history, biographies, even plays and novels"—"such characters as are sketched by a Richardson or a Molière . . . must nevertheless correspond to human nature in kind" (ibid., 233).

15. On conjectural history's "complex and special fusion of fact and fiction, textual observation and speculation," see Bewell, *Wordsworth and the Enlightenment*, 59–60. Thomas Reid characterized Hume's *Treatise* as a Gothic romance. "I find I have only been in an inchanted castle, imposed upon by spectres and apparitions" (*Inquiry into the Human Mind*, 22): a mischievous allusion to Hume's claim that ideas of belief "take firmer hold of my mind, than the ideas of an inchanted castle" (*Treatise on Human Nature*, 673); see Kareem, "Lost in the Castle of Scepticism." Buffon accused Thomas Burnet's *Sacred Theory of the Earth* of being "an elegant romance" ("The Theory of the Earth," in Lyon and Sloan, *From Natural History to the History of Nature*, 148–49).

16. Zalasiewicz et al., "Introduction" to Buffon, *Epochs of Nature*, xxv.

17. Reill, *Vitalizing Nature*, 55–56. See also Bender's qualification: "[Buffon] followed the classic strategy of attempting to maintain the scientificity of his work by associating hypotheses with fictionality and limiting their relevance to his findings, which he stringently contrasted with hypotheses and called 'theories' . . . insisting upon exact analysis of observable physical fact in order to free a space for strictly

controlled speculation in the form of hypotheses that answer the question 'how'" (*Ends of Enlightenment*, 53–54). On the rise of analogical reasoning as method of a new comparative historicism in the Romantic period, see Griffiths, *The Age of Analogy*, especially 14–20, 28–30; on "regulative analogy" in fiction and scientific writing, see Farina, *Everyday Words*, 95–136.

18. Dugald Stewart, "Account of the Life and Writings of Adam Smith, LL.D.," in *Works*, 7: 31. Stewart is referring to David Hume's *Natural History of Religion*: see Palmeri, *State of Nature, Stages of Society*, 6–7, 37–38.

19. Dugald Stewart, ibid., 7: 31.

20. Palmeri, *State of Nature, Stages of Society*, 16; see 34–53; see also Poovey, *History of the Modern Fact*, 219–29.

21. Adam Ferguson, *History of Civil Society*, 8.

22. Rousseau, *The Discourses*, 132. "Commençons donc par écarter tous les faits, car ils ne touchent point à la question. Il ne faut pas prendre les Recherches, dans lesquelles on peut entrer sur ce sujet, pour des vérités historiques, mais seulement pour des raisonnements hypothétiques et conditionnels; plus propres à éclaircir la Nature des choses qu'à montrer la véritable origine, et semblables à ceux que font tous les jours nos Physiciens sur la formation du Monde" (*Discours sur l'origine et les fondements de l'inégalité*, 62–63). By "all the facts," Rousseau primarily means the scriptural account of ancient history, although his reference to "Philosophers who have examined the foundations of society" includes the archives of secular history in the dismissal. See Palmeri, *State of Nature, Stages of Society*, 4–5.

23. De Man, *Allegories of Reading*, 141.

24. Agamben, *The Open*, 29.

25. Ibid., 15.

26. Lovejoy, *The Great Chain of Being*, 198–99.

27. Ibid., 244: "converting the once immutable Chain of Being into the program of an endless Becoming" (259).

28. Buffon, *Natural History*, 2: 2–3, 16; "cette puissance de produire son semblable, cette chaîne d'existences successives d'individus, qui constitue l'existence réelle de l'espèce"; "cette vertu procréatrice qui s'exerce perpétuellement" (*Oeuvres*, 143, 134). See Sloan, "The Gaze of Natural History," 132; Lyon and Sloan, *From Natural History to the History of Nature*, 20–23.

29. Buffon, *Natural History*, 3: 264.

30. Émile Durkheim revived Buffon's nomenclature in the early twentieth century: see, e.g., "The Dualism of Human Nature and Its Social Conditions."

31. Foucault, *Abnormal*, 63.

32. See Zoe Beenstock's summary of the challenge to the Aristotelian foundation of human nature in sociability mounted by Thomas Hobbes's social contract theory and Enlightenment responses, from Shaftesbury and Hutcheson to Kant: *The Politics of Romanticism*, 17–35.

33. Compare Foucault's commentary, in *The Order of Things*, on the "anthropological configuration of modern philosophy" since Kant: "The pre-critical analysis of what man is in his essence becomes the analytic of everything that can, in general, be presented to man's experience" (372–73).

34. De Man, *Allegories of Reading*, 137; Bates, *States of War*, 180.

35. Rousseau, *The Social Contract*, 76. See Bates, ibid., 176–89.

36. Compare Johann Friedrich Blumenbach, *On the Natural Variety of Mankind* (1795): "Man is a domestic animal. . . . Other domestic animals were brought to that state of perfection *through him*. He is the only one who brought *himself* to perfection" (*Anthropological Treatises*, 340).

37. Kant, "Idea for a Universal History with a Cosmopolitan Aim," in *Anthropology, History, and Education*, 111. See Brandt, "The Guiding Idea," 97–99.

38. On the post-Kantian theory of fictions developed by Hans Vaihinger in *The Philosophy of "As If"* (*Die Philosophie des Als-Ob*, 1911), see Iser, *The Fictive and the Imaginary*, 13–17, 130–57. In his autobiographical preface, Vaihinger pays tribute to the formative influence of "Herder's book on the history of mankind": "Herder draws special attention to the evolution of spiritual life out of its first animal origins, and he regards man always as linked up with that Nature from which he has gradually evolved. Thus in 1869, when I first heard Darwin's name and when my school friends told me about the new theory of man's animal ancestry, it was no surprise to me, because through my reading of Herder I was already familiar with the idea" (*The Philosophy of "As If"*, xxiii–xxiv).

39. Beiser, *The Fate of Reason*, 148–49.

40. See Reill, *Vitalizing Nature*, 42–43, 55–56 (on Buffon); 187–88 (on Herder); Zalasiewicz et al., "Introduction" to Buffon, *Epochs of Nature*, xxxiii.

41. Goldstein, "Irritable Figures," 275. See also the reconstructions of Herder's recourse to the poetic and aesthetic as media of philosophical knowledge (via Wolff, Baumgarten, Lessing, and others) by Noyes, "Herder's Unsettling"; and Gaukroger, *The Natural and the Human*, 190–97.

42. Herder, "On Image, Poetry, and Fable," in *Selected Writings on Aesthetics*, 358–59; see Goldstein, ibid., 285–86.

43. Goldstein, ibid., 282.

44. On the "vitalist turn," see Reill, *Vitalizing Nature*; Gigante, *Life*; Mitchell, *Experimental Life*, 1–8. On the distinction between emergent organic-vitalist and epigenetic paradigms—the former positing an autonomous, autotelic life force, the other a "vital materialism" based on Lucretian atomism—see Goldstein, *Sweet Science*, 19–22, 48–50, 56–59.

45. Buffon, *Natural History*, 3: 404–5. "Un individu est un être à part, isolé, détaché, et qui n'a rien de commun avec les autres êtres, sinon qu'il leur ressemble ou bien qu'il en diffère . . . cependant ce n'est ni le nombre ni la collection des individus semblables qui fait l'espèce, c'est la succession constante et le renouvellement non interrompu de ces individus qui la constituent; car un être qui dureroit toûjours ne feroit pas une espèce, non plus qu'un million d'êtres semblables qui dureroient aussi toûjours : l'espèce est donc un mot abstrait et général, dont la chose n'existe qu'en considérant la Nature dans la succession des temps, et dans la destruction constante et le renouvellement tout aussi constant des êtres" (*Oeuvres*, 558–59). See Sloan, "The Gaze of Natural History," 131–33.

46. Buffon, *Natural History*, 8: 67.

47. Rousseau, *The Discourses*, 141. Future references will be given in the text.

48. For the commonplace, compare Addison, *The Spectator* 111 (July 7, 1711): "A Brute arrives at a point of Perfection that he can never pass. In a few Years he has all

the Endowments he is capable of; and were he to live ten thousand more, would be the same thing he is at present" (1: 457). On perfectibility (the term was coined by Turgot and popularized by Rousseau) and its English adherents, especially Godwin, see Davidson, *Breeding*, 165–82.

49. Adam Ferguson, *History of Civil Society*, 10. Future references to this edition will be given in the text.

50. Rousseau, *The Discourses*, 186.

51. See Carrithers, "The Enlightenment Science of Society," 247–49; McDaniel, "Philosophical History and the Science of Man in Scotland."

52. "Après avoir montré que la *perfectibilité*, les vertus sociales et les autres facultés que l'homme Naturel avoit reçues en puissance, ne pouvoient jamais se developper d'elles-mêmes, qu'elles avoient besoin pour cela du concours fortuit de plusieurs causes étrangeres qui pouvoient ne jamais naître, et sans lesquelles il fût demeuré éternellement dans sa condition primitive" (Rousseau, *Discours sur l'origine et les fondements de l'inégalité*, 92).

53. Herder, "Treatise on the Origins of Language," in *Philosophical Writings*, 76. Future references will be given in the text. Although he frames his treatise as a refutation of the Protestant natural theologian Johann Peter Süßmilch, Herder's more subtle engagement is with Rousseau's *Discourse on Inequality*. He had not read Rousseau's "Essay on the Origin of Languages," which was not published until 1781. For a full discussion, see DeSouza, "Language, Reason, and Sociability," 229–39; also Barnard, *Herder on Nationality, Humanity, and History*, 39–46.

54. Herder engages with Hermann Samuel Reimarus's detailed treatment of the question in *General Reflections on the Drives of Animals* (1760); see Zammito, "Herder between Reimarus and Tetens." Reimarus saw instinct as evidence for fixity of type, hence a barrier to evolutionism: see Robert Richards, *Darwin and the Emergence of Evolutionary Theories of Mind and Behavior*, 30. On "the human animal's lack of instinct" and its legacy in Enlightenment theories of aesthetic contemplation, see Von Mücke, *The Practices of Enlightenment*, 27–38 (on Reimarus, 33–36).

55. Herder's Romantic conceptions of the creature's "circle" or "sphere," i.e., its horizon of perception, and of the signifying mark (*Merkmal*) it imprints on the world have a long reach, through Feuerbach's *Essence of Christianity* (see chapter 5) at least as far as the *Umwelt* of Jacob von Uexküll: *A Foray into the Worlds of Animals and Humans*, 43 (the creature's "bubble"), 48–49 (*Wirkmal*, "effect mark," and *Merkmal*, "perception mark"). Heidegger's distinction between the animal, "poor in world" (*weltarm*), and man, "world-forming" (*weltbildend*), is also latent here: see Heidegger's response to Herder, *On the Essence of Language*; also Agamben, *The Open*, 40–42; Canguilhem, *Knowledge of Life*, 111–13.

56. See DeSouza's summary of Herder's account of linguistic development, "Language, Reason, and Sociability," 224–27.

57. "As soon as they are born, brutes bring with them from their mother's womb, as Lucilius says, all that they are going to possess.... The Father infused in man, at his birth, every sort of seed and all sprouts of every kind of life. These seeds will grow and bear fruit in each man who sows them" (Pico della Mirandola, *Oration on the Dignity of Man*, 119–21).

58. On the Enlightenment medico-physiological conception of health as an equilibrium or attunement between the body and its environment, mediated through the senses, see Kevis Goodman's forthcoming *Pathologies of Motion: Medicine, Aesthetics, Poetics*.

59. "Wir wachsen immer aus einer Kindheit, so alt wir sein mögen, sind immer im Gange, unruhig, ungesättigt. Das Wesentliche unsres Lebens ist nie Genuß, sondern immer Progression, und wir sind nie Menschen gewesen, bis wir—zu Ende gelebt haben; dahingegen die Biene Biene war, als sie ihre erste Zelle bauete" (Herder, *Abhandlung über den Ursprung der Sprache*, 84–85). Compare Marx, in the 1844 *Economic and Philosophical Manuscripts*: "[Animals] build themselves nests, dwellings, like the bees, beavers, ants, etc. But an animal only produces what it immediately needs for itself or its young. It produces one-sidedly, whilst man produces universally. It produces only under the dominion of immediate physical need, whilst man produces even when he is free from physical need and only truly produces in freedom therefrom. An animal produces only itself, whilst man reproduces the whole of nature" (*Karl Marx*, 41–42). See also Heidegger on the worker bee versus "world-forming man" (*The Fundamental Concepts of Metaphysics*, 193).

60. See Beiser, *The Fate of Reason*, 136–38.

61. Michael Forster, *Herder's Philosophy*, 253.

62. Herder, "On the Cognition and Sensation of the Human Soul," in *Philosophical Writings*, 188.

63. Herder, *Outlines of a Philosophy of the History of Man*, 63. Future references will be cited in the text. Zammito (*Kant, Herder, and the Birth of Anthropology*, 302–7) defines Herder's commitment to epigenesis as "the scientific effort to discern, to describe, and to account for the *immanent capacity* ('force') of nature to *transform itself*, to construct higher plateaus of order—discontinuously, *emergently*, and thus to preserve the idea that, at least *empirically*, it is possible to conceive nature as inherently lawful" (306).

64. Kant, "Review of Herder's *Ideas*," in *Anthropology, History, and Education*, 131. See Zammito, "Epigenesis." Todd Kontje identifies Herder as "the most influential disseminator of [the new, naturalistic] concept of *Bildung*" in his useful summary of eighteenth-century theories of the concept (*The German Bildungsroman*, 2–3). On Haller, Wolff, Blumenbach, and the *Bildungstrieb*, see Larson, *Interpreting Nature*, 91–92, 138–61; Reill, *Vitalizing Nature*, 160–70; Gigante, *Life*, 16–21.

65. Kant, ibid., 131, 132. See Beiser, *The Fate of Reason*, 147–53; Robert Richards, *The Romantic Conception of Life*, 216–25. Kant is referring to book 5, chapter 1 of *Outlines*: "The more we learn of Nature, the more we observe these indwelling powers . . . and thus all things are full of organically operating omnipotence. We know not where this begins, or where it ends: for, throughout the creation, wherever effect is, there is power, wherever life displays itself, there is internal vitality. Thus there prevails in the invisible realm of creation, not only a *connected chain*, but *an ascending series of powers*; as we perceive these acting before us, in organized forms, in its visible kingdom" (Herder, *Outlines of a Philosophy*, 108).

66. See Lenoir, "Kant, Blumenbach"; *Strategy of Life*, 22–25; Robert Richards, "Kant and Blumenbach"; Van den Berg, *Kant on Proper Science*, 188–206.

67. Robert Richards, "Kant and Blumenbach," 12, 20; see also Van den Berg, *Kant on Proper Science*, 188–89, 204–6. Gigante points out that for Wolff, also, the life force (*vis essentialis*) was a heuristic device, in contrast to Blumenbach's conception of the *Bildungstrieb* (*Life*, 20). On Wolff and Blumenbach, see Larson, *Interpreting Nature*, 159–65. In dialogue with Kant, Blumenbach later moved toward a regulative rather than constitutive conception of the *Bildungstrieb*. Sheehan and Wahrman point to a rhetorical vacillation: even as he invoked the *Bildungstrieb* as an explanatory (i.e., causal) principle, he claimed a merely heuristic function (*Invisible Hands*, 171–72).

68. Robert Richards, "Kant and Blumenbach," 16.

69. Lenoir, *Strategy of Life*, 24–30 (26). See also Beiser, *The Fate of Reason*, 152–58; Brian Jacobs, "Kantian Character and the Problem of a Science of Humanity," 109; Mensch, *Kant's Organicism*, 12–15, 60–66.

70. "When reason saw organic activity in nature, according to Kant, what it was really looking at was itself" (Mensch, *Kant's Organicism*, 144). See also Müller-Sievers on epigenesis "as the generative power of reason as such" (*Self-Generation*, 6); Goldstein, *Sweet Science*, 78–79.

71. "To reach a discussion of the society of his time, Herder *passes through* cosmology, biology, anthropology, and prehistory. He *reaches* man through nature. Nature and society come to be seen as elements of one single—and providential—process of development" (Rossi, *The Dark Abyss of Time*, 113). See also Sheehan and Wahrman: "The grammar of this organization was . . . a reflexive one: materials 'form themselves'; they 'emerge' from chaos" (*Invisible Hands*, 167).

72. Zammito, "Herder on Historicism and Naturalism," 15. On Herder's evolutionary scheme, rehearsing phylogenetic recapitulation, affording wide formal variability, but resisting species mutation, see Reill, *Vitalizing Nature*, 187–89; on *Bildung* in Herder's early writings, see Gjesdal, "Human Nature and Human Science," 181–82.

73. On Herder's account of human historical development "as an integral part of nature," interactive with its geographical and social environments, see Waldow, "Between Nature and History," 153–57.

74. Kant, "Review of Herder's *Ideas*," in *Anthropology, History, and Education*, 142.

75. Gigante, *Life*, 48.

76. Kant, "Review of Herder's *Ideas*," in *Anthropology, History, and Education*, 132. Kant found the implication of transmutation "monstrous," he clarified in a rejoinder to a critic (K. L. Reinhold), not because it offended "*metaphysical orthodoxy*" but because human reason naturally abhors "an idea in which *nothing at all can be thought*" (135). In a footnote to part 2 of the *Anthropology*, Kant coolly (and startlingly) entertains the conjecture of an evolutionary series of "epochs of nature," in the third of which "an orangutan or a chimpanzee formed the organs used for walking, for handling objects, and for speaking, into the structure of a human being, whose innermost part contained an organ for the use of the understanding and which developed gradually through social culture" (423). On Kant's "generic preformationism," an archetypalist conception of species development (via *zweckmäßige Ordnungen*, "plans of organization"), see Lenoir, *Strategy of Life*, 32–35, 84.

77. Goethe to Karl Ludwig Knebel (November 17, 1784), cited in Düntzer, *Life of Goethe*, 342.

78. See Salmon, "Herder's *Essay on the Origin of Language*," 67–68. For Buffon's defence of the barrier between apes and humans, see *Natural History*, 8: 64–76.

79. Subsequent evolutionist natural history converts the former into the latter. For a later iteration of man's state as central and synthetic, see Robert Chambers's recourse, in the first edition of *Vestiges of the Natural History of Creation* (1844), to the "circular" (quinarian) taxonomy of William Sharp MacLeay: "In [man] only is to be found that concentration of qualities from all the other groups of his order.... Man, then, considered zoologically, and without regard to the distinct character assigned to him by theology, simply takes his place as the type of all types of the animal kingdom, the true and unmistakable head of animated nature upon this earth" (271, 272–73).

80. Goldstein, "Irritable Figures," 282–85; Noyes, "Herder's Unsettling," 155–56; Gaukroger, *The Natural and the Human*, 190–97.

81. In *Anthropology from a Pragmatic Point of View*, however, Kant argues that "the shape and organization" of the "*hand, fingers,* and *fingertips*" constitutes a "technical predisposition" of the human as rational animal (*Anthropology, History, and Education*, 165–66). On a different effect of the human erect stance—a "turning around or turning back, a turning from the back or from behind, a dorsal turn"—see Wills, *Dorsality*: "Dorsality will be a name for that which, from behind, from or in the back of the human, turns (it) into something technological, some technological thing" (5).

82. Lamarck, *Zoological Philosophy*, 12.

83. Ibid., 170. "En outre, si les individus dont je parle, mus par le besoin de dominer, et de voir à la fois au loin et au large, s'efforçoient de se tenir debout, et en prenoient constamment l'habitude de génération en génération; il n'est pas douteux encore que leurs pieds ne prissent insensiblement une conformation propre à les tenir dans une attitude redressée.... [I]l n'est pas douteux encore que leur angle facial ne devînt plus ouvert, que leur museau ne se raccourcît de plus en plus, et qu' à la fin étant entièrement effacé, ils n'eussent leurs dents incisives verticales" (*Philosophie zoologique*, 1: 349–50).

84. Cuvier, "Elegy of Lamarck," 1. Lamarck died in 1829; the *Éloge* was read after Cuvier's death, in November 1832. On Lamarck and the Cuvier-Geoffroy debate, see chapter 3.

85. Corsi, *The Age of Lamarck*, 235, 245; see also Burkhardt, "Lamarck, Evolution, and the Politics of Science," 295. The doctrine in question was that of the unity of organic composition.

86. Lyell, *Principles of Geology*, 188.

87. Spencer, *Autobiography*, 1: 201. On Lyell's critique of Lamarck, see Secord, *Visions of Science*, 165–71.

88. For views on either side, see Müller-Sievers, who invokes Kant's superior critical rigor against Herder's tautologous reasoning (*Self-Generation*, 94–100), versus Gaukroger, who contrasts the intellectual dead-end of Kant's anthropology—emptied of useful content for the sake of formal purity—with the fecundity of Herder's, nourishing the Romantic humanism of Schiller and Feuerbach (*The Natural and the Human*, 213–16).

89. "It is equally proved, that, *if a being, or system of beings, be forced out of this permanent condition of its truth, goodness, and beauty, it will again approach it by its internal powers, either in vibrations, or in an asymptote; as out of this state it*

[214] NOTES TO CHAPTER TWO

finds no stability" (Herder, *Outlines of a Philosophy*, 451): "Eben sowohl ists erwiesen, daß *wenn ein Wesen oder ein System derselben aus diesem Beharrungszustande seiner Wahrheit, Güte und Schönheit verrückt worden, es sich demselben durch innere Kraft, entweder in Schwingungen oder in einer Asymptote wieder nähere, weil außer diesem Zustande es keinen Bestand findet*" (*Ideen*, 596).

90. Michael Forster, *Herder's Philosophy*, 253–54.

91. On the absence of realism in nineteenth-century Irish fiction, correlated with the absence of civil society and a national state, see Lloyd, *Anomalous States*, 129–31; Deane, *Strange Country*, 18–20; Eagleton, *Heathcliff and the Great Hunger*, 181–88.

Chapter Two: The Form of the Novel

1. Herder, "This Too a Philosophy for the Formation of Humanity," in *Philosophical Writings*, 288. © Cambridge University Press 2002, reproduced with permission.

2. Staël-Holstein, *Germany*, 2: 54. Future references are given in the text.

3. Goethe presented Staël with a copy of *Wilhelm Meister* in 1797, after translating some of her essays (including the "Essai sur les fictions") for Schiller's *Die Horen*; the two met when Staël visited Weimar in 1803. She acquired reading proficiency in German after her tour, although she may have read Goethe's novel in the 1802 French translation. See Behler, "Madame de Staël and Goethe," 131; Judith Martin, *Germaine de Staël in Germany*, 34; Jaeck, *Madame de Staël and the Spread of German Literature*, 35–36, 44–45.

4. Staël-Holstein, *Germany*, 2: 56–57. A decade later, *Wilhelm Meister*'s English translator, Thomas Carlyle, echoes Staël's enthusiasm: "This mysterious child, at first neglected by the reader, gradually forced on his attention, at length overpowers him with an emotion more deep and thrilling than any poet since the days of Shakspeare [*sic*] has succeeded in producing. . . . It is not tears which her fate calls forth, but a feeling far too deep for tears" (Carlyle, "Translator's Preface" [1824], cited in Cave, *Mignon's Afterlives*, 126–27).

5. Hegel, *Aesthetics*, 2: 1093; 1: 149. Future references will be given in the text.

6. On *Corinne* as *roman philosophique*, see Zanone, "L'esthétique du 'tableau philosophique' dans *Corinne ou l'Italie*," 11. *Corinne* was as popular in Great Britain as elsewhere. Two rival English translations appeared in 1807, the same year as the original: see Garside and Schöwerling, *The English Novel 1770–1829*, 258–59; they were superseded by the 1833 translation for Bentley's *Standard Novels*, with verse renditions of the heroine's rhapsodies by L. E. L. (Letitia Elizabeth Landon).

7. On Corinne as a variant of Mignon, see Cave, *Mignon's Afterlives*, 88, 90, 251–52; Siegel, *Haunted Museum*, 42–43: Staël "gave the most consistent and influential literary form to the sensibility represented in Goethe's *Italian Journeys* and *Wilhelm Meister*" (42).

8. Moers, *Literary Women*, 176.

9. Armstrong, *Desire and Domestic Fiction*, 98–100.

10. See Soare, "The Female Gothic Connoisseur." Soare tracks the reabsorption of the heroine into the Female Quixote archetype in English anti-Corinne satires.

11. Staël, *Corinne*, trans. Raphael, 46. Future references to this edition will be given in the text.

12. "Know you the land where bloom the orange trees" (ibid., 30); see Goethe, *Wilhelm Meisters Lehrjahre*, 148; Staël, *Corinne*, ed. Balayé, 63.

13. Lokke, *Tracing Women's Romanticism*, 4–5. On the German tradition, see also Kontje, "Socialization and Alienation in the Female Bildungsroman"; Judith Martin, *Germaine de Staël in Germany*.

14. The "novelistic revolution" (Moretti, *Distant Reading*, 19–21); the "new genres or sub-genres" (Jameson, *Antinomies of Realism*, 145).

15. Morgenstern, "On the Bildungsroman," 655; Scott, *Waverley*, 5.

16. On Herder as the principal intellectual context for the Bildungsroman, see Beddow, *The Fiction of Humanity*, 64–68 (64); Kontje, *The German Bildungsroman*, 2–3; and, especially, Boes, *Formative Fictions*, 51–60.

17. Campe, "Form and Life in the Theory of the Novel," 53, 55.

18. Laurence Sterne had prepared the type with his deconstruction of the masculine narrator-protagonist, along with the empiricist psychology of the "science of man," in *Tristram Shandy*, a novel beloved by Goethe, Friedrich Schlegel, and other German authors.

19. Moi, " A Woman's Desire to Be Known," 156: referring to Hegel, *The Phenomenology of Spirit*, 274–75.

20. Balzac, *Works*, 1: 2.

21. Ibid., 1: 3.

22. Moretti, *Distant Reading*, 19–21.

23. Buffon, *Epochs of Nature*, 132. On the ideological legacy of Buffon's natural history in "Revolutionary projects of a total transformation of society," see Spary, *Utopia's Garden*, 150–52.

24. Condorcet, *Historical View of the Progress of the Human Mind*, 11.

25. Rousseau, *The Social Contract*, 76–77.

26. Spary, *Utopia's Garden*, 153.

27. Cabanis, *Rapports du physique et du moral de l'homme* (1802), cited in Xavier Martin, *Human Nature and the French Revolution*, 160. Martin S. Staum gives a sympathetic reading: "man appeared as a creature capable of changing his habits rather than slavishly following them. Human nature was transported from the fixity of Being to the flow of Becoming" ("Cabanis and the Science of Man," 143). On revolutionary projects of human regeneration, and the shift from "autonomous regeneration to notions of imposing strength from without" in the 1790s, see also Douthwaite, *The Wild Girl, Natural Man, and the Monster*, 161–81 (178).

28. Staum, *Cabanis*, 207–8, 210; Gaukroger, *The Natural and the Human*, 151–54.

29. Marshall Brown, "Theory of the Novel," 263; Jameson, *Antinomies of Realism*, 203.

30. Jameson, ibid., 203.

31. On the "vast intellectual and formal heterogeneity" of the *Wanderjahre*, see Piper, *Dreaming in Books*, 22–26 (23).

32. Robert Richards, *The Romantic Conception of Life*, 340–42, 368–79. The two met in 1770 in Strasbourg, where Herder wrote the *Treatise on the Origin of Language*. Goethe was an early and appreciative reader of this, as well as of *Ideas for a Philosophy of the History of Mankind*, which Herder was writing concurrently with

the composition of *Wilhelm Meister*: see Lewes, *Life of Goethe*, 2: 286. "Herder in his *Ideen* remarks that one cannot find any differences between man and animal in any particulars," Robert Richards quotes from one of Goethe's letters, and comments: "his interest in the transmutation of species was undoubtedly stimulated by Herder's quasi-evolutionary ideas" (*Romantic Conception of Life*, 375). On Goethe's adherence to principles of epigenesis and vital materialism, in opposition to Kant's teleological organicism, see Goldstein, *Sweet Science*, 75–107, 128–35.

33. Herder, *Philosophical Writings*, 129, 133 ("der Mensch, der Lehrling aller Sinne! der Lehrling der ganzen Welt!").

34. Schlegel, "On Goethe's *Meister*," in Bernstein, *Classic and Romantic German Aesthetics*, 271; Moretti, *The Way of the World*, 5.

35. Morgenstern, "On the Bildungsroman," 655. For a concise account of the cultural contexts of *Bildung* and its emergence in late Enlightenment Germany as "the biological and rational inner development of the human being as a fully developed organism," see Pfau, *Bildungsroman*, 124–32 (126).

36. Dilthey, *Poetry and Experience*, 335–36. The new form, he writes, draws on "the new developmental psychology established by Leibniz," Rousseau's "idea of a natural education in conformity with the inner development of the psyche," and "the ideal of humanity" promoted by Lessing and Herder (336). Thanks to Mark Taylor for steering me here.

37. Bakhtin, "The Bildungsroman and Its Significance in the History of Realism (Towards a Historical Typology of the Novel)," in *Speech Genres*, 19, 21.

38. Swales, *The German Bildungsroman from Wieland to Hesse*, 4.

39. Beddow, *The Fiction of Humanity*, 76.

40. Redfield, *Phantom Formations*, 38, 55.

41. Slaughter, *Human Rights, Inc.*, 111.

42. Humboldt, *The Sphere and Duties of Government*, 11.

43. See Schmitz, *Correspondence between Schiller and Goethe*, vol. 1, passim (for Schiller's general criticism of the novel, 264–65, 420–21); Plath, "Schiller's Influence on *Wilhelm Meisters Lehrjahre*." Goethe resisted Schiller's urging that he make the novel conform more programmatically to his conception of *Bildung*. For judicious summaries, see Swales, *The German Bildungsroman from Wieland to Hesse*, 24–26; Kontje, *The German Bildungsroman*, 10–11.

44. Schiller, *Aesthetic Education*, 17–19. Future references to this edition are cited in the text.

45. Esty, *Unseasonable Youth*, 39–40.

46. Slaughter, *Human Rights, Inc.*, 101. Jameson recapitulates the standard teleological reading: "The plot is thus turned inside out: from a series of chance happenings it is suddenly revealed as a plan and as a deliberately providential design" (*Antinomies of Realism*, 208). See also Friedrich Kittler, "Über der Sozialisation Wilhelm Meisters" (1978), cited in Kontje, "Socialization and Alienation," 221–22. For Moretti, the Tower represents the Hegelian idea of civil society as historical telos (*The Way of the World*, 53–54).

47. Strong teleological readings of the function of the state in Schiller's argument may be influenced, in the English tradition at least, by Matthew Arnold's adaptation of the aesthetic education in *Culture and Anarchy*: putting no faith in an "organic" forum

for the free play of thought (i.e., the market), Arnold invokes the centralized bureaucratic state (drawing on the Prussian model, exemplified by Humboldt, *The Sphere and Duties of Government*, 88–89, 94) as an authoritative "organ of our collective best self, of our national right reason" (72).

48. Compare Hamilton's critical reading of this rhetorical condition as symptomatic and ideological, "Schiller's temporizing," in *Metaromanticism*, 25–43.

49. "By the twenty-seventh Letter the aesthetic state ('Zustand'), which was originally conceived, and as late as the nineteenth Letter is still conceived, as a means to moral freedom, now appears to have subsumed both the physical and the moral and become an end in itself—the 'ästhetischer Staat'" (Nicholas Martin, *Nietzsche and Schiller*, 70). See also Beiser, *Schiller as Philosopher*, for whom the alternating treatments of the aesthetic "sometimes . . . as a means toward morality and sometimes as an end in itself" articulate complementary rather than competing "moral and anthropological perspectives" (166–67).

50. For the critique of Schiller's aesthetic state as an abandonment of the treatise's original political purpose, see Hamilton, *Metaromanticism* (note 48, above); and Pugh, "Schiller as Platonist." For a defense, see Rancière, "Aesthetics as Politics": where the French Revolution's "reign of Law" continued the "reign of free form over servile matter" in another key, Schiller's treatise makes "the aesthetic supremacy of form over matter and of activity over passivity . . . into the principle of a more profound revolution, a revolution of sensible existence itself and no longer only of the forms of State" (31–32).

51. The irony hits at Goethe's own nationalist fervor during his studies in Strasbourg, where he met Herder (characterized, in an early draft of his memoir *Dichtung und Wahrheit*, as *Deutschheit emergierend*, an "awakening of German feeling" [Friedenthal, *Goethe*, 83]).

52. See Slaughter, who views the Tower as a "corporate surrogate for a proto-German bourgeois state" (*Human Rights, Inc.*, 101). "*Wilhelm Meister's Apprenticeship* refuses to find closure in an allegory of the state," writes Boes. "Instead, Goethe's Tower Society reinvents itself as an explicitly cosmopolitan organization" (*Formative Fictions*, 7).

53. See Ngai, "Merely Interesting."

54. See Redfield's exemplary deconstructive reading of *Meister*: "Schillerian *Bildung* is negated and recuperated as the irony of luck: as the lucky chance that only fiction can reliably provide" (*Phantom Formations*, 80).

55. Swales, *The German Bildungsroman from Wieland to Hesse*, 30, 64–65; Lukács, *The Theory of the Novel*, 80.

56. Pfau, "Bildungsspiele," 581, 579. But see Jane Brown, *Goethe's Allegories of Identity*, 95–117, for a subtle account of the novel's "morphological" developmental plan.

57. "Wilhelm is not only led with ease from one thing to another, but is always oscillating in his views of himself. Even his emotions are not persistent" (Lewes, *Life of Goethe*, 2: 396).

58. See Fleming, *Exemplarity and Mediocrity*, 107. Elsewhere, Novalis paid tribute to *Wilhelm Meister* as "the Absolute Novel, without qualification" (Bernstein, *Classic and Romantic German Aesthetics*, 229).

59. Hegel, *Aesthetics*, 2: 593.

60. Ibid., 1: 149–50. Moretti's compelling reading of *Wilhelm Meister* does, however, locate Wilhelm's destiny with the Hegelian vision of civil society, turning the novel toward an "enchantment of everyday life."

61. Hegel, *Aesthetics*, 2: 1092–93.

62. Hegel's term (*schlechte Unendlichkeit*), derived in turn from Schiller's contrast between the "empty infinity" (*leere Unendlichkeit*) of "indetermination through sheer absence of determination" and the "infinity filled with content" (*erfüllte Unendlichkeit*) of "aesthetic freedom from determination" in his twenty-first letter (*Aesthetic Education*, 144–45).

63. Lukács, *Theory of the Novel*, 81, 133. Future references will be given in the text.

64. The work of art "should put before our eyes a content, not in its universality as such, but one whose universality has been absolutely individualized and sensuously particularized" (Hegel, *Aesthetics*, 1: 51).

65. Lukács's synthesis treats Schlegelian irony (79–80) as symptomatic, subsuming it within a larger Hegelian frame as "the highest freedom that can be achieved in a world without God" (92).

66. Or what has recently been called "surface reading": see Best and Marcus, "Surface Reading," especially "surface as a practice of critical description" (11).

67. Schlegel, "On Goethe's *Meister*," in Bernstein, *Classic and Romantic German Aesthetics*, 275.

68. Lacoue-Labarthe and Nancy, *The Literary Absolute*, 12. For a critique of their misappropriation of Schlegelian reflexivity as a technique of the "absolute," see Bode, "Absolut Jena."

69. See, e.g., Duff, *Romanticism and the Uses of Genre*, 164; Arac, "What Kind of History Does a Theory of the Novel Require?," 190–91.

70. Schlegel, "Letter about the Novel," in Bernstein, *Classic and Romantic German Aesthetics*, 293.

71. Marshall Brown, "Theory of the Novel," 256. See also Wellek, "German and English Romanticism: A Confrontation," 43–45; Marshall Brown, *The Shape of German Romanticism*, 199–213.

72. "Die Ironie ist eine permanente Parekbase" (Schlegel, "Zur Philosophie" [1797], in *Kritische Ausgabe*, 18: 85). For a discussion, see Handwerk, "Romantic Irony," 215–17; De Man, "The Concept of Irony," in *Aesthetic Ideology*, 178–81; Behler, "The Theory of Irony," 44–65.

73. "Jede Kunst und jede Wissenschaft die durch die Rede wirkt, wenn sie als Kunst um ihrer selbst willen geübt wird, und wenn sie den höchsten Gipfel erreicht, erscheint als Poesie" (Schlegel, "Gespräch über die Poesie" [1799], in *Kritische Ausgabe*, 2: 304). See Benjamin Dawson, "Science and the Scientific Disciplines," 698. Jonathan Arac suggests "novelistic poesis" as an alternative translation of "*romantische Poesie*" ("What Kind of History Does a Theory of the Novel Require?," 191).

74. Schlegel, "Letter About the Novel," "*Athenaeum* Fragment No. 118," in Bernstein, *Classic and Romantic German Aesthetics*, 293, 249; Schlegel, *Kritische Ausgabe*, 2: 335, 182. See Duff's account of Schlegel's "combinatorial method," based on an aesthetic technique of "rough-mixing," or "juxtaposition rather than synthesis, friction rather than fusion . . . in which formal heterogeneity is foregrounded rather

than concealed"—as opposed to "smooth-mixing," the "seamless fusion of forms" that mimics a totality (*Romanticism and the Uses of Genre*, 165, 178–81).

75. Bernstein, *Classic and Romantic German Aesthetics*, 249. "Die romantische Dichtart ist noch im Werden; ja das ist ihr eigentliches Wesen, daß sie ewig nur werden, nie vollendet sein kann" (*Kritische Ausgabe*, 2: 183).

76. See Boes, "Apprenticeship of the Novel," 273–75. According to Ernst Behler, Condorcet's "infinite perfectibility of the human race" informed Schlegel's idea of poetry as an "infinite becoming, irreducible to a knowable principle with regard to beginning or end," sustained by an ironical praxis that "rejects any type of closure and postpones it to an unrealizable future" (*German Romantic Literary Theory*, 68–70).

77. Bode, "Absolut Jena," 32: "Because of its auto-referentiality and its self-reflexiveness [poesy] displays a feature that is observable in all self-referential systems of sufficient magnitude and complexity—it becomes inexhaustible by finite interpretations so that, provoking ever new ones, it triggers off a series of progressive readings that, however, invariably fail to arrest the 'meaning' of the text. . . . Thus by acknowledging its limitations, literature becomes limitless, inexhaustible" (32). Lukács attempts a synthesis of the Hegelian drive to totality with Schlegelian irony: "the novel, in contrast to other genres whose existence resides within the finished form, appears as something in the process of becoming. . . . Thus the novel, by transforming itself into a normative being of becoming, surmounts itself" (*Theory of the Novel*, 72–73).

78. Lacoue-Labarthe and Nancy, *The Literary Absolute*, 98.

79. Schlegel, "Letter About the Novel," in Bernstein, *Classic and Romantic German Aesthetics*, 293: "das Romantische nicht sowohl eine Gattung ist als auch ein Element der Poesie, das mehr oder weniger herrschen und zurücktreten, aber nie ganz fehlen darf . . . dem Roman aber, insofern er eine besondre Gattung sein will, verabscheue" (*Kritische Ausgabe*, 2: 335).

80. Lacoue-Labarthe and Nancy, *The Literary Absolute*, 92.

81. Lukács, *Theory of the Novel*, 72–73. Thanks to Erin Mackie for this insight.

82. Cf. Saintsbury, *The English Novel*: "Scott, like Miss Austen, at once opened an immense new field to the novelist, and showed how that field was to be cultivated . . . between them they cover almost the entire possible ground of prose fiction" (210).

83. De Man, *Aesthetic Ideology*, 182.

84. Wellek, "German and English Romanticism," 43–44; on *Sartor*'s Scottishness and Carlyle's repudiation of the novel, see Duncan, *Scott's Shadow*, 306–10.

85. Lewes, "Realism in Art," 272. Lewes blamed German transcendental philosophy and its hard dualism between the real and the ideal: "men in general [have] lost sight of the fact that Art is a Representation of Reality" (273).

86. Dilthey, *Poetry and Experience*, 335.

87. Moretti, *The Way of the World*, 15.

88. Musil: "with every true experiment a cultured man educates himself [*bildet sich ein geistiger Mensch*]. This is the organic plasticity of man. In this sense every novel worthy of the name is a *Bildungsroman*" (cited in Redfield, *Phantom Formations*, 42–43).

89. Sammons, "The Mystery of the Missing *Bildungsroman*"; Amrine, "Rethinking the *Bildungsroman*"; Redfield, *Phantom Formations*, 40–43.

90. "Der *Mensch ist allmächtig und allwissend und allgütig*; nur ist d[er] *Mensch* in dem Einzelnen *nicht ganz* sondern nur Stückweise da Der Mensch kann

nie da seyn" (Schlegel, *Kritische Ausgabe*, 18: 506; see Handwerk, "Romantic Irony," 219.

91. Handwerk, ibid., 213, 221.

92. Lacoue-Labarthe and Nancy, *The Literary Absolute*, 98.

93. Armstrong and Tennenhouse, *Novels in the Time of Democratic Writing*.

94. Lukács, *The Historical Novel*, 53.

95. Moretti includes *Waverley* as a Bildungsroman in *The Way of the World*. The most sensitive account of Scott's novel as a Bildungsroman—of the author himself as well as of hero and reader—is Millgate's, *Walter Scott*, 35–57.

96. Lukács, *The Historical Novel*, 35.

97. Scott, *Waverley*, 33–34. Future references will be given in the text.

98. See Duncan, *Modern Romance*, 79–92.

99. On Scott and conjectural history see, e.g., McMaster, *Scott and Society*, 49–77; Chandler, *England in 1819*, 127–35; Duncan, *Scott's Shadow*, 101–4, 135–38.

100. Barthes, *The Fashion System*, cited in Campbell, *Historical Style*, 203.

101. Campbell (ibid.) argues that Scott reclaims fashion as a medium of historical cognition through this ostensible disavowal: "In moving not away from costumery but toward the out-of-date, *Waverley* upholds fashion as a system of meaning more fully coextensive with cultural life than the increasingly stratified literary system" (211).

102. This middle range as the domain of nineteenth-century realism is discussed in chapter 4.

103. Ricoeur, *Memory, History, Forgetting*, 101. See also Bakhtin, *Speech Genres*: the "major task of the modern historical novel" has been to "find an historical aspect of private life" relating "individual life-sequences" to "the life of the nation, the state, mankind" (217).

104. "Homer had the good Fortune to see and learn the Grecian Manners, at their true Pitch and happiest Temper for Verse. Had he been born much sooner, he could have seen nothing but Nakedness and Barbarity: Had he come much later, he had fallen either in *Times of Peace*, when a wide and settled Polity prevailed over *Greece*; or in *General Wars*, when private Passions are buried in the common Order, and established Discipline" (Blackwell, *Enquiry into the Life and Writings of Homer*, 35).

105. See Chandler, *England in 1819*, who situates "the contradiction between the first and last chapters of *Waverley*—turning on the question whether Scott undertook this experiment with historical narrative to preserve evanescent manners or to identify the unchanging features of human nature" (382) within a larger debate in Enlightenment historiography (240–41).

106. Burnett, *On the Origins and Progress of Language*, 1: 262–66n.: "that there are men with tails, such as the ancients gave to their satyrs, is a fact so well attested that I think it cannot be doubted" (262).

107. Thanks to Tanya Llewellyn for this point. On the "anthropological and eclectic" mode of *Waverley*, see Robert Crawford, *Devolving English Literature*, 123–32 (132); on Scott's "autoethnography," Buzard, *Disorienting Fiction*, 68–81, 85–89.

108. "Mit euch zu Roß / durch die Lüfte nicht reitet sie länger; / die magdliche Blume / verblüht der Maid; / ein Gatte gewinnt / ihre weibliche Gunst – / dem herrischen Manne / gehorcht sie fortan, / am Herde sitzt sie und spinnt, / aller Spottenden Ziel und Spiel!" (Wagner, *Der Ring des Nibelungen*, 152–53).

109. Staël, *Politics, Literature, and National Character*, 153. On perfectibility and the Scottish Enlightenment roots of Staël's "science of nations," see Jones, "Madame de Staël and Scotland."

110. See Staum, *Cabanis*, 216.

111. Staël, *Corinne, or Italy*, 246, 247. Future references to this edition will be given in the text. Citations in French are from Balayé's edition. On Staël and women's rights, see Fairweather, *Madame de Staël*, 117.

112. See Judith Martin, *Germaine de Staël in Germany*, 38–40, 48–55; Fairweather, *Madame de Staël*, 299–309.

113. On Corinne as an embodiment of Schlegel's aesthetic of genre mixing, realized in a "choreographic" "art of the moment" (["Künst"] des Augenblicks," 181), see Klettke, "Germaine De Staël: *Corinne ou L'Italie*."

114. Esterhammer, *Romanticism and Improvisation*, 91.

115. Scholars have noted Goethe's description of Emma Hamilton's "attitudes," in his *Italian Journey*, as a precedent for Corinne. "Inside this golden frame dressed in many colors against the black ground, she had sometimes imitated the ancient paintings of Pompeii and even modern master-works" (Goethe, *Italian Journey*, 262). But see Kate Davies, "Pantomime, Connoisseurship, Consumption."

116. On Corinne's performances, see Poulet, "The Role of Improvisation in *Corinne*"; Deneys-Tunney, "*Corinne* by Madame de Staël"; Erik Simpson, *Literary Minstrelsy, 1770–1830*, 51–57; Esterhammer, *Romanticism and Improvisation*, 86–91.

117. See Hirsch, "Spiritual Bildung."

118. Goethe, *Wilhelm Meister*, 330–34, 352–54.

119. Deidre Lynch calls Corinne's last song a "Gothic revenge on Oswald and the paternal literary order" ("The (Dis)Locations of Romantic Nationalism," 215). See also Lynch's discussion of Corinne's "ghostliness": "to be a woman of genius is already to be a ghost" (205).

120. On the eclipse of Corinne's utopian enthusiasm by Oswald's "British melancholy"—a "contagious but prestigious disease that Oswald passes on to Corinne"—see Lokke, *Tracing Women's Romanticism*, 25–53 (47).

121. Francis Jeffrey praised the satirical efficacy of Staël's depictions of English provincial life in his 1807 *Edinburgh Review* article, "Corinne, ou l'Italie."

122. Staël, *Literature Considered in Its Relation to Social Institutions*, 204–5.

123. Ibid., 207.

124. Although impatient with *Pride and Prejudice*—which she found *vulgaire*, i.e., full of "petty things"—Staël ordered *Mansfield Park* for her library at Coppet. Staël had befriended Frances Burney, whose novels she admired, on her visit to England in 1793 (Fairweather, *Madame de Staël*, 163–66).

125. "Le titre du roman se trouve parfaitement justifié par la fiction: Corinne, c'est l'Italie et l'Italie, c'est Corinne . . . véritable métonymie nationale" (Phalèse, *Corinne à la page*, 96).

126. I thus make the opposite argument to Susan Tenenbaum, for whom "Staël described the eclipse of politics by aesthetics as a principal cause of Italy's servitude to Napoleon" ("*Corinne*: Political Polemics and the Theory of the Novel," 157). Among other accounts of literature and the arts as a national medium in *Corinne*, Giulia Pacini argues that Italy's inchoate national condition allows Staël to showcase "the instrumental

function of the arts in the construction and definition of a nation" ("Hidden Politics in Germaine de Staël's *Corinne*," 171). For Patrick Vincent, Staël's "female genius" resolves the "contradiction between Enlightenment universalism's openness and the parochialism of romantic nationalism" by making literature the medium of "an enlightened, liberal form of nationalism" (*The Romantic Poetess*, 76). Suzanne Guerlac argues that Corinne allegorizes, in opposition to the (Napoleonic) militaristic and imperial conception of the nation, "literature, the imagined, and imagining, community of the nation" ("Writing the Nation (Mme de Staël)," 52). In Caroline Franklin's fine account, sensitive to the novel's Napoleonic context, "the tension between Romantic genius and Romantic patriotism produces a fault-line separating art from politics." Staël enthrones her poetic genius in "the decentralized feminine realm of Italy where art flourished undisturbed by civic duties and rivaled religion in nourishing the Italian soul . . . as a fitting opponent to Napoleon, the colossus of public violence, who was crowned 'Rex Totius Italiae' in 1805" (*Female Romantics*, 11–13). In another subtle analysis, Tricia Lootens notes "the symbolic burden of national/sexual shame" that counters the "inspirational, even utopian aspects of Corinne's equation with Italy": "To cast her lot with a politically subjected, and thus corrupt, country; to sacrifice her genius by entering into the spiritual servitude of an English/Scottish wife: these appear to become Corinne's options. To choose either, moreover, from Nelvil's point of view, must be to betray profoundly shocking 'Italian' propensities toward subservience and emotional excess: to confirm, that is, his suspicion of her innate, Italianate 'slavishness'" (*The Political Poetess*, 44).

127. On Corinne's cosmopolitanism, see Catherine Jones's reading of her performance of Ossian, "Madame de Staël and Scotland," 247–48.

128. Tenenbaum, "*Corinne*: Political Polemics and the Theory of the Novel," 158. In a nuanced account of *Corinne*'s nationalist project, Glenda Sluga reads "the story of a nation in discursive gestation," in which the subjection of female liberty by national liberty is a melancholy historical necessity ("Gender and the Nation: Madame de Staël or Italy," 243). Susanne Hillman describes "the exceptional women with lyres who were destined to act as the nation's unifying agents by recreating the gallants' deeds in writing. By imagining the nation these women fulfilled the supreme task of citizenship" ("Men with Muskets, Women with Lyres," 232).

129. See Tracy, "Maria Edgeworth and Lady Morgan"; Trumpener, *Bardic Nationalism*, 131–43; Corbett, *Allegories of Union*, 52–70. It seems unlikely Staël could have read *The Wild Irish Girl* in time for it to have influenced *Corinne*, which she began in 1805 and completed in November 1806. Owenson's 1809 novel *Woman: or Ida of Athens* is clearly indebted to *Corinne*. I thank Claire Connolly and Joep Leerssen for a discussion of the relation between the novels of Staël and Owenson.

130. Armstrong, *Desire and Domestic Fiction*, 4; Moi, "Expressivity and Silence," 157.

131. Brewer, *The Afterlife of Character*, 78–120 (78).

132. Moers, *Literary Women*, 176–78. On writers who followed Staël in assuming a Corinne-like persona, see, e.g., Peterson, "Rewriting *A History of the Lyre*"; Vincent, *The Romantic Poetess*, 18–28, 97–121.

133. George Eliot, *The Mill on the Floss*, 332.

134. Translation adjusted. "De toutes les facultés de l'âme que je tiens de la nature, celle de souffrir est la seule que j'aie exercée tout entière" (*Corinne*, ed. Balayé,

584). On Corinne as Romantic archetype of the "suffering heroine," see Vincent, *The Romantic Poetess*, 17–20.

135. On *Corinne* and *Aurora Leigh*, see Moers, *Literary Women*, 182, 205–6; Lewis, *Germaine de Staël, George Sand, and the Victorian Woman Artist*, 107–16; Kaplan, "Aurora Leigh," 146–51; Chapman, *Networking the Nation*, 101–5, 113–14.

136. Eliot, *Daniel Deronda*, 664. On *Daniel Deronda* and *Corinne*, see Moers, ibid., 196, 199–200; Weliver, "George Eliot and the Prima Donna's 'Script,'" 103–20; Heller, *Literary Sisterhoods*, 3–6, 37–45.

Chapter Three: Lamarckian Historical Romance

1. Lukács, *The Historical Novel*, 19.
2. Ibid., 53.
3. On the satirical "confusion between beasts, man-beasts, and man-made beasts" in *Count Robert of Paris*, see Simmons, "A Man of Few Words," 27.
4. See, e.g., Hobsbaum, "Scott's 'Apoplectic' Novels": "Everyone who has not read *Count Robert of Paris* . . . knows it to be unreadable" (153).
5. See Alexander, "Essay on the Text"; Gamerschlag, "The Making and Unmaking of Sir Walter Scott's *Count Robert of Paris*."
6. Scott, *Count Robert of Paris*, 362. Future references to this edition will be cited in the text.
7. On the affinities between them, see Armstrong, *How Novels Think*, 54–56, 59–61.
8. On Frankenstein's monster as a redaction of Rousseauvian natural man, however, see Marshall, *The Surprising Effects of Sympathy*, 178–227; and Laura Brown, *Homeless Dogs and Melancholy Apes*, 59–63. Brown makes a persuasive case for Shelley's basing her "vividly realized composite being who challenges the distinctiveness of the human" on a "weird collation" of eighteenth-century accounts of the orangutan, notably Lord Monboddo's (62); she identifies a source for the creature's giant stature (59) in Monboddo's citation of a traveler's description of the great ape as "from 7 to 9 feet high" (Burnett, *On the Origins and Progress of Language*, 1: 281). Monboddo is the likely source for Scott's orangutan in *Count Robert of Paris*.
9. Lamarck, *Zoological Philosophy*, 69 ("animal series"), 170–71 (orang-outang).
10. The one-volume *Standard Novels* edition (1833), translated by Frederic Schoberl, came up with the title that stuck: *The Hunchback of Notre-Dame*. Hugo disliked it.
11. Hugo, *Notre-Dame of Paris*, 199; *Notre-Dame de Paris*, ed. Sacy, 251. Future references to Sturrock's English translation will be given, and passages from the original French in endnotes.
12. See Richard Maxwell's discussion of Hugo's and Scott's competing treatments of popular insurgency, focused on the motif of the siege: *The Historical Novel in Europe*, 201–4.
13. See Baldick, *In Frankenstein's Shadow*, 16–20, 60–62. On "the monster as metaphor for the unruly Revolutionary mob" circa 1830, and the teratological context, see Hibberd, "Monsters and the Mob" (35).
14. "C'est le mélange d'un ordre ancien et d'un ordre nouveau, la présence simultanée dé deux états qui, ordinairement, se succèdent l'un à l'autre" (Isidore Geoffroy Saint-Hilaire, *Histoire générale et particulière*, 18).

15. "Ce résultat général et définitif de mes déterminations d'organes, est devenu la conclusion la plus élevée de mes recherches; haute manifestation de l'essence des choses, que j'ai exprimée et proclamée sous le nom d'*Unité de composition organique*" (Etienne Geoffroy Saint-Hilaire, *Philosophie anatomique*, vol. 2, *Des monstruosités humaines* [1822], xxxiv). On a rival Romantic conception of monstrosity as an excess of vitality over form, see Gigante, *Life*, 6, 48.

16. "L'étude des monstres sera donc, pour le physiologiste et pour le philosophe, la recherché des procédés par lesquels la nature opère la génération des espèces" (Geoffroy, ibid., 121: referring to Virey, "Monstre," 133).

17. Appel, *The Cuvier-Geoffroy Debate*, 130–36; Corsi, *The Age of Lamarck*, 233–34.

18. Appel, ibid., 132. Corsi (ibid.) argues that Geoffroy had actually read little Lamarck, and misrepresented key Lamarckian ideas in taking opportunistic advantage of Lamarck's rising popularity in the 1820s. Robert Mitchell draws a distinction between *perfectibility* as evolutionary principle in Lamarck's morphology and *variation* (against the absent ground of the "abstract animal") in Geoffroy's (*Experimental Life*, 172–73).

19. Appel, *The Cuvier-Geoffroy Debate*, 132: citing Geoffroy Saint-Hilaire, "Recherches sur l'organisation des Gavials," 150–51, 151n. For Lamarck's two laws, see *Philosophie zoologique*, 1: 235.

20. Appel, ibid., 158–61 and 276n58; Clubb, "Quasimodo, Quasi-man," 276; Goethe, *Conversations with Eckermann and Soret*, 2: 290–92. Goethe called *Notre-Dame de Paris* "the most abominable book that ever was written," a symptom of Hugo's subjection to "the unhappy Romantic tendency of his time" (2: 403).

21. Corsi, *The Age of Lamarck*, 235, 245; see also Burkhardt, "Lamarck, Evolution, and the Politics of Science," 295. For an account of the debate, see Appel, *The Cuvier-Geoffroy Debate*, 145–55.

22. "Une philosophie qui substitue des métaphores aux raisonnements" (Cuvier, *Discours sur les révolutions de la surface du globe*, 19).

23. Corsi, *The Age of Lamarck*, 260.

24. Appel, *The Cuvier-Geoffroy Debate*, 130–32, 135–36. The point at issue was Geoffroy's argument for the structural homology of anatomical features across different species and orders.

25. See Corsi, *The Age of Lamarck*; Secord, "Edinburgh Lamarckians."

26. On Darwin, Jameson, and Grant, see Desmond, *The Politics of Evolution*, 398–402; Browne, *Charles Darwin*, 69–88.

27. Secord, "Edinburgh Lamarckians," 15–17. See also Patrick Scott, "Comparative Anatomies."

28. Jameson, "Preface to the Fifth Edition": Cuvier, *The Theory of the Earth*, vi.

29. Desmond, *The Politics of Evolution*, 327–28. See also Secord, *Visions of Science*, 141–50, 165–71.

30. See Rudwick, *Bursting the Limits of Time*, 6, 158–60; on Lyell's "reluctant admiration" for Lamarck, 245.

31. Desmond, *The Politics of Evolution*, 328.

32. Lyell, *Principles of Geology*, 193.

33. Buffon, *Natural History*, 8: 76; Rousseau, *The Discourses*, 204–9. "Orang-outang" became the generic name under which traveler's accounts of different species

of great ape were bundled, gorillas and chimpanzees as well as orangutans. See Laura Brown, *Homeless Dogs and Melancholy Apes*, 30–36: "These redactions of the hominoid ape enter the literary culture of the eighteenth century as rough drafts for an ongoing experiment in human identity" (36).

34. Burnett, *Of the Origins and Progress of Language*, 1: 270, 289.

35. Desmond, *The Politics of Evolution*, 288–94. This was not because Owen was anti-evolutionist. In the 1840s, Owen espoused a teleologically regulated evolutionary hypothesis, based on Karl Ernst von Baer's theory of embryological development (also adopted by Robert Chambers in *Vestiges of the Natural History of Creation*): "A preordained divergence from some lower vertebrate form, consistent with the *Vestiges* mechanism, could account for man's animal origin without having him emerge directly from an ape.... It was therefore theoretically consistent for Owen to endorse the *Vestiges* mechanism but dissent from Chambers' speculations on the ape-origin of humanity; and in this case, ideologically essential as well" (Evelleen Richards, "Property Rights," 154–55).

36. Casanova, *The World Republic of Letters*, 77–79.

37. Klancher, "Discriminations, or Romantic Cosmopolitanisms in London"; see also Langan, "Venice."

38. See McCracken-Flesher, *The Doctor Dissected*, 39–46, 53–55.

39. See McMaster, *Scott and Society*, 212; Simmons, "A Man of Few Words," 25. McMaster identifies the orangutan as the terminal figure in a chain of conjectural versions of natural man, next to Hereward the Forester and then Count Robert himself (214–15).

40. Herder, *Outlines of a Philosophy of the History of Mankind*, 71, 72. On the late Enlightenment ascription of sensibility to anthropoid apes, see Laura Brown, *Homeless Dogs and Melancholy Apes*, 53–58.

41. Robb, *Victor Hugo*, 156–57.

42. Ibid., 150, 574n66. On Hazlitt, see Hooker, *The Fortunes of Victor Hugo in England*, 32–34.

43. But see Bartfeld, "Mouvance, mutation et progrès," who argues that the people embody a progressive life force (*élan vital*) oriented toward a future revolutionary horizon: "le peuple est néanmoins l'élément vivant de cette humanité en marche" (38).

44. A decade later, however, Balzac could confidently invoke Geoffroy's "unity of composition" as the organizing formal principle of his zoology of society in the preface to *The Human Comedy: Works*, 1: 2.

45. Clubb, "Quasimodo, Quasi-man," 272–77.

46. "*Préexistence des germes*: ces deux mots, déjà pour moi difficiles à entendre l'un détaché de l'autre, me paraissent, s'ils sont réunis, tout à-fait inintelligibles. Ils doivent naissance à une idée de causalité, à l'explication métaphysique d'un fait qu'on sait très-bien n'avoir été ni observé ni apprécié" (Etienne Geoffroy Saint-Hilaire, *Philosophie anatomique*, 2: 480).

47. Cf., for example, Lewes, *Comte's Philosophy of the Sciences*: "It was then believed,—as, indeed, it is still very generally believed,—that the acorn contained the oak, and the germ contained the man. This Metaphysical conception of primitive germs, potentially containing all that may subsequently be developed from them, naturally led men to argue that a monster was originally a monster—that the

deformation existed potentially in the primitive germ. . . . The third or Positive conception of Epigenesis, or gradual organic development in accordance with conditions, has finally routed the metaphysical conception of 'pre-existent germs' and by considering monsters as simple cases of 'organic deviation,' has, with the aid of Geoffroy St. Hilaire's great law of 'arrested development,' made monstrosity a branch of positive embryology" (33).

48. "Il semblait, avec sa face humaine et sa membrure bestiale, le reptile naturel de cette dalle humide et sombre. . . . C'est ainsi que peu à peu, se développant toujours dans le sens de la cathédrale, y vivant, y dormant, n'en sortant presque jamais, en subissant à toute heure la pression mystérieuse, il arriva à lui ressembler, à s'y incruster, pour ainsi dire, à en faire partie intégrante. Ses angles saillants s'emboîtaient, qu'on nous passé cette figure, aux angles rentrants de l'édifice, et il en semblait, non seulement l'habitant, mais encore le contenu naturel. On pourrait presque dire qu'il en avait pris la forme, comme le colimaçon prend la forme de sa coquille. . . . Il y avait entre la vieille église et lui une sympathie instinctive si profonde, tant d'affinités magnétiques, tant d'affinités matérielles, qu'il y adhérait en quelque sorte comme la tortue à son écaille. La rugueuse cathédrale était sa carapace" (204–5).

49. Cf. Etienne Geoffroy Saint-Hilaire: "Tout cet échafaudage d'un germe primitivement monstrueux, d'un germe préexistant et emboîté de toute éternité avec de vicieuses qualités, reste une pure supposition" (*Philosophie anatomique*, 2: 489).

50. Ibid., 2: 485.

51. Scott may have had an informant in his friend Basil Hall, who accompanied Amherst's expedition and published a *Voyage to Loo-Choo, and Other Places in the Eastern Seas, in the Year 1816* (1817), reprinted as the first title in *Constable's Miscellany* (1826).

52. See, e.g., McMaster, *Scott and Society*, 212–15. On Scott and the historiography of the Crusades, see David Simpson, *Romanticism and the Question of the Stranger*, 89–90; Watt, "Scott, the Scottish Enlightenment, and Romantic Orientalism," 106–8.

53. Brenhilda typifies a parallel thematic of monstrosity to the novel's preoccupation with racial and species difference, namely sexual difference. The "fantastic appearance of her half-masculine garb" (*Count Robert of Paris*, 133) marks her as "[s]omething between man and woman" (180). Her duel with Anna Comnena confirms her opponent's status as a complementary freak, the bluestocking, unsexed by intellectual prowess: "she had somewhat lost the charms of her person as she became enriched in her mind" (37). "A woman who is pitiless, is a worse monster than one who is unsexed," the Empress Irene (herself something of a monster) upbraids her daughter (284).

54. McGann, "Walter Scott's Romantic Postmodernity," 124.

55. Scott, *Waverley*, 5–6. James Chandler notes a source for this Enlightenment commonplace in Hume's *Enquiry* (*England in 1819*, 241).

56. Scott is echoing Fielding's claim that his novel is a true history, although not based on "records," on the grounds that its characters are drawn from "the vast authentic doomsday-book of Nature" (*Tom Jones*, 422–23).

57. "Wire-wove and hot-pressed" refers to the Fourdrinier paper-making machine, operative in Scotland by 1811 (Bell, *Edinburgh History of the Book in Scotland*, 3: 22). See also Garside's note, *Waverley*, 396–97.

58. Scott, *Rob Roy*, 254.

59. Rousseau, *The Discourses*, 186.

60. MacRitchie, *The Testimony of Tradition*, 152–53, 157–58.

61. Scott, *Ivanhoe*, 19–20.

62. See David Simpson's powerful discussion, *Romanticism and the Question of the Stranger*, 98–101, 107–8.

63. See Cagidemetrio, "A Plea for Fictional Histories and Old-Time 'Jewesses.'"

64. See Ragussis, *Figures of Conversion*, 89–93, 116–26.

65. Kant, "On the Use of Teleological Principles in Philosophy," in *Anthropology, History, and Education*, 192–218. On the Kant-Forster debate, see Tucker, *The Moment of Racial Sight*, 55–62.

66. Blumenbach, *Anthropological Treatises*, 57; Lawrence, *Lectures on Physiology*, 108–9, 114, 202–3. Blumenbach multiplied the proofs of human distinction: "The hymen, the guardian of chastity, is adapted to man, who is alone endowed with reason; but the clitoris, the obscene organ of brute pleasure, is given to beasts also" (87).

67. Long, *History of Jamaica*, 351–75 (356, 365).

68. Home, *Sketches of the History of Man*, 1: 3, 20.

69. Ibid., 1: 76.

70. On the rise of polygenesis and racial science between 1830 and 1850, see Stepan, *The Idea of Race in Science*, 29–46; Stocking, *Victorian Anthropology*, 64–69. On Knox, see Desmond, *The Politics of Evolution*, 77–80, 388–89. Kitson, "Coleridge and 'the Ouran utang Hypothesis,'" argues that Knox's polygenetic views were not characteristic of Romantic-period thinking about race.

71. Nancy Armstrong's phrase, referring to late Victorian Gothic romances by H. Rider Haggard and Bram Stoker, flourishing in the imperial heyday of scientific racism: *How Novels Think*, 105–8.

72. See Wilkins, *Species*, 104–9.

73. Sutherland, reading the Scythians and the Orang-Outang as figures of "degeneration," invests *Count Robert of Paris* with a later nineteenth-century preoccupation (*The Life of Walter Scott*, 343–44).

74. Burnett, *Antient Metaphysics*, 3: 254–64. The identification of great apes as "the satyrs of the ancients" goes back to Edward Tyson's 1699 anatomical treatise *Orang-Outang, sive Homo Sylvestrus*; see Nash, *Wild Enlightenment*, 16–41.

75. On language as the gift of humanity and its deprivation as an expulsion into the biopolitical regime of "brute silence," see Seshadri, *HumAnimal*, 23–29, 65–108.

76. See Thomas, *Man and the Natural World*, 30–41; Agamben, *The Open*, 33–37.

77. For a critical account of the Cheselden case, see Kennedy, *Revising the Clinic*, 30–53. On Mitchell, discussed by Dugald Stewart in a paper read to the Royal Society of Edinburgh in 1815 and reprinted in vol. 3 of *Elements of the Philosophy of the Human Mind* (1827), see Coyer, *Literature and Medicine*, 42–43. Stewart cites Mitchell's case as proof of the absolute barrier between mankind and other animals (*Works*, 3: 249).

78. Diderot, *Thoughts on the Interpretation of Nature*, 182. Scott's episode also echoes Diderot's characterization of sight as "a kind of touch which extends to distant objects" (151), as well as his recommendation that a patient restored to sight be confined at first to a dark room lest the unaccustomed glare of sunlight prove traumatic (177).

79. Ibid., 174; for *D'Alembert's Dream*, see 104–5. On Diderot's Lucretianism and transformism, see Staum, *Cabanis*, 28–30.

80. Agamben, *The Open*, 16.

81. See the introduction to this book. On Cullen, see Goodman, *Pathologies of Motion*, chapter 1; on Burke, see Sarafianos, "Pain, Labor, and the Sublime."

82. Burke, *Philosophical Enquiry*, 34.

83. On vision and form, see Schiller, *Aesthetic Education*, Letter 26.

84. Schlegel, "*Athenaeum* Fragment" no. 305, in Bernstein, *Classic and Romantic German Aesthetics*, 257.

85. Hugo, *Dramas*, 11. Future references will be cited in the text.

86. "[U]n étrange centaure moitié homme, moitié cloche" (211); "une fière guêpe," "une espèce de femme abeille" (333). On "hybrid form," see Brombert, *Victor Hugo and the Visionary Novel*, 51–54.

87. "La monstruosité n'est plus un désordre aveugle, mais un autre ordre également régulier, également soumis à des lois; ou, si l'on veut, c'est le mélange d'un ordre ancien et d'un ordre nouveau, la présence simultanée dé deux états qui, ordinairement, se succèdent l'un à l'autre" (Isidore Geoffroy Saint-Hilaire, *Histoire générale et particulière*, 1: 18). See Huet, *Monstrous Imagination*, 108–9.

88. Sarah Hibberd links the grotesque aesthetic of *Cromwell* with Geoffroy's and Isidore's teratology: "Monsters and the Mob," 29.

89. Canguilhem, *Knowledge of Life*, 135–36.

90. On Hugo's innovative representation of the city as labyrinth—necessitating the recourse to allegorical techniques—see Maxwell, *The Mysteries of Paris and London*, 25–54.

91. Moretti, *Atlas of the European Novel*, 44n24.

92. "Par moments, sur le sol, où tremblait la claret des feux, mêlée à de grandes ombres indéfinies, on pouvait voir passer un chien qui ressemblait à un homme, un homme qui ressemblait à un chien. Les limites des races et des espèces semblaient s'effacer dans cette cité comme dans un pandémonium. Hommes, femmes, bêtes, âge, sexe, santé, maladie, tout semblait être en commun parmi ce peuple; tout allait ensemble, mêlé, confondu, superposé; chacun y participait de tout. . . . C'était comme un nouveau monde, inconnu, inouï, difforme, reptile, fourmillant, fantastique" (127–28). Compare the 1833 Shoberl translation: "A dog which looked like a man, or a man who looked like a dog, might be seen from time to time passing over the place on which trembled the reflection of the fires, interspersed with broad ill-defined shadows. The limits between races and species seemed to be done away with in this city, as in a pandemonium. Men, women, brutes, age, sex, health, disease, all seemed to be in common among these people. They were jumbled, huddled together, laid upon one another; each there partook of every thing. . . . It was like a new world, unknown, unheard of, deformed, creeping, crawling, fantastic" (*The Hunchback of Notre Dame*, 74).

93. Compare Bartfeld, "Mouvance, mutation et progrès," for whom the grotesque, the aesthetic state of the urban populace, is an evolutionary stage toward a future sublime horizon: "L'hybridité devient condition du progrès" (37).

94. "[U]ne monstrueuse bête à mille pieds" (527), "un serpent à écailles d'acier" (536).

95. "On eût dit que quelque autre église avait envoyé à l'assaut de Notre-Dame ses gorgones, ses dogues, ses drées, ses démons, ses sculptures les plus fantastiques". C'était

comme un couche de monstres vivants sur les monstres de Pierre de la façade" (539). A "drée" is a "legendary animal with gnashing teeth" (*Notre-Dame de Paris*, "Notes," 693).

96. "Parmi les vieilles églises de Paris une sorte de chimère; elle a la tête d l'une, les membres de celle-là, la croupe de l'autre; quelque chose de toutes" (162); "semblait un énorme sphinx à deux têtes assis au milieu de la ville" (235).

97. "Alors il lui sembla que l'église aussi s'ébranlait, remuait, s'animait, vivait, que chaque grosse colonne devenait une patte énorme qui battait le sol de sa large spatule de pierre, et que la gigantesque cathédrale n'était plus qu'une sorte d'éléphant prodigieux qui soufflait et marchait avec ses piliers pour pieds, ses deux tours pour trompes et l'immense drap noir pour caparaçon" (461).

98. "[L]e dépôt que laisse une nation; les entassements que font les siècles; le résidu des évaporations successives de la société humaine; en un mot, des espèces de formations. Chaque flot du temps superpose son alluvion, chaque race dépose sa couche sur le monument, chaque individu apporte sa pierre. Ainsi font les castors, ainsi font les abeilles, ainsi font les hommes. . . . L'homme, l'artiste, l'individu s'effacent sur ces grandes masses sans nom d'auteur ; l'intelligence humaine s'y résume et s'y totalise. Le temps est l'architecte, le peuple est le maçon" (162–63).

99. Adam Ferguson, *History of Civil Society*, 10.

100. In his preface to the 1832 "definitive edition," Hugo claims that the chapter was one of three originally written but mislaid for the first printing, and now restored.

101. "[L]e grand livre de l'humanité" (238); "la grande écriture du genre humain" (240); "l'écriture principale, l'écriture universelle" (244). See Brière, *Victor Hugo et le roman architecturel*, 65–104.

102. On Hugo's novel as "epic" avatar of the cathedral, see Zarifopol-Johnston, "'Notre-Dame de Paris': The Cathedral in the Book."

103. Brombert, *The Hidden Reader*, 62.

104. Ibid.

105. Buffon's phrase, "le sombre abîme du temps" (recalling Prospero's words in *The Tempest*)—see Rossi, *The Dark Abyss of Time*, 107–8.

106. See Schiebinger, "Skeletons in the Closet."

Chapter Four: Dickens: Transformist

1. Collins, *Dickens: The Critical Heritage*, 470–71.

2. Melville, *Moby-Dick*, 70. Future references to this edition will be given in the text.

3. Samuel Otter links Melville's evocation of the whale's sublime inhuman anatomy with a critique of nineteenth-century ethnological discourses of craniology and anthropometry, "driving a wedge between description and evaluation and rendering the natural strange" (*Melville's Anatomies*, 110).

4. For contrasting views of human and inhuman nature in Melville's novel, see Schultz, "Melville's Environmental Vision in *Moby-Dick*," on the "intrinsic and irresistible interdependency among diverse species of life" figured by an affective kinship between humans and whales (100); and Hugh Crawford, "Networking the Nonhuman," on the novel's distribution of agency across "human and nonhuman actors in shifting and dynamic alliances" (18).

5. Synchronicity: *Harper's New Monthly Magazine* notes a London review of *Moby-Dick* (anon., "Literary Notices") in the same issue (April 1852) in which it begins the American serialization of *Bleak House* (649–68).

6. Dickens, *Bleak House*, 1. Future references to this edition will be given in the text.

7. The allusion to the Noachian deluge is submerged in the modern speculations of catastrophist geology, from Cuvier to "Adhémar and Lehon, distinguished men of science, [who] believe that they have proved that a grand deluge must inevitably devastate the globe every ten thousand five hundred years; that such deluges have regularly occurred during all previous time, and that such will recur again at their stated epochs; and that, although these grand deluges may not be so universal as to desolate the *whole* world, they are cataclysms sufficiently terrific to exterminate the great majority of existing creatures" (Anon., "Natural Selection," in *All the Year Round*, 297). On Dickens's play with the geological debate between Neptunian and Plutonian theories of the earth's formation—by flood or fire—in *Bleak House*, see George Levine, *Darwin and the Novelists*, 122; Wilkinson, "From Faraday to Judgment Day," 225–47.

8. *Quarterly Review* 36 (1827), 434, cited in Zimmerman, *Excavating Victorians*, 1.

9. On the megalosaurus and catastrophist geology in *Bleak House*, see Buckland, "'The Poetry of Science.'" Although increasingly discredited in post-Lyellian geology, catastrophism was sustained in Victorian popular scientific spectacles (dioramas, panoramas, the concrete dinosaur models at the Sydenham Crystal Palace), which tinge Dickens's depiction of the megalosaurus as amusing rather than menacing (679–89).

10. On the entropic vision of *Bleak House*, see Hillis Miller, *Charles Dickens*; George Levine, *Darwin and the Novelists*; MacDuffie, *Victorian Literature, Energy, and the Ecological Imagination*, according to whom the megalosaurus heralds a "Gothic return of the earth's stored energy, released through a wilderness of London chimneys" (94). On "the aggregate of humanity" in *Bleak House* as "a surplus of ungovernable life that exceeds and threatens the survival of the social body," see Steinlight, "Dickens's 'Supernumeraries,'" 246.

11. "Dickens's works abound with bizarre, grotesque, and unlikely creations who seem to say, 'we are artificial nature; we are the unnatural real,'" writes Jesse Oak Taylor, who cites anthropogenic climate change (the London smog) as the phenomenon's historical correlative: "the status of the natural is itself thrown into crisis by an environment in which all facets of existence, including the weather, are materially altered by human action" ("The Novel as Climate Model," 8). Taylor characterizes Dickens's realism as allegorical rather than mimetic: "Dickens does not *represent* a world aspiring to mimetic verisimilitude but rather *creates* one, bringing a fictional world into being that renders legible otherwise invisible dynamics at work in the world" (2).

12. Collins, *Dickens: The Critical Heritage*, 287 (citing *Bentley's Miscellany*, October 1853).

13. Woloch, *The One vs. the Many*, 126.

14. Bowen, *Other Dickens*, 5, 17. Cf. a reviewer in the *Atlantic Monthly* (May 1867), noting "that strong individualizing tendency in [Dickens's] mind which makes him give consciousness even to inanimate things, and which one critic goes so far as to call 'literary Fetishism'" (Collins, *Dickens: The Critical Heritage*, 480); and Andrew Lang (1899) on Dickens's "*Animism* . . . a relapse [into] the early human intellectual

condition" (*Literary Criticism*, 131). For more recent accounts, see Carey, *The Violent Effigy*; David Simpson, *Fetishism and Imagination*.

15. As Richard Maxwell has shown (*The Mysteries of Paris and London*, 54–57, 61–72).

16. Chambers, *Vestiges of the Natural History of Creation*, 251. Kathleen Tillotson calls *Dombey and Son* (1848) "the earliest example of responsible and successful planning" among Dickens's novels: "it has unity not only of action, but of design and feeling" (*Novels of the Eighteen-Forties*, 157).

17. Darwin, *On the Origin of Species*, 303.

18. Nathalie Vanfasse, "'Grotesque, but Not Impossible,'" reads the Dickensian grotesque as "testing [the] limits" of mid-Victorian realism's balancing act between competing realist and idealist imperatives (5).

19. Eliot, *Adam Bede*, 179, 180.

20. Beer, *Darwin's Plots*, 154–68.

21. Balzac, *Works*, 1: 2. Balzac insists on the transcendental anatomists' principle of unity of type.

22. Taylor, "The Novel as Climate Model."

23. Hensley and James, "Soot Moth." Cf. MacDuffie, "Dickens and the Environment": "The idea of an entirely urbanized planet is a staple of the contemporary environmental imaginary, whether in works of dystopian science fiction like *Blade Runner*, or in non-fiction texts like Mike Davis's *Planet of Slums*, but we see it first coming into being in Dickens's fiction" (574). Krook's contribution to the London fog, Elizabeth Vinyard Boyle pointed out in a graduate seminar at Berkeley (October 2018), underscores its general composition from burning animal fat as well as coal and wood.

24. Martin S. Rudwick specifies 1776–1848 as the era of the modern geohistorical "discovery of time" (*Bursting the Limits of Time*, 6).

25. Chambers, *Vestiges of the Natural History of Creation*, 22. Future citations from this edition will be given in the text.

26. Echoing Chambers himself (ibid., 388), scholars have credited *Vestiges* with "[introducing] a developmental cosmology based on natural law to the English-speaking world" (Secord, introduction to *Vestiges of the Natural History of Creation*, ix), with being "the first complete history of the world from its beginnings to the present according to an evolutionary principle" (Stierstofer, "Vestiges of English Literature," 27), and with instantiating what Bernard Lightman calls "the evolutionary epic in its modern form. Chambers took the monad-to-human style cosmic evolutionary narrative, which had its origins in Lucretius, and combined it with the new sciences of the nineteenth century for the first time" (*Victorian Popularizers of Science*, 221). Such claims overlook Herder's *Ideen*, available in English in Thomas Churchill's translation, *Outlines of a Philosophy of the History of Man* (1800, 1803), and extracted in magazines as late as 1840 (Esterhammer, "Continental Literature, Translation, and the Johnson Circle," 102). It is quite likely Chambers would have known it. Herder's conjectural natural history provides a more convincing model for *Vestiges* than the historical novels of Scott, cited by Secord.

27. Chamber relies, in the first edition of *Vestiges*, on the "circular" (or quinarian) taxonomy of the "Macleay system of animated nature" (236): one of the more

vulnerable points in the criticism of his scientific errors, Chambers abandons it in later editions.

28. See Secord, *Victorian Sensation*, 10. Michael Slater estimates some forty-one contributions by Dickens to *The Examiner* between 1837–43 and 1848–49 (*Charles Dickens*, 276–77).

29. Dickens, anonymous review of *The Poetry of Science*, 787.

30. Haight, "Dickens and Lewes on Spontaneous Combustion," 63. For refutations, see Wilkinson, "From Faraday to Judgment Day"; and George Levine, *Darwin and the Novelists*, 122–29. For recent overviews, see Furneaux and Winyard, "Dickens, Science and the Victorian Literary Imagination"; Smith, "Dickens and Astronomy, Geology and Biology."

31. See George Levine, "Dickens and Darwin," 126–29. The articles are almost certainly not by Richard Owen, despite an impression (still current in some quarters) that they might be; Owen's article on the *Origin of Species* for the *Edinburgh Review* is quite different from the article in *All the Year Round*.

32. George Levine, *Darwin and the Novelists*, 129.

33. See, e.g., Gibson, "*Our Mutual Friend* and Network Form."

34. Gowan Dawson, "Dickens, Dinosaurs, and Design."

35. Boehm, *Charles Dickens and the Sciences of Childhood*, 2.

36. Buckland, *Novel Science*, 248–66; see also Buckland, "'The Poetry of Science.'"

37. Lightman compares *Vestiges* to a "spectacle," a "museum of creation," and a "panorama" (*Victorian Popularizers of Science*, 221–22).

38. Boehm, *Charles Dickens and the Sciences of Childhood*, 146. See also Lawrence Frank, who reads *Bleak House* as a paleontological romance in the context of "ongoing controversies over the nebular hypothesis" (*Victorian Detective Fiction and the Nature of Evidence*, 71).

39. See Evelleen Richards, "Property Rights," for a detailed analysis of Owen's version of transmutation and his guarded support for Chambers's project in *Vestiges*. Secord argues that *Vestiges* strategically alludes to the visual culture of early Victorian popular science, while maintaining respectability by refraining from including images (*Victorian Sensation*, 437–48).

40. George Levine, "Dickens and Darwin," 252.

41. Buckland, *Novel Science*, 266.

42. Sedgwick, "Vestiges of the Natural History of Creation," 44: see Frank, *Victorian Detective Fiction*, 76; Secord, *Victorian Sensation*, 14, 90, 203; Buckland, *Novel Science*, 67–69.

43. Dickens, "The Poetry of Science," 787. Dickens's verdict was later amplified by Charles Darwin himself: "In my opinion it has done excellent service in this country in calling attention to the subject [of species transmutation], in removing prejudice, and in thus preparing the ground for the reception of analogous views" ("An Historical Sketch of the Progress of Opinion on the Origin of Species" [1861], in *On the Origin of Species*, ed. Bynum, 433. This edition will be cited throughout *Human Forms* unless otherwise noted.) Darwin's initial reactions were far warier, inflected by tactical concerns for his own work on species transmutation: see Secord, *Victorian Sensation*, 429–33, 508–11.

44. Dickens's popular weekly journals, *Household Words* and its successor *All the Year Round*, were modeled on Robert and William Chambers's Reform-era *Chambers's Edinburgh Journal*. Dickens's subeditor on *Household Words*, William Henry Wills, had served as assistant editor on *Chambers's Journal* and was married to the Chambers's sister Janet.

45. On Dickens, see Ledger, *Dickens and the Popular Radical Imagination*; and Sen, *London, Radical Culture, and the Making of the Dickensian Aesthetic*; on the surge of works of popular science between 1830 and 1855, part of the "industrial book revolution of the nineteenth century" (31), see Lightman, *Victorian Popularizers of Science*, 25–34.

46. See Gowan Dawson, "Literary Megatheriums and Loose Baggy Monsters."

47. Armstrong, *How Novels Think*, 3–10.

48. Maxwell argues that Dickens excelled over more literal imitators (such as Harrison Ainsworth) by adopting Hugo's allegorical technique for representing the urban labyrinth (*The Mysteries of Paris and London*, 71).

49. Gowan Dawson, "Literary Megatheriums," 206. Dawson cites the megatherium (a giant Pleistocene ground-sloth), rather than the megalosaurus (a carnivorous Jurassic sauropod), as the typical monster invoked by reviewers. In his account, the megatherium symbolizes the recuperation of monstrous form under providential design, following Owen's anatomical theory. The present discussion views the megalosaurus as a far less reassuring (scaly rather than furry) type of formal monstrosity: providential design may be at work, but if so, it is remote from local human cognition.

50. Aristotle, "Poetics," chapter 8, in *A New Aristotle Reader*, 546–47. See Gowan Dawson, "Literary Megatheriums," 206.

51. Collins, *Dickens: The Critical Heritage*, 283; see also 281, 295.

52. The novel's totem monster, the megalosaurus, is "forty feet long or so." Gowan Dawson emphasizes the legibility of the dinosaur's anatomy, its availability for rational computation, in the new morphology of Cuvier and Owen, and hence a coded promise that the order of *Bleak House* is finally knowable: "Dickens, Dinosaurs, and Design." (See also Gowan Dawson, "'By a Comparison of Incidents and Dialogue.'") Richard Owen computed the length of the megalosaurus at thirty feet, countering exaggerated earlier estimates (*History of British Fossil Reptiles*, 352–53).

53. James, *Literary Criticism*, 1107. On monstrosity as an effect of "life's relentless fecundity and 'the monstrous' [as] a mode of uncontainable vitality," disrupting formal containment, see Gigante, *Life*, 6, 48.

54. Moretti, *Canon/Archive*, 65–68.

55. Ermarth, *Realism and Consensus in the English Novel*, 65.

56. Moretti, "Serious Century"; George Levine, *The Realistic Imagination*, 5.

57. Levine and Ortiz-Robles, *Narrative Middles*, 6–8.

58. Greiner, *Sympathetic Realism*, 38–46.

59. Griffiths, *Age of Analogy*, 168; Duncan, "The Provincial or Regional Novel," 329–32; for its inauguration in Austen and Scott, see Lee, "Austen's Scale-Making."

60. Lukács, *The Historical Novel*, 33. On Quetelet, the average man, and the realist novel, see Jaffe, *The Affective Life of the Average Man*, 10–13, 30–36; Kolb, "In Search of Lost Causes."

61. [J. A. Strothert], *The Rambler* (January 1854), cited in Collins, *Dickens: The Critical Heritage*, 295.

62. Collins, ibid., 287, 288, citing *Bentley's Miscellany* (October 1853).

63. Collins, ibid., 276, citing [H. F. Chorley], the *Athenaeum* (September 1853). The classic discussion of Dickens's predilection for "flat" rather than "round" characters is E. M. Forster's, in *Aspects of the Novel*, 108-12; the most searching recent discussion is Woloch's, in *The One vs. the Many*, 125-76.

64. Woloch, ibid., 165.

65. James, review in *The Nation* (December 1865), in Collins, *Dickens: The Critical Heritage*, 470-71.

66. Canguilhem, *Knowledge of Life*, 105-6.

67. Boehm, *Dickens and the Sciences of Childhood*, 146-66. On the embryology of *Vestiges*, see Evelleen Richards, "Property Rights," 133-39.

68. Darwin, *On the Origin of Species*, 177.

69. Cited in Secord, *Victorian Sensation*, 433. Compare Darwin's sardonic comment in his introduction to *On the Origin of Species*: "The author of 'Vestiges of Creation' would, I presume, say that, after a certain unknown number of generations, some bird had given birth to a woodpecker, and some plant to the misseltoe, and that these had been produced as perfect as we now see them" (13).

70. Cited by Evelleen Richards, "Property Rights," 147. Owen had been studying the embryological research of Geoffroy and Hunter in the late 1830s, before the publication of *Vestiges*. Richards argues (148-50) that Hunter's version provided Owen with a reassuringly teleological account of embryonic malformation, against the environmental etiology adduced by Geoffroy.

71. Owen, "Darwin on the Origin of Species," 508.

72. Owen, *On the Anatomy of Vertebrates* (1868), 807-8; cited by Evelleen Richards, "Property Rights," 146n63. While endorsing Chambers's teratological scheme for species transmutation, Owen rejected his arguments for the spontaneous generation of life and the descent of humans from apes (Richards, ibid., 154-56). Richards summarizes Owen's transmutationism as an attempt "to reconcile Geoffroy's teratological speculations with conservative teleological concerns" (157).

73. Perhaps Charles Darwin was recalling Dickens's character in one of the controversial passages of *On the Origin of Species* (highlighted, e.g., by Owen, "Darwin on the Origin of Species," 517-19): "In North America the black bear was seen by Hearne swimming for hours with widely open mouth, thus catching, like a whale, insects in the water. . . . I can see no difficulty in a race of bears being rendered, by natural selection, more and more aquatic in their structure and habits, with larger and larger mouths, till a creature was produced as monstrous as a whale" (169).

74. The dejected ape recalls Herder's orangutan, with its melancholy intuition of a proximate but inaccessible horizon of human reason: see chapter 3.

75. Thanks to Deborah Nord for this example.

76. As Lamarck had argued (*Zoological Philosophy*, 68-72).

77. See MacDuffie, *Victorian Literature, Energy, and the Ecological Imagination*; Parham, "Dickens in the City"; Taylor, "The Novel as Climate Model."

78. Buzard, *Disorienting Fiction*, 156.

79. "The one great principle of the English law is, to make business for itself.... Viewed by this light it becomes a coherent scheme, and not the monstrous maze the laity are apt to think it" (Dickens, *Bleak House*, 621). Compare, with a different emphasis, Caroline Levine: the novel's network structure of interconnections "[casts] narrative persons less as powerful or symbolic agents in their own right than as moments in which complex and invisible social forces cross" (*Forms*, 126).

80. Compare D. A. Miller's Foucauldian account: "If Chancery thus names an organization of power that is total but not totalizable, total because it is not totalizable, then what is most radically the matter with being 'in Chancery' is not that there may be no way out of it ... but, more seriously, that the binarisms of inside/outside, here/elsewhere become meaningless and the ideological effects they ground impossible" ("Discipline in Different Voices," 61).

81. On Jarndyce's Bleak House (ii) within Bleak House (i) within *Bleak House*, compare Stewart, *On Longing*: "the dollhouse is a materialized secret; what we look for is the dollhouse within the dollhouse.... The dollhouse has two predominant motifs: wealth and nostalgia" (161).

82. Buzard, *Disorienting Fiction*, 107; on the operation of the double narrative, 121–23. See also, e.g., Harvey, "*Bleak House*: The Double Narrative"; Newsom, *Dickens on the Romantic Side of Familiar Things* (affording a double perspective on characters), 54–55; Herbert, "The Occult in *Bleak House*" ("[bringing] to the surface the inherently fantastic property of omniscient narration"), 103; Welsh, *Dickens Redressed* (a modern rather than postmodern mode of "discontinuous narration, comic versus satiric, with comic prevailing"), 24–127. Woloch discusses the failure of Victorian commentators to note either of the novel's formal innovations, the double narrative and the present-tense impersonal narration, and the implications for our own critical act ("*Bleak House*: 19, 20, 21").

83. On Esther, "the affective center of the novel" (19), see Zwerdling, "Esther Summerson Rehabilitated"; Welsh, *Dickens Redressed*, 20–58, 77–81; Jordan, *Supposing Bleak House*, 2–25; Rajan, "The Epistemology of Trust," 82–89.

84. Caroline Levine, *Forms*, 129 (treating the novel, however, as though it consisted solely of the impersonal narrative). See also Gibson, "*Our Mutual Friend* and Network Form": "Dickens's characters never exist all at once, autonomously, but only as parts of an open, interactive, and distributive network" (66).

85. Harvey, "The Double Narrative," 228.

86. On the impersonal narrative in *Bleak House*, see Arac, *Impure Worlds*, 82–84. I am concurring here with those critics who read omniscience as a rhetorical performance rather than as the expression or symptom of a theological condition: in Audrey Jaffe's formulation, "a fantasy: of unlimited knowledge and mobility; of transcending the boundaries imposed by physical being and by an ideology of unitary identity" (*Vanishing Points*, 6). On the technique, see also Culler, "Omniscience"; Nelles, "Omniscience for Atheists"; Paul Dawson, *The Return of the Omniscient Narrator*.

87. Jaffe, *Vanishing Points*, 128.

88. See Stanzel, *A Theory of Narrative*, 47–62, 66–68, 190–92. On Stanzel's "figural narrative situation," see also Bode, *The Novel*, 143–63; on its relation to free indirect discourse, 158–60.

89. On Charles Bally's *style indirect libre* and Etienne Lorck's *erlebte Rede*, and early taxonomic and linguistic debates around the terms and usage, see Pascal's useful account: *The Dual Voice*, 8–30; on "represented speech and thought, that style for representing consciousness which has become almost synonymous with novelistic style," Banfield, *Unspeakable Sentences*, 16–17, 65–108; on "psychonarration" and "narrated monologue," Dorrit Cohn, *Transparent Minds*, 26–57, 13–14, 100–115. While Cohn says "narrated monologue" corresponds to *style indirect libre*, she notes that the "insensible shading of narrated monologue into psycho-narration, or vice versa, is very frequent in figural narrative situations" (137). For a descriptive analysis of different accounts circa 1978, see McHale, "Free Indirect Discourse," 250–64.

90. Pascal, *The Dual Voice*, 26.

91. Genette, *Narrative Discourse*, 174.

92. Banfield, *Unspeakable Sentences*, 257–58.

93. See, e.g., McHale, "Free Indirect Discourse"; Fludernik, *The Fictions of Language and the Languages of Fiction*, 11–12, 280.

94. See Banfield, *Unspeakable Sentences*, 228–31; compare Hamburger, *The Logic of Literature*, 63; McHale, "Free Indirect Discourse," 257–58, 282–84. Frances Ferguson goes so far as to call free indirect discourse "the novel's one and only formal contribution to literature" ("Jane Austen, *Emma*, and the Impact of Form," 159). Many commentators follow Pascal (*The Dual Voice*, 99–102) in seeing a decisive mid-century establishment of the technique by Flaubert. Dissenting, Monika Fludernik finds the technique used in earlier periods, and argues that it is not specifically literary (*The Fictions of Language and the Languages of Fiction*, chapter 8; also, "The Linguistic Illusion of Alterity," 92–93).

95. Ermarth, *Realism and Consensus*, 65, 67, 70.

96. See Greiner, *Sympathetic Realism*, 37–44.

97. Eliot, "The Natural History of German Life," in *Essays of George Eliot*, 296; Lewes, *Comte's Philosophy of the Sciences*, 33, 264; Spencer, "The Social Organism."

98. Arendt, *The Human Condition*, 39.

99. Moretti, "Serious Century," 395–96.

100. Greiner, *Sympathetic Realism*, 42–43.

101. See Frances Ferguson's account of free indirect discourse in *Emma* as the "representation of a general will or social character": "Jane Austen, *Emma*, and the Impact of Form," 163–65, 170–80 (175).

102. Austen, *Emma*, 183. See also my discussion of this passage in *Scott's Shadow*, 117–19.

103. Plotz, *Semi-Detached*, 126–27. Compare Leavis, *The Great Tradition*: Austen's *Emma* manifests "a vital capacity for experience, a kind of reverent openness before life, and a marked moral intensity" (9), in contrast to the Flaubertian disdain for "life." Frederic Jameson argues that Flaubertian free indirect discourse signals a "nihilistic" destabilizing of the realist aesthetic (*Antinomies of Realism*, 178–81).

104. Pascal, *Dual Voice*, 75–76; Dorrit Cohn, *Transparent Minds*, 21–26. George Eliot's well-known criticism of Dickens for failing to combine his depiction of "the external traits of our town population" ("idiom and manners," "gestures and phrases") with an equivalent representation of "psychological character" (people's "conceptions

of life, and their emotions") diagnoses, in effect, a technical absence of free indirect discourse: "The Natural History of German Life," in *Essays of George Eliot*, 271.

105. In an alternative account, Amanda Anderson makes a case for a functional integration of the two narratives "into an implied complementary relation in which Esther's narrative expresses the voice of moral aspiration while the third-person narrative presents a world dominated by negative psychological and sociological conditions," in "a complex formal enactment of the estranging but also enabling moral consequences of attempting to think the moral and the sociological perspective in relation to one another" (*Bleak Liberalism*, 51, 56).

106. "Flaubert is the first author systematically to use the past imperfect tense (the 'imparfait') for the free indirect form" (Pascal, *The Dual Voice*, 102).

107. Jordan, *Supposing Bleak House*, 22.

108. See Knoepflmacher and Tennyson, *Nature and the Victorian Imagination*, 422. For Garrett Stewart, the echo of *The Prelude* affords a "lyric inwardness of style . . . Dickens's secret prose" (*Dickens and the Trials of the Imagination*, 239–40).

109. Wordsworth, *The Prelude*, 479.

110. Ibid., 481.

111. See Litvack, "What Books Did Dickens Buy and Read?," 94–95. *The Prelude* was printed by Dickens's own publishers, Bradbury and Evans.

112. Dickens revisits the "spots of time" scenario in *Great Expectations*: see Knoepflmacher and Tennyson, *Nature and the Victorian Imagination*, 422–25.

113. Most conspicuously, Dickens bases two of his secondary characters on literary figures, still alive at time of writing, associated with the earlier generation, Leigh Hunt (Harold Skimpole) and Walter Savage Landor (Lawrence Boythorn). Unusual in Dickens's fiction, the "personality" is a satirical device typical of magazines of the 1820s to the 1830s, such as *Blackwood's* and *Fraser's*. As has often been noted, *Bleak House* is a kind of historical novel—one, however, that systematically blurs the genre's foremost topos, "the period" (as Dickens calls it in the opening chapter of *A Tale of Two Cities*), with topical allusions and hints that run from the late 1820s to the eve of the present. A late reference to the coming of the railways in Lincolnshire (839) brings a jolt of chronological precision, as though the novel is suddenly accelerating to its close.

114. Lamarck, *Zoological Philosophy*, 170. The classic account of the imperial prospect in locodescriptive poetry, in which the privileged viewer assumes aesthetic (i.e., disinterested) rather than literal possession of the scene, is Barrell's, *English Literature in History, 1730-80*, 56–79.

115. See Price, *The Anthology and the Rise of the Novel*, 93; Edward Jacobs, "Ann Radcliffe and Romantic Print Culture," 57–63.

116. See Duncan, *Modern Romance*, 38–43.

117. Jameson, *Antinomies of Realism*, 11, 28–38.

118. Scott, *Waverley*, 84.

119. Abrams, *Natural Supernaturalism*, 385. "Greater Romantic lyric" is Abrams's coinage.

120. Ibid., 418–22: Wordsworth's "visionary gleam," Shelley's "evanescent visitations," Hölderlin's *der Augenblick*.

121. Woolf, *The Common Reader*, 148.

122. See, e.g., Christensen, *Romanticism at the End of History*, 107–28, 153–75. On lyric versus narrative, see Culler, *Theory of the Lyric*, 226, 283–95.

123. Eliot, *Middlemarch*, 788.

124. James, "The Art of Fiction," in *The Art of Criticism*, 174.

125. As an audience member pointed out at the Boston Romanticism Colloquium, April 2017, Dorothea is shortsighted: a deft realist alibi for the archetypal vagueness of what she is seeing (or thinks she is seeing).

126. Meillassoux, *After Finitude*, 5–26.

127. Wordsworth, *Poems*, 2: 39.

128. Vermeule, *Why Do We Care about Literary Characters?*, 71–81. Free indirect discourse is developed to "put pressure upon" and "stimulate our mind-reading capacities" (72–73, 77).

129. See Wilkinson, "From Faraday to Judgment Day," 242–47; Hack, "Sublimation Strange," 132–48.

130. Esther opens her narrative, in chapter 2, with an acknowledgment of the book's division of labor: "I have a great deal of difficulty in beginning to write my portion of these pages" (27). The impersonal narrative acknowledges her, in turn, when it resumes at the opening of chapter 7: "While Esther sleeps, and while Esther wakes, it is still wet weather down at the place in Lincolnshire" (103). Neither recurs to the other once the double narrative is established. We never learn who has commissioned Esther to write her portion, or who is compiling the remainder—in contrast, e.g., to Franklin Blake's editorial as well as narrative role in Wilkie Collins's *The Moonstone*.

131. See Puckett, *Bad Form*, 7. Buzard reads the exhibition of "an obsolete omniscience" in this passage (*Disorienting Fiction*, 121).

132. See Henchman's discussion of point of view and parallax in Victorian fiction, radicalized in the division between narratives in *Bleak House* (*The Starry Sky Within*, 128).

133. Storey, *Charles Dickens: Bleak House*, 35.

134. Wordsworth, *Poems*, 1: 574–75. Compare, more prosaically, the opening of chapter 47 of *Bleak House*: "the high church spires and the distances are so near and clear in the morning light that the city itself seems renewed by rest" (719).

135. Buzard, *Disorienting Fiction*, 154–56; Tracy, "Lighthousekeeping"; Ledger, *Dickens and the Popular Radical Imagination*, 193–95. See also Taylor, "The Novel as Climate Model," 4–8.

136. George Levine, *Darwin and the Novelists*, 142.

Chapter Five: George Eliot's Science Fiction

1. J. M. Bernstein, ed., *Classic and Romantic German Aesthetics*, 236. © Cambridge University Press 2003, reproduced with permission.

2. Eliot, *Middlemarch*, 3. Future references will be cited in the text.

3. Fielding, *Tom Jones*, 30.

4. Hume, *A Treatise of Human Nature*, 43.

5. E.g., among contemporary reviewers, R. H. Hutton: "one of the great books of the world" ("George Eliot's Moral Anatomy," 302); Simcox: "an epoch in the history of fiction" ("Review," 323).

6. See Meisel, "On the Age of the Universe." Martin S. Rudwick sets 1848 as a terminus for the modern scientific "discovery of time" (*Bursting the Limits of Time*, 6). On Romantic-period anthropological analogues of "deep time," see Heringman, *Sciences of Antiquity*.

7. Goodlad, "The *Longue Durée* of Political *Bildungsromane*," 104.

8. Eliot, *George Eliot Letters*, 3: 227; see also 3: 214.

9. On Eliot's reading of Darwin, see Beer, *Darwin's Plots*, 146–47. On the distinction between Darwinian natural selection and Victorian models of development, see Bowler, *The Non-Darwinian Revolution*, who argues that "Darwin's theory should be seen not as the central theme in nineteenth-century evolutionism but as a catalyst that helped to bring about the transition to an evolutionary viewpoint within an essentially non-Darwinian conceptual framework" (5). Dov Ospovat argues that Darwin's theory became a developmental theory in response to the work of his contemporaries, including Owen and Spencer (*The Development of Darwin's Theory*, 216, 228, 232–33). See also Allen MacDuffie's claim that the "vision of evolution as progressive development," widely adopted in Darwin's wake, obscured the radical implications of the theory of natural selection: "Charles Darwin and the Victorian Pre-History of Climate Denial," 547–48.

10. Ospovat, ibid., 116; Bowler, ibid., 51. On Von Baer's developmental morphology, see Lenoir, *Strategy of Life*, 76–95.

11. Postlethwaite, "George Eliot and Science," 107. The Google Engram viewer shows the phrase "Development Hypothesis" emerging c. 1841, rising steeply between 1845 and 1852, and then falling off.

12. Bowler, *The Non-Darwinian Revolution*, 5–14, 47–51 (51). See also Ospovat, *The Development of Darwin's Theory*, 232–33.

13. Lewes, "Mr. Darwin's Hypotheses," *Fortnightly Review* 3: 357. Bowler characterizes Haeckel as "pseudo-Darwinian," bound to the development paradigm despite his declared allegiance to Darwin's theory (*The Non-Darwinian Revolution*, 83–90).

14. Lewes, ibid., 4: 75–79.

15. Beer, *Darwin's Plots*, 275n12; see also 148.

16. For a summary, see Griffiths, *The Age of Analogy*, 206–10.

17. Spencer, "Progress: Its Law and Cause," 465; Lamarck, *Zoological Philosophy*, 10–11, 126–30 (127).

18. See Ospovat, *The Development of Darwin's Theory*, 225–28 (227).

19. Darwin, *On the Origin of Species*, 426.

20. On Wallace's 1864 lecture "The Origin of Human Races and the Antiquity of Man" and Reade's 1872 Social-Darwinist universal history *The Martyrdom of Man*, see Hesketh, "The Future Evolution of 'Man,'" 195–201.

21. Darwin, *On the Origin of Species*, 426.

22. Grosz, *Becoming Undone*, 24.

23. Beer, *Darwin's Plots*, 146.

24. See Prum, *The Evolution of Beauty*; Evelleen Richards, *Darwin and the Making of Sexual Selection*.

25. Darwin, *On the Origin of Species*, 423.

26. Sprat, *History of the Royal Society*, 113. See, however, Tita Chico's demonstration of the pervasiveness of figuration within Royal Society members' disavowals (*The Experimental Imagination*, 19–43).

27. As Beer and others argue; see, for an extended treatment, Young, *Darwin's Metaphor*, 92–120.

28. Darwin, *The Variation of Animals and Plants under Domestication*, 1: 6. Darwin inserted an earlier version of this passage into the third edition (1861) of *On the Origin of Species*, as the second paragraph of chapter 4, "Natural Selection," 84–85.

29. See Anger's discussion, *Victorian Interpretation*, 89–91: Whewell adapted "the idea of progressive revelation in biblical exegesis" to scientific discovery (91).

30. Whewell, *The Philosophy of the Inductive Sciences*, 2: 479.

31. Ibid., 2: 480–81.

32. Eliot, "The Natural History of German Life," in *Essays of George Eliot*, 287–88.

33. R. H. Hutton, unsigned review (*The Spectator*, March 30, 1872), in David Carroll, *George Eliot: The Critical Heritage*, 296. See Beer, *Darwin's Plots*, 139–41.

34. See, e.g., Shuttleworth, *George Eliot and Nineteenth-Century Science*; Rothfield, *Vital Signs*; Gallagher, *The Body Economic*; Matus, *Shock, Memory and the Unconscious in Victorian Fiction*; Ryan, *Thinking without Thinking in the Victorian Novel*; Henchman, *The Starry Sky Within*; Jenkins, "George Eliot, Geometry and Gender." Before *Darwin's Plots*, Michael York Mason reads *Middlemarch* through the debate between Whewell and J. S. Mill over scientific method ("*Middlemarch* and Science").

35. Beer, *Darwin's Plots*, 140.

36. Compare Daniel Wright's argument, from the angle of Victorian language philosophy, in "George Eliot's Vagueness": "she investigates the power of figurative language: those moments in which the imaginative precision of metaphor affords an alternative to a precision that is lost to language itself" (627).

37. Beer, *Darwin's Plots*, 159.

38. Griffiths, *The Age of Analogy*, 24.

39. On *The Mill on the Floss* and the Bildungsroman tradition, see Fraiman, *Unbecoming Women*, 124–35; Esty, *Unseasonable Youth*, 53–64. On the novel's repudiation of "Spencer's assumptions about the biological fixity of gender differences and their implications for human morality" (75) and George Eliot's "resolutely feminist resistance to . . . Spencer's biological determinism" more broadly (5), see Paxton, *George Eliot and Herbert Spencer*, 71–75, 90.

40. On George Eliot and the social evolutionism of Comte, Spencer, and Lewes, see Graver (who argues for a "community of interests" and "shared affinities" rather than influence), *George Eliot and Community*, 4–10, 40–48; and Shuttleworth, *George Eliot and Nineteenth-Century Science*, 8–23; on Eliot and Spencer, see Paxton, ibid., who cites Spencer's admission (in his *Autobiography*) that it was Eliot who got him to read Comte (18).

41. Comte, *The Positive Philosophy*, 91.

42. Ibid., 346. Cf. Lewes on Comte: "He decides against Lamarck's celebrated development hypothesis. Although his admiration of Lamarck, and appreciation of his influence on philosophical zoology, is such as may be expected from so great and liberal a thinker, he does not, as it appears to me, fully appreciate the immense value of this hypothesis if merely treated as a philosophic artifice, let its truth be what it may" (*Comte's Philosophy of the Sciences*, 190). On the Lamarckism of Comte, Spencer, and Lewes, see Graver, *George Eliot and Community*, 47, 64–65.

43. Spencer, "Progress: Its Law and Cause," 446.

44. Spencer, *Social Statics*, 439, 440. For the principle, Spencer and Lewes cite Coleridge's late essay "Theory of Life," derived in turn from Schelling: "I define life as *the principle of individuation*, the power which unites a given *all* into a *whole* that is presupposed by its parts" (Lewes, *Comte's Philosophy of the Sciences*, 167–68). On Spencer's sources in Von Baer and the German Romantic life sciences, see Ospovat, *The Development of Darwin's Theory*, 216; Lightman, "Life," 29–30.

45. Lewes, ibid., 167. On Comte's "dialectical conception of the relations between the organism and the milieu," limited however to the human species, see Canguilhem, *Knowledge of Life*, 101–3. Robert Mitchell suggests that "the vitalist dimension of Romantic-era understandings of media" began to dissipate with the mid-nineteenth-century separation of "life" and "culture" into "different fields of knowledge production" (*Experimental Life*, 145).

46. Eliot, "Notes on Form in Art," in *Essays of George Eliot*, 433.

47. Eliot, "Prospectus of the *Westminster and Foreign Quarterly Review*," in *Selected Essays*, 4–5.

48. Eliot, "The Natural History of German Life," in *Essays of George Eliot*, 286, 289–90.

49. Compare Shuttleworth's somewhat different analysis of a tension between contradictory senses of "natural history" (*George Eliot and Nineteenth-Century Science*, 25–30).

50. On Spencer's "obsessively literal and baroque elaboration of the widespread metaphor of society as an organism," see Palmeri, *State of Nature, Stages of Society*, 113.

51. See Griffiths, *Age of Analogy*, 169, 183–89.

52. Eliot, "The Natural History of German Life," in *Essays of George Eliot*, 274–75. Eliot is freely translating Riehl's text here (see Fleishman, *George Eliot's Intellectual Life*, 80).

53. Shuttleworth, *George Eliot and Nineteenth-Century Science*, 24–25, 49–50, 64–68, 76–77.

54. On Morgenstern, the Bildungsroman, and the female protagonist, see chapter 2.

55. Eliot, *The Mill on the Floss*, 31. Future references to this edition will be given in the text.

56. For analytical summaries, see Anger, *Victorian Interpretation*, 100–105; Nazar, "The Continental Eliot," 414–21.

57. Feuerbach, *The Essence of Christianity*, 1–2. Future references will be cited in the text.

58. See chapter 1. "The leaf on which the caterpillar lives is for it a world, an infinite space" (Feuerbach, ibid., 7); Uexküll, *A Foray into the Worlds of Animals and Humans*, 46–52; on the Herderian "sphere" as the creature's "bubble," 43.

59. Darwin, *On the Origin of Species*, 427.

60. Kucich, *Repression and Victorian Fiction*, 123. See Marx, "Theses on Feuerbach," No. 6: "Feuerbach resolves the religious essence into the human essence. But the human essence is no abstraction inherent in each single individual. In its reality it is the ensemble of the social relations. Feuerbach, who does not enter upon a criticism

of this real essence, is consequently compelled: 1.) To abstract from the historical process and to fix the religious sentiment as something by itself and to presuppose an abstract—isolated—human individual. 2.) Essence, therefore, can be comprehended only as 'genus' [*Gattung*, species], as an internal, dumb generality which naturally unites the many individuals" (Marx, *Karl Marx*, 22).

61. See Marx's *Economic and Philosophical Manuscripts* (1844): "Man is a species-being, not only because in practice and in theory he adopts the species (his own as well as those of other things) as his object, but—and this is only another way of expressing it—also because he treats himself as the actual, living species; because he treats himself as a *universal* and therefore a free being" (ibid., 40).

62. Ibid., 22. Compare the opposition between general category and solitary individual that sustained Buffon's definition of species: see chapter 1. Nazar cites recent scholarship on the continuity of Marx's emphasis on social relations and practice with Feuerbach's critical project ("The Continental Eliot," 416).

63. Kucich, *Repression and Victorian Fiction*, 122, 123.

64. Ibid., 117.

65. "Music is the language of feeling; melody is audible feeling—feeling communicating itself" (Feuerbach, *The Essence of Christianity*, 3); cf. *The Mill on the Floss*, 305, 385; "her sensibility to the extreme excitement of music was only one form of that passionate sensibility which belonged to her whole nature" (401).

66. "In love, the reality of the species, which otherwise is only a thing of reason, an object of mere thought, becomes a matter of feeling, a truth of feeling . . . [man] declares the life which he has through love to be the truly human life, corresponding to the idea of man, i.e., of the species" (Feuerbach, ibid., 155).

67. These passages in *The Mill on the Floss* may be a source for Darwin's account of the evolution of morality—via a conflict between instincts, one appetitive and immediate, the other social and temporally "enduring"—in *The Descent of Man*, 122–23.

68. Feuerbach, *The Essence of Christianity*, 59–60.

69. Compare Armstrong, *How Novels Think*: "Maggie becomes not only less of an individual but also more of a human being than Lucy, who lacks the capacity to be caught up in something larger than herself and be carried away on the current of an impersonal desire that appears to have something more like the interests of the species in mind" (95).

70. Elisha Cohn, *Still Life*, 66–67.

71. Beer, *Darwin's Plots*, 139–40.

72. Müller, *Lectures on the Science of Language* (1861); Tylor, *Primitive Culture* (1871), vol. 1, chapter 1, "The Science of Culture." On comparative philology and mythology in the genealogy of early anthropology, see Stocking, *Victorian Anthropology*, 23–25, 55–62, 157–58.

73. Eliot lists Schwann and Schleiden in her "quarry" for *Middlemarch*, along with T. H. Huxley's essay "The Cell-Theory" (*Quarry for* Middlemarch, 612, 617–18). On Eliot and Virchow, see Blair, "A Change in the Units."

74. The phrase is Hillis Miller's: "Optic and Semiotic in *Middlemarch*," 125.

75. Eliot, "The Progress of the Intellect," in *Essays of George Eliot*, 35.

76. For George Eliot's summary of Bryant's argument in her "Folger notebook," c. 1869, see Eliot, *George Eliot's* Middlemarch *Notebooks*, 48, 135; on Bryant as a model

for Casaubon, xlvii; on Eliot's reading of Creuzer, 48–51, 135. Colin Kidd argues that Creuzer's work and the attendant controversy constitute the "new discoveries" of which Casaubon is ignorant (*The World of Mr Casaubon*, 23–25); on Mackay, 26–27; on Bryant and the old regime of apologetic mythography (a nuanced account), 111–23. See also Harvey, "The Intellectual Background of the Novel," 33–35.

77. Creuzer, *Symbolism and Mythology of Ancient Peoples* (1810), in Feldman and Richardson, *The Rise of Modern Mythology*, 395.

78. Mackay, *The Progress of the Intellect*, 1: 62–63.

79. Ibid., 1: 65.

80. Ibid., 1: 142.

81. Müller, *Comparative Mythology*, 12; *Lectures on the Science of Language*, 21.

82. Müller, *Comparative Mythology*, 124, 135.

83. Eliot, "The Progress of the Intellect," in *Essays of George Eliot*, 39.

84. Hertz, *George Eliot's Pulse*, 23.

85. For the former view, see Harvey, "The Intellectual Background to the Novel," 36; McCarthy, "Lydgate, 'The New, Young Surgeon' of *Middlemarch*"; Rothfield, *Vital Signs*, 92–99; Otis, *Networking*, 111; Blair, "Contagious Sympathies," 147; Fleishman, *George Eliot's Intellectual Life*, 168–70. For the latter view, see, e.g., Beer, *Darwin's Plots*: "There is not one 'primitive tissue,' just as there is not one 'key to mythologies'" (143).

86. Jacyna, "The Romantic Programme and the Reception of Cell Theory in Britain," 19–20, 15. Mason notes the ambiguity as to whether Lydgate's "search for a 'Primitive Tissue' is for the origin of all tissue (embryology and, perhaps, comparative anatomy) or for its basic form (histology)" ("*Middlemarch* and Science," 162).

87. Jacyna, ibid., 20n28 (citing Rather, "Some Relations between Eighteenth-Century Fiber Theory and Nineteenth-Century Cell Theory").

88. Eliot, *Quarry for* Middlemarch, 617–18.

89. T. H. Huxley, "The Cell-Theory," 253–54, 261.

90. Richmond, "Thomas Huxley's Developmental View of the Cell," 64. See also Richmond's more detailed statement of her argument, "T. H. Huxley's Criticism of German Cell Theory."

91. Otis, *Networking*, 81–119. In the 1860s, Eliot and Lewes were reading Hermann von Helmholtz, Wilhelm Wundt, and Claude Bernard on the nervous system, as well as Virchow; later, Lewes would repudiate the "fetichistic deification of the [nerve-]cell" that "invested it with miraculous powers" in the early development of neurology (*On the Physical Basis of Mind*, 248).

92. Lydgate's failure has been characterized in various ways. Otis sees him "entangled in his own metaphor" of the "independent tissue," "mistakenly believing that he can act as an autonomous element in a small provincial town" (*Networking*, 101); Blair diagnoses an inability (despite his own reforming ambition) to grasp the "contagious" paradigm of medical pathology as a social condition that would be promoted by Virchow ("Contagious Sympathies," 152–54).

93. Mason, "*Middlemarch* and Science," 160–61 (citing Lewes, *The Principles of Success in Literature*). Fleishman argues, however, that Lydgate's research fails through insufficient imagination: "having learned about tissues from Bichat, he seeks yet another tissue, a primary one, instead of going outside or beyond that framework to seek a more elementary constituent or unit of tissue" (*George Eliot's Intellectual Life*, 170).

94. Whewell, *The Philosophy of the Inductive Sciences*, 2: 42. The quotation from *Middlemarch* echoes Tyndall's "The Scientific Use of the Imagination," quoted by Lewes in *Problems of Life and Mind* (261), which summarizes Whewellian principles; see also Beer, *Darwin's Plots*, 141; George Levine (on Eliot and Lewes), "George Eliot's Hypothesis of Reality," 11–14. On Baconian induction, hermeneutics, and the imagination in Victorian scientific inquiry, see Smith, *Fact and Feeling*, 16–22, 37–41; Anger, *Victorian Interpretation*, 87–94. Shuttleworth aligns Lydgate with Claude Bernard, whom Eliot and Lewes were reading in the late 1860s (*George Eliot and Nineteenth-Century Science*, 145).

95. Eliot takes the example from Tyndall's 1868 lecture "On the Methods and Tendencies of Physical Investigation": "in the eye of science the animal body is just as much the product of molecular force as the stalk or ear of corn, or as the crystal of salt or sugar. . . . Take the human heart, for example, with its exquisite system of valves" (*Scientific Addresses*, 14).

96. Eliot, "The Progress of the Intellect," in *Essays of George Eliot*, 35; Mackay, *The Progress of the Intellect*, 1: 62.

97. De Witt, *Moral Authority, Men of Science, and the Victorian Novel*, 74. Compare Barbara Hardy: "Casaubon's failure, interestingly enough, lies both in ardour and intellect, George Eliot's concept of knowledge being one of integrated ardour and learning. The integration is there in Lydgate, as far as his science is concerned, but it is imperfectly sustained in his human relations" (*The Novels of George Eliot*, 64).

98. Whewell, *The Philosophy of the Inductive Sciences*, 2: 65.

99. Eliot, "Notes on Form in Art," in *Essays of George Eliot*, 433.

100. Fleishman calls the novel "a preeminent example of superorganic form, in which a great diversity of characters and social contexts is allowed to develop partially overlapping interrelations, while developing a wholeness that accords with Eliot's idea of form in art" (*George Eliot's Intellectual Life*, 170).

101. Whewell, *The Philosophy of the Inductive Sciences*, 2: 74.

102. Puckett, *Bad Form*, 116.

103. Bechtel, *Discovering Cell Mechanisms*, 67–68. Lydgate may have one: we learn that he makes diligent use of the microscope, "which research had begun to use again with new enthusiasm of reliance" (148), and that his instrument is superior to amateur naturalist Mr Farebrother's (chapter 36). Lilian R. Furst reckons the new technology arrives "just a little too late" for him: *Between Doctors and Patients*, 60. Reviewing the new advances in microscopic technology, with which Eliot was well informed via Lewes, Mark Wormald argues that Lydgate's condescension to Farebrother exposes the deductive rigidity of his own approach (a view I do not share): "Microscopy and Semiotic in *Middlemarch*," 508–10, 518–22.

104. Lewes, "Mr. Darwin's Hypothesis," *Fortnightly Review* 4: 61.

105. Huxley, *On the Physical Basis of Life*, 5. Involuntary, palpitating: "When viewed with a sufficiently high magnifying power, the protoplasmic layer of the nettle hair is seen to be in a condition of unceasing activity. Local contractions of the whole thickness of its substance pass slowly and gradually from point to point, and give rise to the appearance of progressive waves, just as the bending of successive stalks of corn by a breeze produces the apparent billows of a corn-field" (7: like the fields Dorothea views from her window). Hugo von Mohl, who gave the term "protoplasm"

scientific currency in 1846, described it as "an opaque, viscid fluid, having granules intermingled in it" (Bechtel, *Discovering Cell Mechanisms*, 73). On viscid physiology in *Middlemarch*, see Brilmyer, "Plasticity, Form, and the Matter of Character."

106. "I have translated the term 'Protoplasm,' which is the scientific name of the substance of which I am about to speak, by the words 'the physical basis of life'" (*On the Physical Basis of Life*, 5). Huxley's lecture popularized the term in English: see Brain, "Protoplasmania." In the same year (1868), Huxley named what he supposed to be a specimen of deep-sea protoplasm *Bathybius haeckellii*, after Haeckel's account of "an entirely homogeneous and structureless substance, a living particle of albumen, capable of nourishment and reproduction" (Rehbock, "Huxley, Haeckel and the Oceanographers," 508, 515). *Bathybius* subsequently turned out to be an inorganic precipitate.

107. Ruskin, *Unto This Last*, 156. On the nineteenth-century ascendancy of life as value, see Gallagher's major discussion, *The Body Economic*, 90–91.

108. Coleridge, *The Statesman's Manual*, 37.

109. Banfield, *Unspeakable Sentences*, 257.

110. Dorothea's epiphany exhibits the sublation of metonymy into metaphor—protoplasm into life, matter into form—that Elaine Freedgood has described as *Middlemarch's* totalizing rhetorical technique: "To render the metonymic ground of the novel selectively and intensively metaphoric guarantees its reproducible symbolic coherence and allows for consensual, predictable interpretations of it as a literary work" (*The Ideas in Things*, 115–16). Hillis Miller argues that the novel undoes that "reproducible symbolic coherence" by pitching incompatible metaphoric systems against each other: "Optic and Semiotic in *Middlemarch*," 143–44.

111. Darwin, *On the Origin of Species*, 427.

112. Lewes, "Mr. Darwin's Hypotheses," 4 (1868), 494; Rehbock (quoting Haeckel's "Beiträge zur Plastidentheorie," 1870), "Huxley, Haeckel and the Oceanographers," 522. Lewes revives the Lucretian hypothesis, proposed in the previous century by Bonnet and Blumenbach, that the earth was seeded with germs of life that evolved into distinct species: see Robert Richards, *The Romantic Conception of Life*, 222–25. Jesse Cordes Selbin, in a private communication, has pointed to a possible source for the phrase "involuntary, palpitating life" in Claude Bernard's *Introduction à l'étude de la médecine expérimentale* (1865): "Pourtant on n'arrivera jamais à des généralisations vraiment fécondes et lumineuses sur les phénomènes vitaux, qu'autant qu'on aura expérimenté soi-même et remué dans l'hôpital, l'amphithéâtre ou le laboratoire, le terrain fétide ou palpitant de la vie" (28: "Yet we will never achieve truly fertile and luminous generalizations about vital phenomena until we have ourselves experimented in the hospital, operating theater or laboratory, and stirred the fetid or palpitating ground of life"). Lewes owned and read the book, so it is likely George Eliot read it too: see Shuttleworth, *George Eliot and Nineteenth-Century Science*, 226n5.

113. See Eliot's source, Herodotus's *Histories*, book 1, chapter 189.

114. On the organic figure of the web in *Middlemarch*, see Beer, *Darwin's Plots*, 156–61; on the developmental plan of branching lines of descent, see Ospovat, *The Development of Darwin's Theory*, 144–65; Bowler, *The Non-Darwinian Revolution*, 11–13.

115. Gallagher, "George Eliot, Immanent Victorian," 71.

116. See Beer, *Darwin's Plots*, 167; Shuttleworth, *George Eliot and Nineteenth-Century Science*, 166. The first reader to pick up on this appears to have been James

Clerk Maxwell, attending a lecture on solar mythology in April 1873: "I think Middlemarch is not a mere unconscious myth, as the Odyssey was to its author, but an elaborately conscious one, in which all the characters are intended to be astronomical or meteorological.... The whole thing is, and is intended to be, a solar myth from beginning to end" (quoted in Beer, *Open Fields*, 233–34). On Ladislaw as (also) a Dionysian figure, see Wiesenfarth, *George Eliot's Mythmaking*, 190, 195–96; Hillis Miller, "A Conclusion in Which Almost Nothing Is Concluded," 145–46.

117. *OED*, "metamorphosis," 3.b. and 3.c. Given broad scientific currency by Goethe in *Die Metamorphose der Pflanzen*, "metamorphosis" enters English as a term for biological transformation via Robert Edmond Grant, the most influential British follower of Lamarck and Geoffroy, around the time that the action of *Middlemarch* takes place, in the late 1820s.

118. On Ladislaw and Rigg as analogous figures "of shape-shifting, of metamorphosis," see Trotter, "Space, Movement and Sexual Feeling in *Middlemarch*," 52–55.

119. Lovecraft, "The Shadow over Innsmouth," in *Tales*, 596, 597–98.

120. See Kidd, *The World of Mr Casaubon*, 177–80 (177).

121. "Undoubtedly a debased, quasi-pagan thing imported from the East a century before" (Lovecraft, "The Shadow over Innsmouth," in *Tales*, 596).

122. This is the case, also, for the other outsiders to Middlemarch, Bulstrode and Lydgate. A fifth, Raffles, returns only to die there.

123. Austen, *Northanger Abbey*, 160. On the topos, see Duncan, "The Regional or Provincial Novel"; Plotz, *Portable Property*, 96–98.

124. "The world of the realistic novel is irrecoverably in fragments" (George Levine, "George Eliot's Hypothesis of Reality," 16). On *Daniel Deronda* as a "novel haunted by the future," see Beer, *Darwin's Plots*, 169–75; on Jewish "ancestrality" in *Daniel Deronda*, see Heffernan, "The Stamp of Rarity."

125. On *Daniel Deronda* and provincial life, see Plotz, *Portable Property*, 75–77.

126. Darwin, *On the Origin of Species*, 423.

127. Ibid., 409.

128. See Eliot, *George Eliot's* Daniel Deronda *Notebooks*, 19, 308, 312–14.

129. Eliot, *Daniel Deronda*, 389; Darwin, *On the Origin of Species*, 201, 420.

130. Eliot, ibid., 397. Future references to this edition will be given in the text.

131. Tylor, *Primitive Culture*, 1: 19. Eliot also uses the term to characterize the archery tournament at Brackenshaw Park, where Gwendolen and Grandcourt first meet (*Daniel Deronda*, 101).

132. See Rothfield's discussion of the tension "between pathological and embryological-evolutionary perspectives" within the novel's biological register (*Vital Signs*, 119).

133. Recent discussions of the novel's zoological metaphors tend to stabilize their figural volatility: Pielak, "Hunting Gwendolen"; Kingstone, "Human-Animal Elision."

134. Darwin opens *The Descent of Man* by designating "malconformations, the result of arrested development" and "reversion to some former and ancient type of structure" as key criteria in his evolutionary argument (21). The scheme of phylogenetic recapitulation in embryonic development, formulated by Friedrich Tiedemann, was popularized in English in the preceding generation by Perceval Lord and Robert Chambers: "The brain of man, which exceeds that of all other animals in complexity

of organization and fulness of development, is, at one early period, only 'a simple fold of nervous matter, with difficulty distinguishable into three parts, while a little tail-like prolongation towards the hinder parts, and which had been the first to appear, is the only representation of a spinal marrow.... In a short time, however, the structure is become more complex, the parts more distinct, the spinal marrow better marked; it is now the brain of a reptile'" (Chambers, *Vestiges of the Natural History of Creation*, 199–200, quoting Lord, *Popular Physiology*, 347).

135. See, e.g., Bryant, *A New System*, 2: 451–52, 531–32; Mackay, *The Progress of the Intellect*, 1: 421–34; Tylor, *Primitive Culture*, 2: 240–42. For a discussion, see Kidd, *The World of Mr Casaubon*, 180–90 and (on its resurgence as a controversial theme in the new anthropology of the 1860s and 1870s) 194–99.

136. Tylor, *Primitive Culture*, 1: 106.

137. On Gwendolen's name, and Eliot's reading in Celtic sources, see Eliot, *George Eliot's* Daniel Deronda *Notebooks*, 445–50.

138. Darwin, *The Descent of Man*, 147.

139. Ibid., 151.

140. Plotz, *Portable Property*, 73.

141. Amanda Anderson, *The Powers of Distance*, 121, 129–38: reiterating, in the key of liberal humanism, the Christian sublation of the old law (letter) into the new (spirit). See also Sebastian Lecourt, who argues that Deronda's access to his Jewish heritage via reading (scripture, his father's writings, etc.) effects a liberal aeration of the organic racial attachment promoted by Mordecai (*Cultivating Belief*, 102–4, 116–17).

142. Chase, "The Decomposition of the Elephants." See also Beer, *Darwin's Plots*: "In *Daniel Deronda* causal sequence is disturbed and pressed upon by resurgence, synchronicity, the miraculous, the hermeneutic, and by unassuageable human need" (169). As for technical terms: I use "metalepsis" in the rhetorical sense used by Chase, "a reversal of the temporal status of effect and cause" (218), rather than in the narratological sense, denoting a vertical transgression of narrative frames, used by Gérard Genette (corresponding to Friedrich Schlegel's term "parabasis," derived from Aristophanic drama). See Cohn and Gleich, "Metalepsis and *Mise en Abyme*"; Handwerk, "Romantic Irony," 215–17. On romance and realism, see Cave's introduction to *Daniel Deronda*, xxiv–xxxiii.

143. See Shaffer's discussion of Gwendolen as a penitent Magdalen (*"Kubla Khan" and the Fall of Jerusalem*, 278–81).

144. Chase, "'The Decomposition of the Elephants,'" 222–23.

145. Lewes, *Comte's Philosophy of the Sciences*, 27; see Comte, *The Positive Philosophy*, 25–26.

146. Lewes, ibid., 28. On Mordecai's relation with Daniel as one of scientific discovery, see George Levine, "George Eliot's Hypothesis of Reality," 25–27.

147. "In highlighting the notion of 'the beloved ideas made flesh,' Eliot reconfigures the Christian scriptural paradigm of Jewish identity that severs the Jewish body from the those 'beloved ideas' and texts that Christianity appropriates" (Scheinberg, "'The Beloved Ideas Made Flesh,'" 833).

148. Plotz, *Portable Property*, 73.

149. Lewes, "Mr. Darwin's Hypotheses," *Fortnightly Review* 4: 504.

150. Darwin, *The Variation of Animals and Plants under Domestication*, 2: 373. Future references will be given in the text.

151. See Trower, "Nerves, Vibrations and the Aeolian Harp."

152. Lewes, "Mr. Darwin's Hypotheses," *Fortnightly Review* 4: 507.

153. Kant, "Conjectural Beginning of Human History," in *Anthropology, History, and Education*, 163.

154. Darwin, "Recollections of the Development of My Mind and Character," 395.

155. Eliot, *Impressions of Theophrastus Such*, 199. Eliot is developing the conceit of Samuel Butler's newspaper sketch "Darwin Among the Machines" (1863), revised as "The Book of the Machines" (chapters 23–25) in *Erewhon* (1872).

156. See Lightman, "Life," 30–36: Tyndall's romantic evolutionism (hylozoism) "led him to conceive of all nature as a living, organic being, and evolution as full of purpose" (31). On the theme's currency in mid-Victorian culture, see Mershon, "Ruskin's Dust." On Owen and the "polarizing force" of crystalline and organic formation (resumed by Lewes in "Mr. Darwin's Hypotheses"), see Ospovat, *The Development of Darwin's Theory*, 133. On its likely source in Von Baer's embryology, see Lenoir, *Strategy of Life*, 87. But compare Herder, *Outlines of a Philosophy for the History of Man*: "The inflammable matter of the air probably converted silica into calcareous Earth, and in this the first living creatures of the sea, shellfish, were formed, for throughout all nature the materials appear before the organized animated structure. A still more powerful and pure action of fire and of cold was requisite to crystallization, which inclines not to the shelly form, exhibited by silica in its fractures, but to geometrical Angles. These, too, vary according to the component parts of each individual, until they approach the semi-metals, metals, and ultimately the germs of plants" (27).

157. Grosz, *Becoming Undone*, 19, 27, 34.

158. Eliot, *Impressions of Theophrastus Such*, 201.

BIBLIOGRAPHY

Sources Published before 1900

Addison, Joseph, and Richard Steele. *The Spectator.* Ed. Donald F. Bond. 5 vols. Oxford: Clarendon Press, 1965.
[Anon.] "Literary Notices." *Harper's New Monthly Magazine* 4: 23 (April 1852), 711.
[Anon.] "Natural Selection." *All the Year Round* 63 (July 7, 1860), 293–99.
[Anon.] "Species." *All the Year Round* 58 (June 2, 1860), 174–78.
[Anon.] "Transmutation of Species." *All the Year Round* 98 (March 9, 1861), 519–21.
Aristotle. *A New Aristotle Reader.* Ed. J. L. Ackrill. Princeton: Princeton University Press, 1987.
Arnold, Matthew. *Culture and Anarchy.* Ed. Jane Garnett. Oxford: Oxford University Press, 2006.
Austen, Jane. *Emma.* Ed. James Kinsley. Oxford: Oxford University Press, 2008.
Austen, Jane. *Northanger Abbey, Lady Susan, The Watsons, Sanditon.* Ed. John Davie. Oxford: Oxford University Press, 1998.
Balzac, Honoré de. *The Works of Honoré de Balzac.* Vol. 1, *Introduction to Comedy, etc.* Ed. George Saintsbury. New York: McKinlay, Stone and Mackenzie, 1915.
Bernard, Claude. *Introduction à l'étude de la médecine expérimentale.* Paris: Baillière, 1865.
Bernstein, J. M., ed. *Classic and Romantic German Aesthetics.* Cambridge: Cambridge University Press, 2003.
Blackwell, Thomas. *An Enquiry into the Life and Writings of Homer.* Hildesheim and New York: Georg Olms, 1976.
Blumenbach, Johann Friedrich. *The Anthropological Treatises of Johann Friedrich Blumenbach, and the Inaugural Dissertation of John Hunter, M.D. on the Varieties of Man.* Trans. and ed. John Bendyshe. London: Longmans, 1865.
Browning, Elizabeth Barrett. *Aurora Leigh.* Ed. Kerry McSweeney. Oxford: Oxford University Press, 1993.
Bryant, Jacob. *A New System; Or, an Analysis of Antient Mythology.* 3rd ed. 6 vols. London: Walker, 1807.
Buffon, Georges-Louis Leclerc, Comte de. *The Epochs of Nature.* Trans. and ed. Jan Zalasiewicz, Anne-Sophie Milon, and Mateusz Zalasiewicz. Chicago: University of Chicago Press, 2018.
Buffon, Georges-Louis Leclerc, Comte de. *Natural History, General and Particular.* Trans. William Smellie. 2nd ed. 9 vols. London: Cadell, 1785.
Buffon, Georges-Louis Leclerc, Comte de. *Oeuvres.* Ed. S. Schmitt. Paris: Gallimard, 2007.
Burke, Edmund. *A Philosophical Enquiry into the Origin of Our Ideas of the Sublime and Beautiful.* Ed. Adam Phillips. Oxford: Oxford University Press, 1998.
Burke, Edmund. *Selected Letters of Edmund Burke.* Ed. Harvey V. Mansfield Jr. Chicago: University of Chicago Press, 1984.

Burnett, James, Lord Monboddo. *Antient Metaphysics*. Vol. 3, *Containing the History and Philosophy of Men*. London: Cadell, 1784.
Burnett, James, Lord Monboddo. *Antient Metaphysics*. Vol. 4, *Containing the History of Man*. Edinburgh: Bell & Bradfute, Cadell, 1795.
Burnett, James, Lord Monboddo. *Of the Origins and Progress of Language*. 2nd ed. 2 vols. Edinburgh: Balfour, Cadell, 1774.
Butler, Samuel. *Erewhon; or, Over the Range*. London: Trübner, 1872.
Carlyle, Thomas. *Sartor Resartus*. Ed. Kerry McSweeney and Peter Sabor. Oxford: Oxford University Press, 1987.
Chambers, Robert. *Vestiges of the Natural History of Creation and Other Evolutionary Writings*. Ed. James A. Secord. Chicago: University of Chicago Press, 1994.
Coleridge, Samuel Taylor. *The Statesman's Manual; Or, The Bible the Best Guide to Political Skill and Foresight: A Lay Sermon*. London: Gale and Fenner, 1816.
Collins, Wilkie. *The Moonstone*. Ed. John Sutherland. Oxford: Oxford University Press, 2000.
Comte, Auguste. *The Positive Philosophy*. Trans. Harriet Martineau. New York: Blanchard, 1855.
Condorcet, Marie-Jean-Antoine-Nicolas Caritat, Marquis de. *Outlines of an Historical View of the Progress of the Human Mind*. Philadelphia: Carey, 1796.
Creuzer, Georg Friedrich. *Symbolik und Mythologie der alten Völker, besonders der Griechen*. 3rd ed. 4 vols. Leipzig: Leske, 1837–42.
Cuvier, Baron Georges. *Discours sur les révolutions de la surface du globe, et sur les changemens qu'elles ont produits dans le règne animal*. 3rd ed. Paris, 1825.
Cuvier, Baron Georges. "Elegy of Lamarck." *Edinburgh New Philosophical Journal* 20 (1836), 1–22.
Cuvier, Baron Georges. *Essay on the Theory of the Earth: With Geological Illustrations by Professor Jameson*. Edinburgh: Blackwood, 1827.
Cuvier, Baron Georges. *Recherches sur les ossemens fossiles de quadrupèdes: où l'on rétablit les caractères de plusieurs espèces d'animaux que les révolutions du globe paroissent avoir détruites*. 4 vols. Paris: Deterville, 1812.
D'Alembert, Jean le Rond. *Preliminary Discourse to the Encyclopedia of Diderot*. Trans. Richard N. Schwab. Chicago: University of Chicago Press, 1963.
Darwin, Charles. *The Descent of Man, and Selection in Relation to Sex*. Ed. James Moore and Adrian Desmond. London: Penguin, 2004.
Darwin, Charles. *The Expression of Emotions in Man and Animals*. London: John Murray, 1872.
Darwin, Charles. *On the Origin of Species*. Ed. William Bynum. London: Penguin, 2009.
Darwin, Charles. *On the Origin of Species by Means of Natural Selection, or the Preservation of Favoured Races in the Struggle for Life*. 3rd ed. London: John Murray, 1861.
Darwin, Charles. "Recollections of the Development of My Mind and Character." In *Evolutionary Writings*. Ed. James A. Secord, 354–425. Oxford: Oxford University Press, 2008.
Darwin, Charles. *The Variation of Animals and Plants under Domestication*. 2 vols. London: John Murray, 1868.

Dickens, Charles. *Bleak House*. Ed. Nicola Bradbury. London: Penguin, 2006.
[Dickens, Charles.] "The Poetry of Science." *The Examiner* (9 December, 1848), 787–88.
Diderot, Denis. *Thoughts on the Interpretation of Nature and Other Philosophical Works*. Ed. David Adams. Manchester: Clinamen Press, 1999.
Dilthey, Wilhelm. *Poetry and Experience*. Ed. Rudolf A. Makkreel and Frithjof Rodi. Princeton: Princeton University Press, 1985.
Eliot, George. *Adam Bede*. Ed. Stephen Gill. Harmondsworth: Penguin, 1985.
Eliot, George. *Daniel Deronda*. Ed. Terence Cave. Harmondsworth: Penguin, 1995.
Eliot, George. *Essays of George Eliot*. Ed. Thomas Pinney. New York: Columbia University Press, 1963.
Eliot, George. *The George Eliot Letters*. Ed. Gordon S. Haight. 9 vols. New Haven: Yale University Press, 1954–55.
Eliot, George. *George Eliot's* Daniel Deronda *Notebooks*. Ed. Jane Irwin. Cambridge: Cambridge University Press, 1996.
Eliot, George. *George Eliot's* Middlemarch *Notebooks: A Transcription*. Ed. John Clark Pratt and Victor A. Neufeldt. Berkeley: University of California Press, 1979.
Eliot, George. *Impressions of Theophrastus Such*. London: Blackwood, 1879.
Eliot, George. *Middlemarch*. Ed. Rosemary Ashton. Harmondsworth: Penguin, 1994.
Eliot, George. *The Mill on the Floss*. Ed. Gordon S. Haight. Oxford: Oxford University Press, 1981.
Eliot, George. *Quarry for* Middlemarch. Ed. Anna Theresa Kitchel. 1950. Reprinted in George Eliot, *Middlemarch*. Ed. Bert G. Hornback, 607–42. New York: Norton, 1977.
Eliot, George. *Selected Essays, Poems and Other Writings of George Eliot*. Ed. A. S. Byatt and Nicholas Warren. London: Penguin, 1990.
Feldman, Burton, and Robert D. Richardson, eds. *The Rise of Modern Mythology 1680–1860*. Bloomington: Indiana University Press, 1972.
Ferguson, Adam. *An Essay on the History of Civil Society*. Ed. Fania Oz-Salzberger. Cambridge: Cambridge University Press, 1995.
Feuerbach, Ludwig. *The Essence of Christianity*. Trans. Marian Evans. London: Chapman, 1854.
Fielding, Henry. *Joseph Andrews and Shamela*. Ed. Douglas Brooks-Davies and Thomas Keymer. Oxford: Oxford University Press, 1999.
Fielding, Henry. *Tom Jones, the History of a Foundling*. Ed. Simon Stern and John Bender. Oxford: Oxford University Press, 1996.
Goethe, Johann Wolfgang. *Conversations with Eckermann and Soret*. Trans. John Oxenford. 2 vols. London: Smith, Elder, 1850.
Goethe, Johann Wolfgang. *Italian Journey*. Ed. Thomas P. Saine and Jeffrey L. Sammons, trans. Robert R. Heitner. Princeton: Princeton University Press, 1989.
Goethe, Johann Wolfgang. *Wilhelm Meister's Apprenticeship*. Trans. E. Blackall. Princeton: Princeton University Press, 1995.
Goethe, Johann Wolfgang. *Wilhelm Meister's Apprenticeship, A Novel. From the German of Goethe*. Trans. Thomas Carlyle. 3 vols. Edinburgh: Oliver & Boyd, 1824.
Goethe, Johann Wolfgang. *Wilhelm Meisters Lehrjahre*. Ed. Erhard Bahr. Stuttgart: Philipp Reclam, 1982.
Hall, Basil. *Voyage to Loo-Choo, and Other Places in the Eastern Seas, in the Year 1816*. Edinburgh: Constable, 1826.

Hartley, David. *Observations on Man, his Frame, his Duty, and his Expectations.* 2 vols. London: Charles Hitch and Stephen Austen, 1749.
Hegel, G. W. F. *Aesthetics: Lectures on Fine Art.* Trans. T. M. Knox. 2 vols. Oxford: Oxford University Press, 1975.
Hegel, G. W. F. *The Phenomenology of Spirit.* Trans. A. V. Miller. Oxford: Oxford University Press, 1977.
Herder, Johann Gottfried. *Abhandlung über den Ursprung der Sprache.* Ed. Hans Dietrich Irmscher. Stuttgart: Philipp Reclam, 1966.
Herder, Johann Gottfried. *Ideen zur Philosophie der Geschichte der Menschheit.* Ed. Wolfgang Pross. 2 vols. Munich: Carl Hanser Verlag, 2002.
Herder, John Godfrey. *Outlines of a Philosophy of the History of Man.* Trans. T. Churchill. London: Joseph Johnson, 1800.
Herder, Johann Gottfried. *Philosophical Writings.* Ed. Michael N. Forster. Cambridge: Cambridge University Press, 2002.
Herder, Johann Gottfried. *Selected Writings on Aesthetics.* Princeton: Princeton University Press, 2006.
Herodotus. *The Histories.* Trans. Tom Holland. London: Penguin, 2015.
Home, Henry, Lord Kames. *Sketches of the History of Man.* 2nd ed. 4 vols. Edinburgh: Strahan, Cadell, Creech, 1788.
Hugo, Victor. *Dramas.* Vol. 3, *Cromwell, Mary Tudor, Esmeralda.* Boston: Little, Brown, 1909.
Hugo, Victor. *The Hunchback of Notre-Dame.* Trans. Frederic Schoberl. London: Bentley, 1833.
Hugo, Victor. *Notre-Dame de Paris: 1482.* Ed. S. de Sacy. Paris: Gallimard, 2002.
Hugo, Victor. *Notre-Dame: A Tale of the "Ancien Régime."* Trans. William Hazlitt Jr. 3 vols. London: E. Wilson, 1833.
Hugo, Victor. *Notre-Dame of Paris.* Trans. John Sturrock. Harmondsworth: Penguin, 1978.
Humboldt, Wilhelm von. *The Sphere and Duties of Government.* Trans. Joseph Coulthard. London: Chapman, 1854.
Hume, David. *An Enquiry Concerning Human Understanding and Other Writings.* Ed. Stephen Buckle. Cambridge: Cambridge University Press, 2007.
Hume, David. "Of National Characters." In *Essays: Moral, Political and Literary.* Ed. Eugene F. Miller, 197–215. Indianapolis: Liberty Fund, 1985.
Hume, David. "Of the Standard of Taste." In *Essays: Moral, Political and Literary.* Ed. Eugene F. Miller, 226–49. Indianapolis: Liberty Fund, 1985.
Hume, David. *A Treatise of Human Nature.* Ed. Ernest C. Mossner. Harmondsworth: Penguin, 1984.
Hunt, Robert. *The Poetry of Science, or Studies of the Physical Phenomena of Nature.* London: Bohn, 1848.
Hutton, James. "Theory of the Earth; or an Investigation of the Laws Observable in the Composition, Dissolution, and Restoration of Land upon the Globe." *Transactions of the Royal Society of Edinburgh* 1 (1788), 209–304.
Hutton, R. H. "George Eliot's Moral Anatomy." *The Spectator*, 45 (October 15, 1872), 1262–64. In *George Eliot: The Critical Heritage.* Ed. David Carroll, 302–5. New York: Barnes and Noble, 1971.

Huxley, T. H. "The Cell-Theory." *British and Foreign Medico-Chirurgical Review* 12 (July–October 1853), 285–314.
Huxley, T. H. *On the Physical Basis of Life*. New Haven: Yale University Press, 1869.
James, Henry. *The Art of Criticism: Henry James on the Theory and Practice of Fiction*. Ed. William M. Veeder and Susan Griffin. Chicago: University of Chicago Press, 1986.
James, Henry. *Literary Criticism: French Writers, Other European Writers, the Prefaces to the New York Edition*. Ed. Leon Edel. New York: Library of America, 1984.
[Jeffrey, Francis.] "*Corinne, ou l'Italie*." *Edinburgh Review* 11: 21 (1807), 183–95.
Kant, Immanuel. *Anthropology, History, and Education*. Ed. Günter Zöller and Robert B. Louden. Cambridge: Cambridge University Press, 2007.
Lamarck, Jean Baptiste. *Philosophie zoologique ou exposition des considérations relatives à l'histoire naturelle des animaux, à la diversité de leur organisation et des facultés qu'ils en obtiennent*. 2 vols. Paris: Dentu, 1809.
Lamarck, Jean Baptiste. *Zoological Philosophy: An Exposition with Regard to the Natural History of Animals*. Trans. Hugh Elliott. London: Macmillan, 1914.
Lang, Andrew. *The Selected Writings of Andrew Lang*. Vol. 3, *Literary Criticism*. Ed. Tom Hubbard. Oxford and New York: Routledge, 2017.
Lawrence, William. *Lectures on Physiology, Zoology, and the Natural History of Man*. London: Benbow, 1822.
Lewes, G. H. *Comte's Philosophy of the Sciences*. London: H. G. Bohn, 1853.
Lewes, G. H. "The Development Hypothesis of the 'Vestiges.'" *The Leader* 4 (August–September 1853), 784–85, 812–14, 832–34, 883–84.
Lewes, G. H. "Dickens in Relation to Criticism." In *Dickens: The Critical Heritage*. Ed. Philip Collins, 569–77. London: Routledge and Kegan Paul, 1971.
Lewes, G. H. *The Life of Goethe*. 2 vols. London: Smith, Elder, 1864.
Lewes, G. H. "Lyell and Owen on Development." *The Leader* 2 (October 1851), 996–97.
Lewes, G. H. "Mr. Darwin's Hypotheses." *Fortnightly Review* (New Series) 3: 16 (1868), 353–73; 3: 18 (1868), 611–28; 4: 19 (1868), 61–80; 4: 23 (1868), 492–509.
Lewes, G. H. *On the Physical Basis of Mind*. London, 1877.
Lewes, G. H. *The Principles of Success in Literature*. Ed. Fred N. Scott. Boston: Allyn and Bacon, 1894.
Lewes, G. H. *Problems of Life and Mind. First Series*. London: Trübner, 1874.
Lewes, G. H. "Realism in Art: Recent German Fiction." *Westminster Review* 70: 138 (October 1858), 271–87.
Lewes, G. H. "Von Baer on the Development Hypothesis." *The Leader* 4 (June 1853), 617–18.
Long, Edward. *The History of Jamaica*. Vol. 2, *Reflections on Its Situation, Settlements, Inhabitants, Climate, Products, Commerce, Laws, and Government*. Montreal: McGill-Queen's University Press, 2000.
Lord, Perceval B. *Popular Physiology: Being a Familiar Explanation of the Most Interesting Facts Connected with the Structure and Function of Animals, and Particularly of Man*. London: Parker, 1834.
Lyell, Charles. *Principles of Geology*. Ed. James A. Secord. London: Penguin, 1997.
Lyon, John, and Phillip R. Sloan, eds. *From Natural History to the History of Nature: Readings from Buffon and His Critics*. Notre Dame: University of Notre Dame Press, 1981.

Lytton, Edward Bulwer. *The Coming Race*. London: Blackwood, 1871.
Mackay, Robert William. *The Progress of the Intellect, as Exemplified in the Religious Development of the Greeks and Hebrews*. 2 vols. London: Chapman, 1850.
MacRitchie, David. *The Testimony of Tradition*. London: Kegan Paul, 1890.
Marx, Karl. *Karl Marx: A Reader*. Ed. John Elster. Cambridge: Cambridge University Press, 1986.
Melville, Herman. *Moby-Dick, or, the Whale*. Ed. Andrew Delbanco and Tom Quirk. London: Penguin, 2003.
Morgenstern, Karl. "On the Nature of the Bildungsroman." 1820. Trans. Tobias Boes. *PMLA* 124: 2 (2009), 647–59.
Müller, F. Max. *Comparative Mythology: An Essay*. London: Routledge, 1856.
Müller, F. Max. *Lectures on the Science of Language, Delivered at the Royal Institution of Great Britain, in April, May, and June 1861*. London: Longmans, 1861.
Müller, F. Max. *Lectures on the Science of Religion*. New York: Scribner, 1874.
[Owen, Richard.] "Darwin on the Origin of Species." *Edinburgh Review* 111: 226 (1860), 487–532.
Owen, Richard. *A History of British Fossil Reptiles*. Vol. 1. London: Cassell, 1849.
Owen, Richard. *On the Anatomy of Vertebrates*. Vol. 3, *Mammals*. London: Longmans, 1868.
Owenson, Sydney. *The Wild Irish Girl: A National Tale*. Ed. Claire Connolly and Stephen Copley. London: Pickering and Chatto, 2000.
Owenson, Sydney (Lady Morgan). *Woman: or Ida of Athens*. London: Longmans, 1809.
Pico della Mirandola, Giovanni. *Oration on the Dignity of Man: A New Translation and Commentary*. Ed. and trans. Francesco Borghesi, Michael Papio, and Massimo Riva. Cambridge: Cambridge University Press, 2012.
Quetelet, M. A. *A Treatise on Man and the Development of His Faculties*. Trans. Robert Knox. Edinburgh: W. & R. Chambers, 1842.
Radcliffe, Ann. *The Mysteries of Udolpho: A Romance*. Ed. Bonamy Dobrée. Oxford: Oxford University Press, 1998.
Reade, William Winwood. *The Martyrdom of Man*. London: Kegan Paul, 1872.
Reid, Thomas. *An Inquiry into the Human Mind, On the Principles of Common Sense*. Ed. Derek R. Brookes. Edinburgh: Edinburgh University Press, 1997.
Riehl, Wilhelm Heinrich. *Die Naturgeschichte des Volkes als Grundlage einer deutschen Social-Politik*. 4 vols. Stuttgart: Cotta, 1851–69.
Rousseau, Jean-Jacques. *Discourse on Political Economy and The Social Contract*. Trans. Christopher Betts. Oxford: Oxford University Press, 1994.
Rousseau, Jean-Jacques. *The Discourses and Other Early Political Writings*. Ed. Victor Gourevitch. Cambridge: Cambridge University Press, 1997.
Rousseau, Jean-Jacques. *Discours sur l'origine et les fondements de l'inégalité parmi les hommes*. Ed. Jean Starobinski. Paris: Gallimard, 1969.
Ruskin, John. *Fiction—Fair and Foul*. New York: John B. Alden, 1885.
Ruskin, John. *Unto This Last: Four Essays on the First Principles of Political Economy*. London: Smith, Elder, 1862.
Saint-Hilaire, Etienne Geoffroy. "Considérations d'où sont déduites des règles pour l'observation des monstres et pour leur classification." In *Philosophie anatomique*. Vol. 2, *Des monstruosités humaines*, 103–23. Paris: Chez l'auteur, 1822.

Saint-Hilaire, Etienne Geoffroy. "Monstre." In *Dictionnaire classique d'histoire naturelle*, vol. 11, 108-51. Paris: Rey et Gravier, 1827.
Saint-Hilaire, Etienne Geoffroy. *Philosophie anatomique*. Vol. 2, *Des monstruosités humaines*. Paris: Chez l'auteur, 1822.
Saint-Hilaire, Etienne Geoffroy. "Recherches sur l'organisation des Gavials, sur leurs affinités naturelles, desquelles résulte la nécessité d'une autre distribution générique." *Mémoires du Muséum d'Histoire naturelle* 12 (1825), 97-155.
Saint-Hilaire, Isidore Geoffroy. *Histoire générale et particulière des anomalies de l'organisation chez l'homme et les animaux*. Paris: J. B. Baillière, 1832-37.
Schiller, Friedrich. *On the Aesthetic Education of Man in a Series of Letters*. Ed. and trans. E. M. Wilkinson and L. A. Willoughby. Oxford: Clarendon Press, 1982.
Schlegel, Friedrich. "From 'Athenaeum Fragments.'" In *Classic and Romantic German Aesthetics*. Ed. J. M. Bernstein, 246-60. Cambridge: Cambridge University Press, 2003.
Schlegel, Friedrich. *Kritische Ausgabe*. Vol. 2, *Schriften aus dem Nachlass / Vorlesungen und Fragmente zur Literatur*. Ed. Ernst Behler, Hans Eichner, and Jean J. Anstett. Munich: Schöningh, 1981.
Schlegel, Friedrich. *Kritische Ausgabe*. Vol. 18, *Philosophische Lehrjahre, I (1796-1806)*. Ed. Ernst Behler. Paderborn, Vienna, and Munich: Schöningh, 1963.
Schlegel, Friedrich. "Letter about the Novel." In *Classic and Romantic German Aesthetics*. Ed. J. M. Bernstein, 287-96. Cambridge: Cambridge University Press, 2003.
Schlegel, Friedrich. "On Goethe's *Meister*." In *Classic and Romantic German Aesthetics*. Ed. J. M. Bernstein, 269-86.
Schmitz, L. Dora, trans. and ed. *Correspondence between Schiller and Goethe, from 1794 to 1805*. 2 vols. London: G. Bell, 1877.
Scott, Walter. *Count Robert of Paris*. Ed. J. H. Alexander. Edinburgh: Edinburgh University Press, 2006.
[Scott, Walter.] "Frankenstein." *Blackwood's Edinburgh Magazine* 2 (March 20/April 1, 1818), 613-20.
Scott, Walter. *Ivanhoe*. Ed. Ian Duncan. Oxford: Oxford University Press, 1996.
Scott, Walter. *Rob Roy*. Ed. Ian Duncan. Oxford: Oxford University Press, 1998.
Scott, Walter. *Waverley*. Ed. P. D. Garside. Edinburgh: Edinburgh University Press, 2007.
[Sedgwick, Adam.] "Vestiges of the Natural History of Creation." *Edinburgh Review* 82: 165 (July 1845), 1-85.
Shelley, Mary. *Frankenstein*. Ed. Marilyn Butler. Oxford: Oxford University Press, 1998.
Shelley, Mary. *Frankenstein: Or, the Modern Prometheus*. 3 vols. London: Colburn and Bentley, 1831.
Shelley, Mary. *The Last Man*. Ed. Morton D. Paley. Oxford: Oxford University Press, 1998.
Simcox, Edith. Review of *Middlemarch*. *The Academy* 4 (1 January 1873), 1-4. In *George Eliot: The Critical Heritage*. Ed. David Carroll, 322-30. New York: Barnes and Noble, 1971.
Spencer, Herbert. *Autobiography*. 2 vols. New York: Appleton, 1904.
Spencer, Herbert. "The Development Hypothesis." In *Essays Scientific, Political & Speculative*. Vol. 1, 1-7. London: Williams and Norgate, 1891.

Spencer, Herbert. "Progress: Its Law and Cause." *Westminster Review* (New Series) 11 (April 1857), 445–85.
Spencer, Herbert. "The Social Organism." *Westminster Review* (New Series) 17 (January 1860), 90–121.
Spencer, Herbert. *Social Statics; Or, the Conditions Essential to Human Happiness.* London: Chapman, 1851.
Sprat, Thomas. *History of the Royal Society.* Ed. Jackson I. Cope and Harold Whitmore Jones. St. Louis: Washington University Studies, 1958.
Staël, Madame de. *Corinne.* London: Bentley, 1833.
Staël, Germaine de. *Corinne, or Italy.* Trans. S. Raphael. Oxford: Oxford University Press, 1998.
Staël, Germaine de. *Corinne ou l'Italie.* Ed. Simone Balayé. Paris: Gallimard, 1985.
Staël, Germaine de. *Literature Considered in Its Relation to Social Institutions.* In *Politics, Literature, and National Character.* Ed. Morroe Berger, 139–256. New Brunswick: Transaction, 2000.
Staël-Holstein, Germaine de. *Germany.* Trans. O. W. Wright. 2 vols. Cambridge, MA: Riverside Press, 1871.
Sterne, Lawrence. *The Life and Opinions of Tristram Shandy, Gentleman.* Ed. Ian Campbell Ross. Oxford: Oxford University Press, 2009.
Stewart, Dugald. *Works.* 7 vols. Cambridge, MA: Hillard and Brown, 1829.
Tylor, Edward Burnett. *Primitive Culture: Researches into the Development of Mythology, Philosophy, Religion, Language, Art and Customs.* 2 vols. London: John Murray, 1871.
Tyndall, John. *Scientific Addresses.* New Haven: Yale University Press, 1870.
Tyson, Edward. *Orang-Outang, sive Homo Sylvestrus: or, The Anatomy of a Pygmy Compared with that of a Monkey, an Ape, and a Man.* London: Bennett and Brown, 1699.
Vaihinger, Hans. *The Philosophy of "As If": A System of the Theoretical, Practical and Religious Fictions of Mankind.* Trans. C. K. Ogden. New York: Harcourt, Brace, 1925.
Virey, Julien-Joseph. *Histoire naturelle du genre humain.* 2 vols. Paris: Dupart, 1801.
Virey, Julien-Joseph. "Monstre." In *Dictionaire des sciences médicales.* Vol. 34, 131–48. Paris: Panckoucke, 1819.
Wagner, Richard. *Der Ring des Nibelungen. Das Rheingold / Die Walküre / Siegfried / Götterdämmerung.* Berlin: Hofenberg, 2016.
Wallace, Alfred Russel. "The Origin of the Human Races and the Antiquity of Man Deduced from the Theory of 'Natural Selection.'" *Journal of the Anthropological Society of London* 2 (1864), 158–70.
Whewell, William. *The Philosophy of the Inductive Sciences: Founded upon Their History.* 2nd ed. 2 vols. London: J. W. Parker, 1847.
Williams, Ioan, ed. *Novel and Romance 1700–1800: A Documentary Record.* London: Routledge & Kegan Paul, 1970.
Wordsworth, William. *The Poems.* Ed. John O. Hayden. 2 vols. New Haven: Yale University Press, 1981.
Wordsworth, William. *The Prelude: A Parallel Text.* Ed. J. C. Maxwell. New Haven: Yale University Press, 1981.

Sources Published after 1900

Abrams, M. H. *Natural Supernaturalism: Tradition and Revolution in Romantic Literature*. New York: Norton, 1971.
Ackerman, Diane. *The Human Age: The World Shaped by Us*. New York: Norton, 2014.
Agamben, Giorgio. *The Open: Man and Animal*. Trans. Kevin Attell. Stanford: Stanford University Press, 2004.
Alexander, J. H. "Essay on the Text." In Walter Scott, *Count Robert of Paris*, ed. J. H. Alexander, 381–439. Edinburgh: Edinburgh University Press, 2006.
Allen, Richard C. *David Hartley on Human Nature*. Albany: SUNY Press, 1999.
Amend-Söchting, Anne. "Corinne ou l'Italie / Corinne et l'Italie: Stratégies autour d'une allégorie." In *Madame de Staël/Corinne ou l'Italie*. Ed. Balayé and Perchellet, 35–51. Paris: Klincksieck, 1999.
Amrine, Frederick. "Rethinking the *Bildungsroman*." *Michigan Germanic Studies* 13: 2 (1987), 119–39.
Anderson, Amanda. *Bleak Liberalism*. Chicago: University of Chicago Press, 2016.
Anderson, Amanda. *The Powers of Distance: Cosmopolitanism and the Cultivation of Detachment*. Princeton: Princeton University Press, 2001.
Anderson, Benedict. *Imagined Communities: Reflections on the Origin and Spread of Nationalism*. London: Verso, 1983.
Anger, Suzy. *Victorian Interpretation*. Ithaca: Cornell University Press, 2005.
Appel, Toby. *The Cuvier-Geoffroy Debate: French Biology in the Decades before Darwin*. New York: Oxford University Press, 1987.
Arac, Jonathan. *Impure Worlds: The Institution of Literature in the Age of the Novel*. New York: Fordham University Press, 2010.
Arac, Jonathan. "What Kind of History Does a Theory of the Novel Require?" *Novel* 42: 2 (2009), 190–95.
Arendt, Hannah. *The Human Condition*. Chicago: University of Chicago Press, 1998.
Armstrong, Nancy. *Desire and Domestic Fiction: A Political History of the Novel*. New York: Oxford University Press, 1987.
Armstrong, Nancy. *How Novels Think: The Limits of British Individualism from 1719–1900*. New York: Columbia University Press, 2005.
Armstrong, Nancy, and Leonard Tennenhouse. *Novels in the Time of Democratic Writing: The American Example*. Philadelphia: University of Pennsylvania Press, 2017.
Bakhtin, M. M. *Speech Genres and Other Late Essays*. Ed. C. Emerson and M. Holquist, trans. Vern W. McGee. Austin: University of Texas Press, 1986.
Balayé, Simone, and Jean-Pierre Perchellet, ed. *Madame de Staël/Corinne ou l'Italie*. Paris: Klincksieck, 1999.
Baldick, Chris. *In Frankenstein's Shadow: Myth, Monstrosity, and Nineteenth-Century Writing*. Oxford: Clarendon Press, 1987.
Banfield, Ann. *Unspeakable Sentences: Narration and Representation in the Language of Fiction*. London: Routledge, 1982.
Barnard, F. M. *Herder on Nationality, Humanity, and History*. Toronto: McGill-Queen's University Press, 2003.
Barrell, John. *English Literature in History, 1730–80: An Equal, Wide Survey*. London: Hutchinson, 1980.

Bartfeld, Fernande. "Mouvance, mutation et progrès dans *Notre-Dame de Paris*." *Les Lettres Romanes* 47: 1–2 (1993), 33–39.
Bataille, Georges. *Visions of Excess: Selected Writings, 1927–1939*. Trans. Allan Stoekl, Carl R. Lovitt, and Donald M. Leslie Jr. Minneapolis: University of Minnesota Press, 1989.
Bates, David. *States of War: Enlightenment Origins of the Political*. New York: Columbia University Press, 2012.
Bechtel, William. *Discovering Cell Mechanisms: The Creation of Modern Cell Biology*. Cambridge: Cambridge University Press, 2006.
Beddow, Michael. *The Fiction of Humanity: Studies in the Bildungsroman from Wieland to Thomas Mann*. Cambridge: Cambridge University Press, 1982.
Beenstock, Zoe. *The Politics of Romanticism: The Social Contract and Literature*. Edinburgh: Edinburgh University Press, 2016.
Beer, Gillian. *Darwin's Plots: Evolutionary Narrative in Darwin, George Eliot and Nineteenth-Century Fiction*. 3rd ed. Cambridge: Cambridge University Press, 2009.
Beer, Gillian. *Open Fields: Science in Cultural Encounter*. Oxford: Oxford University Press, 1996.
Behler, Ernst. *German Romantic Literary Theory*. Cambridge: Cambridge University Press, 1993.
Behler, Ernst. "Madame de Staël and Goethe." In *The Spirit of Poesy: Essays on Jewish and German Literature and Thought in Honor of Géza von Molnar*. Ed. Richard A. Block and Peter David Fenves, 131–49. Evanston: Northwestern University Press, 2000.
Behler, Ernst. "The Theory of Irony in German Romanticism." In *Romantic Irony*. Ed. Frederick Garber, 43–81. Amsterdam: Benjamins, 1988.
Beiser, Frederick C. *The Fate of Reason: German Philosophy from Kant to Fichte*. Cambridge, MA: Harvard University Press, 1987.
Beiser, Frederick C. *Schiller as Philosopher*. Oxford: Oxford University Press, 2005.
Bell, Bill, ed. *The Edinburgh History of the Book in Scotland.*, Vol. 3, *Ambition and Industry, 1800–80*. Edinburgh: Edinburgh University Press, 2007.
Bender, John. *Ends of Enlightenment*. Stanford: Stanford University Press, 2012.
Best, Stephen, and Sharon Marcus. "Surface Reading: An Introduction." *Representations* 108: 1 (2009), 1–21.
Bevir, Mark, ed. *Historicism and the Human Sciences in Victorian Britain*. Cambridge: Cambridge University Press, 2017.
Bewell, Alan. "Jefferson's Thermometer: Colonial Biogeographical Constructions of the Climate of America." In *Romantic Science: The Literary Forms of Natural History*. Ed. Noah Heringman, 111–38. Albany: State University of New York Press, 2003.
Bewell, Alan. *Wordsworth and the Enlightenment: Nature, Man and Society in the Experimental Poetry*. New Haven: Yale University Press, 1989.
Blair, Kirstie. "A Change in the Units: *Middlemarch*, G. H. Lewes, and Rudolf Virchow." *George Eliot–George Henry Lewes Studies* 39–40 (September 2001), 9–24.
Blair, Kirstie. "Contagious Sympathies: George Eliot and Rudolf Virchow." In *Unmapped Countries: Biological Visions in Nineteenth-Century Literature and Culture*. Ed. Anne-Julia Zwierlein, 145–54. London: Anthem Press, 2005.
Bode, Christoph. "Absolut Jena: A Second Look at Lacoue-Labarthe's and Nancy's Representations of the Literary Theory of *Frühromantik*." In *Romanticism and*

Philosophy: Thinking with Literature. Ed. Sophie Laniel-Musitelli and Thomas Constantinesco, 19–39. Oxford: Routledge, 2015.

Bode, Christoph. *The Novel*. Trans. James Vigus. Oxford: Wiley-Blackwell, 2011.

Boehm, Katharina. *Charles Dickens and the Sciences of Childhood: Popular Medicine, Child Health and Victorian Culture*. New York: Palgrave Macmillan, 2013.

Boes, Tobias. "Apprenticeship of the Novel: The Bildungsroman and the Invention of History, ca. 1770–1820." *Comparative Literature Studies* 45: 3 (2008), 269–88.

Boes, Tobias. *Formative Fictions: Nationalism, Cosmopolitanism, and the Bildungsroman*. Ithaca: Cornell University Press, 2012.

Bowen, John. *Other Dickens: Pickwick to Chuzzlewit*. Oxford: Oxford University Press, 2000.

Bowler, Peter J. *The Non-Darwinian Revolution: Reinterpreting a Historical Myth*. Baltimore: Johns Hopkins University Press, 1988.

Brain, R. M. "Protoplasmania: Huxley, Haeckel, and the Vibratory Organism in Fin de Siècle Visual Cultures." In *The Art of Evolution: Darwin, Darwinisms, and Visual Cultures*. Ed. F. Brauer and B. Larson, 92–123. Lebanon, NH: University Presses of New England, 2009.

Brandt, Reinhardt. "The Guiding Idea of Kant's Anthropology and the Vocation of the Human Being." In *Essays on Kant's Anthropology*. Ed. Brian Jacobs and Patrick Kain, 85–104. Cambridge: Cambridge University Press, 2003.

Brantlinger, Patrick. *Rule of Darkness: British Literature and Imperialism, 1830–1914*. Ithaca: Cornell University Press, 1990.

Brassier, Ray. *Nihil Unbound: Enlightenment and Extinction*. Basingstoke: Palgrave Macmillan, 2007.

Brewer, David. *The Afterlife of Character, 1726–1825*. Philadelphia: University of Pennsylvania Press, 2011.

Brière, Chantal. *Victor Hugo et le roman architectural*. Paris: Champion, 2007.

Brilmyer, S. Pearl. "Plasticity, Form, and the Matter of Character in *Middlemarch*." *Representations* 130: 1 (2015), 60–83.

Brombert, Victor. *The Hidden Reader: Stendhal, Balzac, Hugo, Baudelaire, Flaubert*. Cambridge, MA: Harvard University Press, 1988.

Brombert, Victor. *Victor Hugo and the Visionary Novel*. Cambridge, MA: Harvard University Press, 1984.

Brooks, Douglas. *Number and Pattern in the Eighteenth-Century Novel: Defoe, Fielding, Smollett, and Sterne*. London: Routledge and Kegan Paul, 1973.

Brown, Jane K. *Goethe's Allegories of Identity*. Philadelphia: University of Pennsylvania Press, 2014.

Brown, Laura. *Homeless Dogs and Melancholy Apes: Humans and Other Animals in the Modern Literary Imagination*. Ithaca: Cornell University Press, 2010.

Brown, Marshall. *The Shape of German Romanticism*. Ithaca: Cornell University Press, 1979.

Brown, Marshall. "Theory of the Novel." In *The Cambridge History of Literary Criticism*. Vol. 5, *Romanticism*. Ed. Marshall Brown, 250–71. Cambridge: Cambridge University Press, 2000.

Browne, Janet. *Charles Darwin: Voyaging*. Princeton: Princeton University Press, 1995.

Buckland, Adelene. *Novel Science: Fiction and the Invention of Nineteenth-Century Geology.* Chicago: University of Chicago Press, 2013.

Buckland, Adelene. "'The Poetry of Science': Charles Dickens, Geology, and Visual and Material Culture in Victorian London." *Victorian Literature and Culture* 35: 2 (2007), 679–94.

Burkhardt, Richard, Jr. "Lamarck, Evolution, and the Politics of Science." *Journal of the History of Biology* 3: 2 (1970), 275–98.

Buzard, James. *Disorienting Fiction: The Autoethnographic Work of Nineteenth-Century British Novels.* Princeton: Princeton University Press, 2005.

Cagidemetrio, Alide. "A Plea for Fictional Histories and Old-Time 'Jewesses.'" In *The Invention of Ethnicity.* Ed. Werner Sollors, 13–43. Oxford: Oxford University Press, 1989.

Calè, Luisa, and Adriana Craciun. "The Disorder of Things." *Eighteenth-Century Studies* 45: 1 (2011), 1–13.

Campbell, Timothy. *Historical Style: Fashion and the New Mode of History, 1740–1830.* Philadelphia: University of Pennsylvania Press, 2016.

Campe, Rüdiger. "Form and Life in the Theory of the Novel." *Constellations* 18: 1 (2011), 53–66.

Canguilhem, Georges. *Knowledge of Life.* Ed. Paola Marrati and Todd Meyers, trans. Stefanos Geroulanos and Daniela Ginsberg. New York: Fordham University Press, 2008.

Carey, John. *The Violent Effigy: A Study of Dickens's Imagination.* London: Faber, 1973.

Carrithers, David. "The Enlightenment Science of Society." In *Inventing Human Science: Eighteenth-Century Domains.* Ed. Christopher Fox, Roy Porter, and Robert Wokler, 232–70. Berkeley: University of California Press, 1995.

Carroll, David. *George Eliot: The Critical Heritage.* New York: Barnes and Noble, 1971.

Carroll, Siobhan. *An Empire of Air and Water: Uncolonizable Space in the British Imagination, 1750–1850.* Philadelphia: University of Pennsylvania Press, 2015.

Casanova, Pascale. *The World Republic of Letters.* Trans. M. B. DeBevoise. Cambridge, MA: Harvard University Press, 2004.

Cave, Terence. *Mignon's Afterlives: Crossing Cultures from Goethe to the Twenty-First Century.* Oxford: Oxford University Press, 2011.

Chai, Leon. "Life and Death in Paris: Medical and Life Sciences in the Romantic Era." In *The Oxford Handbook of European Romanticism.* Ed. Paul Hamilton, 712–29. Oxford: Oxford University Press, 2016.

Chakrabarty, Deepak. "The Climate of History: Four Theses." *Critical Inquiry* 35: 2 (2009), 197–222.

Chandler, James. *England in 1819: The Politics of Literary Culture and the Case of Romantic Historicism.* Chicago: University of Chicago Press, 1998.

Chapman, Alison. *Networking the Nation: British and American Women's Poetry and Italy, 1840–1870.* Oxford: Oxford University Press, 2015.

Chase, Cynthia. "The Decomposition of the Elephants: Double-Reading *Daniel Deronda.*" *PMLA* 93: 2 (1978), 215–27.

Chico, Tita. *The Experimental Imagination: Literary Knowledge and Science in the British Enlightenment.* Stanford: Stanford University Press, 2018.

Christensen, Jerome. *Romanticism at the End of History.* Baltimore: Johns Hopkins University Press, 2000.

Clayton, Jay. *Romantic Vision and the Novel*. Cambridge: Cambridge University Press, 1987.
Cleary, Joe. "Realism after Modernism and the Literary World System." In "Peripheral Realisms," ed. Joe Cleary, Jed Esty, and Colleen Lye. Special issue, *Modern Language Quarterly* 73: 3 (2012), 255–68.
Clubb, William Graham. "Quasimodo, Quasi-man: A Man of the Woods in Victor Hugo's *Notre Dame de Paris*." *Italian Quarterly* 37: 143–46 (2000), 267–80.
Cohen, Margaret, and Carolyn Dever. *The Literary Channel: The Inter-National Invention of the Novel*. Princeton: Princeton University Press, 2002.
Cohn, Dorrit. *Transparent Minds: Narrative Modes for Presenting Consciousness in Fiction*. Princeton: Princeton University Press, 1978.
Cohn, Dorrit, and Lewis S. Gleich. "Metalepsis and *Mise en Abyme*." *Narrative* 20: 1 (2012), 105–14.
Cohn, Elisha. *Still Life: Suspended Development in the Victorian Novel*. Oxford: Oxford University Press, 2015.
Coleman, Patrick. *Reparative Realism: Mourning and Modernity in the French Novel, 1730–1830*. Geneva: Librarie Droz, 1998.
Collins, Philip, ed. *Dickens: The Critical Heritage*. London: Routledge and Kegan Paul, 1971.
Colvin, Howard. Review of *Space and the Eighteenth-Century English Novel*, by Simon Varey. *Review of English Studies* (New Series) 44: 174 (1993), 261–62.
Corbett, Mary Jean. *Allegories of Union in Irish and English Writing, 1790–1870: Politics, History, and the Family from Edgeworth to Arnold*. Cambridge: Cambridge University Press, 2000.
Corsi, Pietro. *The Age of Lamarck: Evolutionary Theories in France, 1790–1830*. Berkeley: University of California Press, 1988.
Corsi, Pietro. "Before Darwin: Transformist Concepts in European Natural History." *Journal of the History of Biology* 38: 1 (2005), 67–83.
Coyer, Megan. *Literature and Medicine in the Nineteenth-Century Periodical Press: Blackwood's Edinburgh Magazine, 1817–1858*. Edinburgh: Edinburgh University Press, 2017.
Crawford, Robert. *Devolving English Literature*. Oxford: Clarendon Press, 1992.
Crawford, Robert, ed. *The Scottish Invention of English Literature*. Cambridge: Cambridge University Press, 1998.
Crawford, T. Hugh. "Networking the Nonhuman: *Moby-Dick*, Mathew Fontaine Maury, and Bruno Latour." *Configurations* 5: 1 (1997), 1–21.
Culler, Jonathan. "Omniscience." *Narrative* 12: 1 (2004), 22–34.
Culler, Jonathan. *Theory of the Lyric*. Cambridge, MA: Harvard University Press, 2015.
Curran, Andrew. "Logics of the Human in Diderot's *Supplément au Voyage de Bougainville*." In *New Essays on Diderot*. Ed. James Fowler, 158–72. Cambridge: Cambridge University Press, 2011.
Dames, Nicholas. *The Physiology of the Novel: Reading, Neural Science, and the Form of Victorian Fiction*. Oxford: Oxford University Press, 2007.
Daudin, H. *Cuvier et Lamarck: Les Classes zoologiques et l'idée de série animale*. 2 vols. Paris: Alcan, 1926–27.

Davidson, Jenny. *Breeding: A Partial History of the Eighteenth Century*. New York: Columbia University Press, 2008.
Davies, Jeremy. *The Birth of the Anthropocene*. Oakland: University of California Press, 2016.
Davies, Kate. "Pantomime, Connoisseurship, Consumption: Emma Hamilton and the Politics of Embodiment." *CW3* (Corvey Women Writers on the Web), http://www2.shu.ac.uk/corvey/cw3journal/issue%20two/davies.html.
Dawson, Benjamin. "Science and the Scientific Disciplines." In *The Oxford Handbook of European Romanticism*. Ed. Paul Hamilton, 684–709. Oxford: Oxford University Press, 2016.
Dawson, Gowan. "'By a Comparison of Incidents and Dialogue': Richard Owen, Comparative Anatomy and Victorian Serial Fiction." *19: Interdisciplinary Studies in the Long Nineteenth-Century* 11 (2010), http://doi.org/10.16995/ntn.577.
Dawson, Gowan. "Dickens, Dinosaurs, and Design." *Victorian Literature and Culture* 44: 4 (2016), 761–78.
Dawson, Gowan. "Literary Megatheriums and Loose Baggy Monsters: Paleontology and the Victorian Novel." *Victorian Studies* 53: 2 (2011), 203–30.
Dawson, Paul. *The Return of the Omniscient Narrator: Authorship and Authority in Twenty-First Century Fiction*. Columbus: The Ohio State University Press, 2013.
Deane, Seamus. *Strange Country: Modernity and Nationhood in Irish Writing since 1790*. Oxford: Clarendon Press, 1997.
De Man, Paul. *Aesthetic Ideology*. Ed. Andrzej Warminski. Minneapolis: University of Minnesota Press, 1996.
De Man, Paul. *Allegories of Reading: Figural Language in Rousseau, Nietzsche, Rilke, and Proust*. New Haven: Yale University Press, 1982.
DeJean, Joan. "Staël's *Corinne*: The Novel's Other Dilemma." In *The Novel's Seductions: Staël's Corinne in Critical Inquiry*. Ed. Karyna Szmurlo, 117–26. Lewisburg: Bucknell University Press, 1999.
Deneys-Tunney, Anne. "*Corinne* by Madame de Staël: The Utopia of Feminine Voice as Music within the Novel." *Dalhousie French Studies* 28 (1994), 55–63.
Derrida, Jacques. "The Ends of Man." In *Margins of Philosophy*. Trans. Alan Bass, 111–36. Chicago: University of Chicago Press, 1972.
Desmond, Adrian. *The Politics of Evolution: Morphology, Medicine and Reform in Radical London*. Chicago: University of Chicago Press, 1989.
DeSouza, Nigel. "Language, Reason, and Sociability: Herder's Critique of Rousseau." *Intellectual History Review* 22: 2 (2012), 221–40.
De Witt, Anne. *Moral Authority, Men of Science, and the Victorian Novel*. Cambridge: Cambridge University Press, 2013.
Dolin, Tim. "Eliot and Victorian Science." In *George Eliot: Authors in Context*, 190–215. Oxford: Oxford University Press, 2005.
Douthwaite, Julia V. *The Wild Girl, Natural Man, and the Monster: Dangerous Experiments in the Age of Enlightenment*. Chicago: University of Chicago Press, 2010.
Duff, David. *Romanticism and the Uses of Genre*. Oxford: Oxford University Press, 2009.
Duff, David, and Catherine Jones, eds. *Scotland, Ireland, and the Romantic Aesthetic*. Lewisburg: Bucknell University Press, 2007.

Duncan, Ian. *Modern Romance and Transformations of the Novel: The Gothic, Scott, Dickens.* Cambridge: Cambridge University Press, 1992.
Duncan, Ian. "The Regional or Provincial Novel." In *A Companion to the Victorian Novel.* Ed. Patrick Brantlinger and William B. Thesing, 318–35. Oxford: Blackwell, 2002.
Duncan, Ian. *Scott's Shadow: The Novel in Romantic Edinburgh.* Princeton: Princeton University Press, 2007.
Düntzer, Heinrich. *Life of Goethe.* Trans. Thomas W. Lyster. London: Unwin, 1908.
Durkheim, Émile. "The Dualism of Human Nature and Its Social Conditions." Trans. I. Eulriet and W. Watts Miller. *Durkheimian Studies/Études Durkheimiennes* 11 (2005), 35–45.
Eagleton, Terry. *Heathcliff and the Great Hunger: Studies in Irish Culture.* London: Verso, 1995.
Ermarth, Elizabeth Deeds. *Realism and Consensus in the English Novel: Time, Space and Narrative.* Edinburgh: Edinburgh University Press, 1998.
Esterhammer, Angela. "Continental Literature, Translation, and the Johnson Circle." *The Wordsworth Circle* 33: 3 (2002), 101–4.
Esterhammer, Angela. *Romanticism and Improvisation, 1750–1850.* Cambridge: Cambridge University Press, 2008.
Esty, Jed, and Colleen Lye. "Peripheral Realisms Now." In "Peripheral Realisms," ed. Joe Cleary, Jed Esty, and Colleen Lye. Special issue, *Modern Language Quarterly* 73: 3 (2012), 269–87.
Esty, Jed. *Unseasonable Youth: Modernism, Colonialism and the Fiction of Development.* Oxford: Oxford University Press, 2004.
Fabian, Johannes. *Time and the Other: How Anthropology Makes Its Object.* New York: Columbia University Press, 1983.
Fairweather, Maria. *Madame de Staël.* London: Constable, 2005.
Farina, Jonathan. *Everyday Words and the Character of Prose in Nineteenth-Century Britain.* Cambridge: Cambridge University Press, 2017.
Faull, Katherine M., ed. *Anthropology and the German Enlightenment: Perspectives on Humanity.* Lewisburg: Bucknell University Press, 1995.
Feltes, N. N. *Modes of Production of Victorian Novels.* Chicago: University of Chicago Press, 1989.
Ferguson, Frances. "Jane Austen, Emma, and the Impact of Form." *MLQ: Modern Language Quarterly* 61: 1 (2000), 157–80
Ferris, Ina. *The Achievement of Literary Authority: Gender, History and the Waverley Novels.* Ithaca: Cornell University Press, 1991.
Festa, Lynn. *Sentimental Figures of Empire in Eighteenth-Century Britain and France.* Baltimore: Johns Hopkins University Press, 2007.
Fleishman, Avrom. *George Eliot's Intellectual Life.* Cambridge: Cambridge University Press, 2010.
Fleming, Paul. *Exemplarity and Mediocrity: The Art of the Average from Bourgeois Tragedy to Realism.* Stanford: Stanford University Press, 2011.
Fludernik, Monika. *The Fictions of Language and the Languages of Fiction: The Linguistic Representation of Speech and Consciousness.* London and New York: Routledge, 1993.

Fludernik, Monika. "The Linguistic Illusion of Alterity: The Free Indirect as Paradigm of Discourse Representation." *Diacritics* 25: 4 (1995), 89–115.
Forster, E. M. *Aspects of the Novel*. New York: Harcourt, Brace, 1927.
Forster, Michael N. *Herder's Philosophy*. Oxford: Oxford University Press, 2018.
Foucault, Michel. *Abnormal: Lectures at the Collège de France, 1974–1975*. Ed. Valerio Marchetti and Antonella Salomoni, trans. Graham Burchell. London: Picador, 2004.
Foucault, Michel. *Introduction to Kant's Anthropology*. Ed. Roberto Nigro. Trans. Kate Briggs and Roberto Nigro. Los Angeles: semiotext(e), 2008.
Foucault, Michel. *The Order of Things: An Archaeology of the Human Sciences*. New York: Vintage Books, 1994.
Fox, Christopher, Roy Porter, and Robert Wokler, eds. *Inventing Human Science: Eighteenth-Century Domains*. Berkeley: University of California Press, 1995.
Fraiman, Susan. *Unbecoming Women: British Women Writers and the Novel of Development*. New York: Columbia University Press, 1993.
Frank, Lawrence. *Victorian Detective Fiction and the Nature of Evidence: The Scientific Investigations of Poe, Dickens, and Doyle*. Basingstoke: Palgrave, 2003.
Franklin, Caroline. *Female Romantics: Nineteenth-Century Women Novelists and Byronism*. London: Routledge, 2012.
Fredrickson, Kathleen. *The Ploy of Instinct: Victorian Sciences of Nature and Sexuality in Liberal Governance*. New York: Fordham University Press, 2014.
Freedgood, Elaine. *The Ideas in Things: Fugitive Meaning in the Victorian Novel*. Chicago: University of Chicago Press, 2006.
Friedenthal, Richard. *Goethe: His Life and Times*. London: Routledge, 2007.
Fulford, Tim, Debbie Lee and Peter J. Kitson. *Literature, Science and Exploration in the Romantic Era*. Cambridge: Cambridge University Press, 2004.
Furneaux, Holly, and Ben Winyard. "Dickens, Science and the Victorian Literary Imagination." *19: Interdisciplinary Studies in the Long Nineteenth Century* 10 (April 2010), http://doi.org/10.16995/ntn.572.
Furst, Lilian R. *Between Doctors and Patients: The Changing Balance of Power*. Charlottesville: University of Virginia Press, 1998.
Gallagher, Catherine. *The Body Economic: Life, Death, and Sensation in Political Economy and the Victorian Novel*. Princeton: Princeton University Press, 2006.
Gallagher, Catherine. "Formalism and Time." *MLQ: Modern Language Quarterly* 61: 1 (2000), 229–51.
Gallagher, Catherine. "George Eliot, Immanent Victorian." *Representations* 90: 1 (2005), 61–74.
Gallagher, Catherine. "The Rise of Fictionality." In *The Novel*. Vol. 1, *History, Geography, and Culture*. Ed. Franco Moretti, 336–63. Princeton: Princeton University Press, 2006.
Gamerschlag, Kurt. "The Making and Unmaking of Sir Walter Scott's *Count Robert of Paris*." *Studies in Scottish Literature* 15: 1 (1980), 95–123.
Garside, P. D., and Rainer Schöwerling. *The English Novel 1770–1829: A Bibliographical Survey of Prose Fiction Published in the British Isles*. Vol. 2, *1800–1829*. Oxford: Oxford University Press, 2000.
Gaukroger, Stephen. *The Natural and the Human: Science and the Shaping of Modernity, 1739–1841*. Oxford: Oxford University Press, 2016.

Genette, Gérard. *Narrative Discourse: An Essay in Method*. Trans. Jane E. Lewin. Ithaca: Cornell University Press, 1980.
Ghosh, Amitav. *The Great Derangement: Climate Change and the Unthinkable*. Chicago: University of Chicago Press, 2016.
Gibson, Anna. "*Our Mutual Friend* and Network Form." *Novel: A Forum on Fiction* 48:1 (2015), 63–84.
Gigante, Denise. "Facing the Ugly: The Case of *Frankenstein*." *ELH* 67: 2 (2000), 565–87.
Gigante, Denise. *Life: Organic Form and Romanticism*. New Haven: Yale University Press, 2009.
Gjesdal, Kristin. "Human Nature and Human Science: Herder and the Anthropological Turn in Hermeneutics." In *Herder: Philosophy and Anthropology*. Ed. Anik Waldow and Nigel DeSouza, 166–84. Oxford: Oxford University Press, 2017.
Goldstein, Amanda Jo. "Irritable Figures: Herder's Poetic Empiricism." In *The Relevance of Romanticism: Essays on German Romantic Philosophy*. Ed. Dalia Nassar, 273–95. Oxford: Oxford University Press, 2014.
Goldstein, Amanda Jo. *Sweet Science: Romantic Materialism and the New Logics of Life*. Chicago: University of Chicago Press, 2017.
Golinski, Jan. *Making Natural Knowledge: Constructivism and the History of Science*. Cambridge: Cambridge University Press, 1998.
Goodlad, Lauren M. E. "The *Longue Durée* of Political *Bildungsromane*: Putting George Eliot into Dialogue with Danish Television." In *Replotting Marriage in Nineteenth-Century British Literature*. Ed. Jill Galvan and Elsie Michie, 100–126. Columbus: Ohio State University Press, 2018.
Goodlad, Lauren M. E. *The Victorian Geopolitical Aesthetic: Realism, Sovereignty, and Transnational Experience*. Oxford: Oxford University Press, 2015.
Goodman, Kevis. *Pathologies of Motion: Medicine, Aesthetics, Poetics*. New Haven: Yale University Press, forthcoming.
Graver, Suzanne. *George Eliot and Community: A Study in Social Theory and Fictional Form*. Berkeley: University of California Press, 1984.
Greif, Mark. *The Age of the Crisis of Man: Thought and Fiction in America, 1933–1973*. Princeton: Princeton University Press, 2015.
Greiner, Rae. *Sympathetic Realism in Nineteenth-Century British Fiction*. Baltimore: Johns Hopkins University Press, 2012.
Griffiths, Devin. *The Age of Analogy: Science and Literature between the Darwins*. Baltimore: Johns Hopkins University Press, 2016.
Grosz, Elizabeth. *Becoming Undone: Darwinian Reflections on Life, Politics, and Art*. Durham: Duke University Press, 2011.
Grosz, Elizabeth. *The Nick of Time: Politics, Evolution, and the Untimely*. Durham: Duke University Press, 2004.
Guerlac, Suzanne. "Writing the Nation (Mme. de Staël)." *French Forum* 30: 3 (2005), 43–56.
Hack, Daniel. "Sublimation Strange: Allegory and Authority in *Bleak House*." *ELH* 66: 1 (1999), 129–56.
Haight, Gordon S. "Dickens and Lewes on Spontaneous Combustion." *Nineteenth-Century Fiction* 10: 1 (1955), 53–63.

Hamburger, Käte. *The Logic of Literature*. 2nd ed. Trans. Marilynne J. Rose. Bloomington: Indiana University Press, 1973.

Hamilton, Paul. *Metaromanticism: Aesthetics, Literature, Theory*. Chicago: University of Chicago Press, 2008.

Handwerk, Gary. "Romantic Irony." In *The Cambridge History of Literary Criticism*. Vol. 5, *Romanticism*. Ed. Marshall Brown, 203–25. Cambridge: Cambridge University Press, 2000.

Hardy, Barbara. *The Novels of George Eliot: A Study in Form*. London: Athlone Press, 1959.

Harvey, W. J. "*Bleak House*: The Double Narrative." In *Bleak House: A Casebook*. Ed. A. E. Dyson, 224–34. London: Macmillan, 1969.

Harvey, W. J. "The Intellectual Background of the Novel: Casaubon and Lydgate." In *"Middlemarch": Critical Approaches to the Novel*. Ed. Barbara Hardy, 25–37. London: Athlone Press, 1967.

Heffernan, Julián Jiménez. "The Stamp of Rarity: Ancestrality and Extinction in *Daniel Deronda*." *Representations* 144: 1 (2018), 90–123.

Heidegger, Martin. *The Fundamental Concepts of Metaphysics*. Trans. William McNeill and Nicholas Walker. Bloomington: Indiana University Press, 1995.

Heidegger, Martin. *On the Essence of Language: The Metaphysics of Language and the Essencing of the Word Concerning Herder's Treatise on the Origin of Language*. Trans. Wanda Torres Gregory and Yvonne Unna. Albany: SUNY Press, 2004.

Heller, Deborah. *Literary Sisterhoods: Imagining Women Artists*. Montreal: McGill-Queen's University Press, 2005.

Henchman, Anna. *The Starry Sky Within: Astronomy and the Reach of the Mind in Victorian Literature*. Oxford: Oxford University Press, 2014.

Hensley, Nathan K., and John Patrick James. "Soot Moth: *Biston Betularia* and the End of Nature." In *BRANCH: Britain, Representation and Nineteenth-Century History*. Ed. Dino Franco Felluga. Extension of *Romanticism and Victorianism on the Net*. Uploaded August 27, 2018. http://www.branchcollective.org/?ps_articles=nathan-k-hensley-and-john-patrick-james-soot-moth-biston-betularia-and-the-victorian-end-of-nature.

Herbert, Christopher. "The Occult in *Bleak House*." *Novel* 17: 2 (1984), 101–15.

Heringman, Noah. "The Commerce of Literature and Natural History." In *Romantic Science: The Literary Forms of Natural History*. Ed. Noah Heringman, 1–19. Albany: State University of New York Press, 2003.

Heringman, Noah. "Deep Time at the Dawn of the Anthropocene." *Representations* 129: 1 (2015), 56–85.

Heringman, Noah. *Sciences of Antiquity: Romantic Antiquarianism, Natural History and Knowledge Work*. Oxford: Oxford University Press, 2013.

Hertz, Neil. *George Eliot's Pulse*. Stanford: Stanford University Press, 2005.

Hesketh, Ian. "The Future Evolution of 'Man.'" In *Historicizing Humans: Deep Time, Evolution, and Race in Nineteenth-Century British Sciences*. Ed. Efram Sera-Shriar, 193–217. Pittsburgh: University of Pittsburgh Press, 2018.

Hibberd, Sarah. "Monsters and the Mob: Depictions of the Grotesque on the Parisian Stage, 1826–1836." In *Textual Intersections: Literature, History and the Arts in Nineteenth-Century Europe*. Ed. Rachael Langford, 29–40. Amsterdam: Rodopi, 2009.

Hilles, Frederick W. "Art and Artifice in *Tom Jones*." In Henry Fielding, *Tom Jones*, ed. Sheridan Baker, 916–32. New York: Norton, 1973.

Hillman, Susanne. "Men with Muskets, Women with Lyres: Nationality, Citizenship, and Gender in the Writings of Germaine de Staël." *Journal of the History of Ideas* 72: 2 (2011), 231–54.

Hirsch, Marianne. "Spiritual Bildung: The 'Beautiful Soul' as Paradigm." In *The Voyage In: Fictions of Female Development*. Ed. Elizabeth Abel, M. Hirsch and Elizabeth Langland, 23–48. Hanover: University Press of New Hampshire, 1983.

Hobsbaum, Philip. "Scott's 'Apoplectic' Novels." In *Scott and His Influence*. Ed. J. H. Alexander and David Hewitt, 149–56. Aberdeen: Association for Scottish Literary Studies, 1983.

Hooker, Kenneth W. *The Fortunes of Victor Hugo in England*. New York: Columbia University Press, 1938.

Huet, Marie. *Monstrous Imagination*. Cambridge, MA: Harvard University Press, 1993.

Iser, Wolfgang. *The Fictive and the Imaginary: Charting Literary Anthropology*. Baltimore: Johns Hopkins University Press, 1993.

Jacobs, Brian. "Kantian Character and the Problem of a Science of Humanity." In *Essays on Kant's Anthropology*. Ed. Brian Jacobs and Patrick Kain, 105–34. Cambridge: Cambridge University Press, 2003.

Jacobs, Brian, and Patrick Kain, eds. *Essays on Kant's Anthropology*. Cambridge: Cambridge University Press, 2003.

Jacobs, Edward. "Ann Radcliffe and Romantic Print Culture." In *Ann Radcliffe, Romanticism and the Gothic*. Ed. Dale Townshend and Angela Wright, 49–66. Cambridge: Cambridge University Press, 2014.

Jacyna, L. S. "The Romantic Programme and the Reception of Cell Theory in Britain." *Journal of the History of Biology* 17: 1 (1984), 13–48.

Jaeck, Emma Gertrude. *Madame de Staël and the Spread of German Literature*. New York: Oxford University Press, 1915.

Jaffe, Audrey. *The Affective Life of the Average Man: The Victorian Novel and the Stock-Market Graph*. Columbus: Ohio State University Press, 2010.

Jaffe, Audrey. *Vanishing Points: Dickens, Narrative, and the Subject of Omniscience*. Berkeley: University of California Press, 1991.

Jameson, Fredric. *The Antinomies of Realism*. London: Verso, 2013.

Jenkins, Alice. "George Eliot, Geometry and Gender." In *Literature and Science*. Ed. Sharon Ruston, 72–90. Woodbridge: D. S. Brewer, 2008.

Jones, Catherine. "Madame de Staël and Scotland: *Corinne*, Ossian and the Science of Nations." *Romanticism* 15: 3 (2009), 239–53.

Jordan, John. *Supposing "Bleak House."* Charlottesville: University of Virginia Press, 2012.

Joshi, Priya. *In Another Country: Colonialism, Culture and the English Novel in India*. New York: Columbia University Press, 2002.

Kaplan, Cora. "Aurora Leigh." In *Feminist Criticism and Social Change: Sex, Class and Race in Literature and Culture*. Ed. Judith Lowder Newton and Deborah Silverton Rosenfelt, 146–51. London: Routledge, 2013.

Kareem, Sarah. "Lost in the Castle of Scepticism: Sceptical Philosophy as Gothic Romance." In *Fictions of Knowledge: Fact, Evidence, Doubt*. Ed. Yota Batsaki,

Subha Mukherji, and Jan-Melissa Schramm, 152–73. New York: Palgrave Macmillan, 2011.
Kennedy, Meegan. *Revising the Clinic: Vision and Representation in Victorian Medical Narratives and the Novel*. Columbus: Ohio State University Press, 2010.
Kidd, Colin. *The World of Mr. Casaubon: Britain's Wars of Mythography*. Cambridge: Cambridge University Press, 2016.
Kingstone, Helen. "Human-Animal Elision: A Darwinian Universe in George Eliot's Novels." *Nineteenth-Century Contexts* 40: 1 (2018), 87–103.
Kitson, Peter J. "Coleridge and 'the Ouran utang Hypothesis': Romantic Theories of Race." In *Coleridge and the Science of Life*. Ed. Nicholas Roe, 91–116. Oxford: Oxford University Press, 2001.
Klancher, Jon. "Discriminations, or Romantic Cosmopolitanisms in London." In *Romantic Metropolis: The Urban Scene of British Culture, 1780–1840*. Ed. James Chandler and Kevin Gilmartin, 65–82. Cambridge: Cambridge University Press, 2005.
Klancher, Jon. *Transfiguring the Arts and Sciences: Knowledge and Cultural Institutions in the Romantic Age*. Cambridge: Cambridge University Press, 2013.
Klettke, Cornelia. "Germaine De Staël: *Corinne ou L'Italie*: Grenzüberschreitung und Verschmelzung der Künste im Sinne der frühromantischen Universalpoesie." *Romanische Forschungen* 115: 2 (2003), 171–93.
Knoepflmacher, U. C., and G. B. Tennyson. *Nature and the Victorian Imagination*. Berkeley: University of California Press, 1977.
Kolb, Margaret. "In Search of Lost Causes: Walter Scott and Adolphe Quetelet's Revolutions." *Configurations* 27: 1 (2019), 59–85.
Kontje, Todd. *The German Bildungsroman: History of a National Genre*. Columbia: Camden House, 1993.
Kontje, Todd. "Socialization and Alienation in the Female Bildungsroman." In *Impure Reason: Dialectic of Enlightenment in Germany*. Ed. Daniel W. Wilson and Robert Holub, 221–41. Detroit: Wayne State University Press, 1993.
Kucich, John. *Repression and Victorian Fiction: Charlotte Bronte, George Eliot and Charles Dickens*. Berkeley: University of California Press, 1987.
Lacoue-Labarthe, Philippe, and Jean-Luc Nancy. *The Literary Absolute: The Theory of Literature in German Romanticism*. Trans. P. Barnard and C. Lester. Albany: SUNY Press, 1988.
Langan, Celeste. "Venice." In *Romantic Metropolis: The Urban Scene of British Culture, 1780–1840*. Ed. James Chandler and Kevin Gilmartin, 261–69. Cambridge: Cambridge University Press, 2005.
Laqueur, Thomas. *Making Sex: Body and Gender from the Greeks to Freud*. Cambridge, MA: Harvard University Press, 1990.
Larson, James L. *Interpreting Nature: The Science of Living Form from Linnaeus to Kant*. Baltimore: Johns Hopkins University Press, 1994.
Leavis, F. R. *The Great Tradition: George Eliot, Henry James, Joseph Conrad*. London: Chatto & Windus, 1962.
Lecourt, Sebastian. *Cultivating Belief: Victorian Anthropology, Liberal Aesthetics, and the Secular Imagination*. Oxford: Oxford University Press, 2018.
Ledger, Sally. *Dickens and the Popular Radical Imagination*. Cambridge: Cambridge University Press, 2007.

Lee, Yoon Sun. "Austen's Scale-Making." *Studies in Romanticism* 52: 3 (2013), 171–95.
Lehleiter, Christine. *Fact and Fiction: Literary and Scientific Cultures in Germany and Britain*. Toronto: University of Toronto Press, 2016.
Lenoir, Timothy. "Kant, Blumenbach and Vital Materialism in German Biology." *Isis* 71: 256 (1980), 77–108.
Lenoir, Timothy. *The Strategy of Life: Teleology and Mechanics in Nineteenth-Century German Biology*. Chicago: University of Chicago Press, 1989.
Levine, Caroline. *Forms: Whole, Rhythm, Hierarchy, Network*. Princeton: Princeton University Press, 2015.
Levine, Caroline, and Mario Ortiz-Robles, eds. *Narrative Middles: Navigating the Nineteenth-Century British Novel*. Columbus: Ohio State University Press, 2011.
Levine, George. *Darwin and the Novelists: Patterns of Science in Victorian Fiction*. Cambridge, MA: Harvard University Press, 1988.
Levine, George. "Dickens and Darwin, Science, and Narrative Form." *Texas Studies in Literature and Language* 28: 3 (1986), 250–80.
Levine, George. "George Eliot's Hypothesis of Reality." *Nineteenth-Century Fiction* 35: 1 (1980), 1–28.
Levine, George. *The Realistic Imagination: English Fiction from Frankenstein to Lady Chatterley*. Chicago: University of Chicago Press, 1984.
Lewis, Linda M. *Germaine de Staël, George Sand, and the Victorian Woman Artist*. Columbia: University of Missouri Press, 2003.
Lewis, Wyndham. *The Human Age*. 2 vols. London: Methuen, 1955–56.
Lightman, Bernard. "Life." In *Historicism and the Human Sciences in Victorian Britain*. Ed. Mark Bevir, 21–47. Cambridge: Cambridge University Press, 2017.
Lightman, Bernard. *Victorian Popularizers of Science: Designing Nature for New Audiences*. Chicago: University of Chicago Press, 2007.
Litvack, Leon. "What Books Did Dickens Buy and Read? Evidence from the Book Accounts with His Publishers." *The Dickensian* 94: 2 (1998), 85–130.
Lloyd, David. *Anomalous States: Irish Writing and the Post-Colonial Moment*. Durham: Duke University Press, 1993.
Lokke, Kari. *Tracing Women's Romanticism: Gender, History, and Transcendence*. London & New York: Routledge, 2004.
Lootens, Tricia. *The Political Poetess: Victorian Femininity, Race, and the Legacy of Separate Spheres*. Princeton: Princeton University Press, 2016.
Lovecraft, H. P. *Tales*. New York: Library of America, 2005.
Lovejoy, Arthur O. *The Great Chain of Being: A Study in the History of an Idea*. Cambridge, MA: Harvard University Press, 1936.
Lukács, Georg. *The Historical Novel*. Trans. Hannah Mitchell and Stanley Mitchell. Lincoln: University of Nebraska Press, 1983.
Lukács, Georg. *Studies in European Realism: A Sociological Survey of the Writings of Balzac, Stendhal, Zola, Tolstoy, Gorki and Others*. Trans. Edith Bone. New York: Grosset & Dunlap, 1964.
Lukács, Georg. *The Theory of the Novel: A Historico-Philosophical Essay on the Forms of Great Epic Literature*. Trans. Anna Bostock. Cambridge: MIT Press, 1971.
Lynch, Deidre Shauna. "The (Dis)Locations of Romantic Nationalism: Shelley, Staël, and the Home-Schooling of Monsters." In *The Literary Channel: The Inter-National*

Invention of the Novel. Ed. Margaret Cohen and Carolyn Dever, 194–224. Princeton: Princeton University Press, 2002.
Lyotard, Jean-François. *The Inhuman: Reflections on Time*. Trans. Geoffrey Bennington and Rachel Bowlby. Stanford: Stanford University Press, 1991.
MacDuffie, Allen. "Charles Darwin and the Victorian Pre-History of Climate Denial." *Victorian Studies* 60: 4 (2018), 543–64.
MacDuffie, Allen. "Dickens and the Environment." In *The Oxford Handbook of Charles Dickens*. Ed. John Jordan, Robert Patten, and Catherine Waters, 566–79. Oxford: Oxford University Press, 2018.
MacDuffie, Allen. *Victorian Literature, Energy, and the Ecological Imagination*. Cambridge: Cambridge University Press, 2014.
Marcus, Sharon. *Between Women: Marriage, Desire, and Friendship in Victorian England*. Princeton: Princeton University Press, 2007.
Marshall, David. *The Surprising Effects of Sympathy: Marivaux, Diderot, Rousseau, and Mary Shelley*. Chicago: University of Chicago Press, 1988.
Martin, Judith E. *Germaine de Staël in Germany: Gender and Literary Influence (1800–1850)*. Madison: Farleigh Dickinson University Press, 2011.
Martin, Nicholas. *Nietzsche and Schiller: Untimely Aesthetics*. Oxford: Clarendon Press, 1996.
Martin, Xavier. *Human Nature and the French Revolution: From the Enlightenment to the Napoleonic Code*. New York: Berghahn, 2001.
Mason, Michael York. "*Middlemarch* and Science: Problems of Life and Mind." *Review of English Studies* (New Series) 22: 86 (1971), 151–69.
Matus, Jill L. *Shock, Memory and the Unconscious in Victorian Fiction*. Cambridge: Cambridge University Press, 2009.
Maxwell, Richard. *The Historical Novel in Europe, 1650–1950*. Cambridge: Cambridge University Press, 2009.
Maxwell, Richard. *The Mysteries of Paris and London*. Charlottesville: University Press of Virginia, 1992.
McCarthy, Patrick J. "Lydgate, 'The New, Young Surgeon' of *Middlemarch*." *SEL: Studies in English Literature* 10: 4 (1970), 805–16.
McCracken-Flesher, Caroline. *The Doctor Dissected: A Cultural Autopsy of the Burke and Hare Murders*. Oxford: Oxford University Press, 2012.
McDaniel, Iain. "Philosophical History and the Science of Man in Scotland: Adam Ferguson's Response to Rousseau." *Modern Intellectual History* 10: 3 (2013), 543–68.
McGann, Jerome. "Walter Scott's Romantic Postmodernity." In *Scotland and the Borders of Romanticism*. Ed. Leith Davis, Ian Duncan, and Janet Sorensen, 113–29. Cambridge: Cambridge University Press, 2004.
McHale, Brian. "Free Indirect Discourse: A Survey of Recent Accounts." *PTL: A Journal of Descriptive Poetics and Theory of Literature* 3: 2 (1978), 249–87.
McKibben, Bill. *The End of Nature*. New York: Anchor, 1989.
McLane, Maureen. *Romanticism and the Human Sciences: Poetry, Population, and the Discourse of the Species*. Cambridge: Cambridge University Press, 2000.
McMaster, Graham. *Scott and Society*. Cambridge: Cambridge University Press, 1981.
McNeil, Ken. *Scotland, Britain, Empire: Writing the Highlands, 1760–1860*. Columbus: Ohio State University Press, 2007.

Meillassoux, Quentin. *After Finitude: An Essay on the Necessity of Contingency*. Trans. Ray Brassier. London: Bloomsbury, 2008.
Meisel, Martin. "On the Age of the Universe." In *BRANCH: Britain, Representation and Nineteenth-Century History*. Ed. Dino Franco Felluga. Extension of *Romanticism and Victorianism on the Net*. Uploaded October 2011. http://www.branchcollective.org/?ps_articles=martin-meisel-on-the-age-of-the-universe.
Mensch, Jennifer. *Kant's Organicism: Epigenesis and the Development of Critical Philosophy*. Chicago: University of Chicago Press, 2014.
Mershon, Ella. "Ruskin's Dust." *Victorian Studies* 58: 3 (2016), 464–92.
Miller, D. A. "Discipline in Different Voices: Bureaucracy, Police, Family, and *Bleak House*." *Representations* 1: 1 (1983), 59–89.
Miller, J. Hillis. *Charles Dickens: The World of His Novels*. Cambridge, MA: Harvard University Press, 1958.
Miller, J. Hillis. "A Conclusion in Which Almost Nothing Is Concluded: *Middlemarch's* 'Finale.'" In *Middlemarch in the 21st Century*. Ed. Karen Chase, 133–56. Oxford: Oxford University Press, 2006.
Miller, J. Hillis. "Optic and Semiotic in *Middlemarch*." In *The Worlds of Victorian Fiction*. Ed. Jerome Buckley, 125–45. Cambridge, MA: Harvard University Press, 1975.
Millgate, Jane. *Walter Scott: The Making of the Novelist*. Edinburgh: Edinburgh University Press, 1984.
Mitchell, Robert. *Experimental Life: Vitalism in Romantic Science and Literature*. Baltimore: Johns Hopkins University Press, 2013.
Moers, Ellen. *Literary Women*. New York: Doubleday, 1976.
Moi, Toril. "A Woman's Desire to Be Known: Expressivity and Silence in *Corinne*." In "Untrodden Regions of the Mind: Romanticism and Psychoanalysis," ed. Ghislaine McDayter. Special Issue, *Bucknell Review* 45: 2 (2002), 143–75.
Moretti, Franco. *Atlas of the European Novel 1800–1900*. London: Verso, 1998.
Moretti, Franco, ed. *Canon/Archive: Studies in Quantitative Formalism from the Stanford Literary Lab*. New York: n+1 Books, 2017.
Moretti, Franco. *Distant Reading*. London: Verso, 2013.
Moretti, Franco. *Graphs, Maps, Trees: Abstract Models for a Literary History*. New York: Verso, 2007.
Moretti, Franco. "Serious Century." In *The Novel*. Vol. 1, *History, Geography and Culture*. Ed. Franco Moretti, 364–400. Princeton: Princeton University Press, 2006.
Moretti, Franco. *The Way of the World: The Bildungsroman in European Culture*. London: Verso, 1987.
Morgan, Benjamin. *The Outward Mind: Materialist Aesthetics in Victorian Science and Literature*. Chicago: University of Chicago Press, 2017.
Mücke, Dorothea von. *The Practices of Enlightenment: Aesthetics, Authorship, and the Public*. New York: Columbia University Press, 2015.
Müller-Sievers, Helmut. *Self-Generation: Biology, Philosophy and Literature around 1800*. Stanford: Stanford University Press, 1997.
Muthu, Sankar. *Enlightenment against Empire*. Princeton: Princeton University Press, 2003.
Nash, Richard. *Wild Enlightenment: The Borders of Human Identity in the Eighteenth Century*. Charlottesville: University of Virginia Press, 2003.

Nazar, Hina. "The Continental Eliot." In *A Companion to George Eliot*. Ed. Amanda Anderson and Harry E. Shaw, 413–27. Oxford: Wiley-Blackwell, 2013.
Neill, Anna. *Primitive Minds: Evolution and Spiritual Experience in the Victorian Novel*. Columbus: Ohio State University Press, 2013.
Nelles, William. "Omniscience for Atheists: Or, Jane Austen's Infallible Narrator." *Narrative* 14: 2 (2006), 118–31.
Newsom, Robert. *Dickens on the Romantic Side of Familiar Things:* Bleak House *and the Novel Tradition*. New York: Columbia University Press, 1977.
Ngai, Sianne. "Merely Interesting." *Critical Inquiry* 34: 4 (2008), 777–817.
Noyes, John K. "Herder's Unsettling of the Distinction between Fact and Fiction." In *Fact and Fiction: Literary and Scientific Cultures in Germany and Britain*. Ed. Christine Lehleiter, 155–74. Toronto: University of Toronto Press, 2016.
Oppenlander, Ella Ann. *Dickens's* All the Year Round: *Descriptive Index and Contributor List*. Troy, NY: Whitston Publishing Co., 1984.
Ospovat, Dov. *The Development of Darwin's Theory: Natural History, Natural Theology, and Natural Selection, 1838-1859*. Cambridge: Cambridge University Press, 1981.
Otis, Laura, ed. *Literature and Science in the Nineteenth Century*. Oxford: Oxford University Press, 2002.
Otis, Laura. *Networking: Communicating with Bodies and Machines in the Nineteenth Century*. Ann Arbor: University of Michigan Press, 2001.
Otter, Samuel. *Melville's Anatomies*. Berkeley: University of California Press, 1999.
Pacini, Giulia. "Hidden Politics in Germaine de Staël's *Corinne ou l'Italie*." *French Forum* 24: 2 (1999), 163–77.
Palmeri, Frank. *State of Nature, Stages of Society: Enlightenment Conjectural History and Modern Social Discourse*. New York: Columbia University Press, 2016.
Parham, John. "Dickens in the City: Science, Technology, Ecology in the Novels of Charles Dickens." *19: Interdisciplinary Studies in the Long Nineteenth Century* 10 (2010), http://doi.org/10.16995/ntn.529.
Pascal, Roy. *The Dual Voice: Free Indirect Speech and Its Functioning in the Nineteenth-Century European Novel*. Manchester: Manchester University Press, 1977.
Paxton, Nancy. *George Eliot and Herbert Spencer*. Princeton: Princeton University Press, 1991.
Peterson, Linda H. "Rewriting *A History of the Lyre*: Letitia Landon, Elizabeth Barrett Browning and the (Re)Construction of the Nineteenth-Century Woman Poet." In *Women's Poetry, Late Romantic to Late Victorian: Gender and Genre, 1830–1900*. Ed. Isobel Armstrong and Virginia Blain, 115–32. New York: Macmillan-St. Martin's, 1999.
Pfau, Thomas. "*Bildung*: Etiology, Function, Structure (with Some Reflections on Beethoven)." In *Die Romantik: Ein Gründungsmythos der Europäischen Moderne*. Ed. Paul Geyer and Anja Ernst, 132–41. Bonn: Bonn University Press, 2010.
Pfau, Thomas. "*Bildungsroman*." In *The Encyclopedia of Romantic Literature*. Vol. 1, *A–G*. Ed. Frederick Burwick, 124–32. Oxford: Wiley Blackwell.
Pfau, Thomas. "*Bildungsspiele*: Vicissitudes of Socialization in *Wilhelm Meister's Apprenticeship*." *European Romantic Review* 21: 5 (2010), 567–87.
Phalèse, Hubert de, ed. *Corinne à la page*. Paris: Nizet, 1999.

Pielak, Chase. "Hunting Gwendolen: Animetaphors in *Daniel Deronda*." *Victorian Literature and Culture* 40: 1 (2012), 99–115.
Piper, Andrew. *Dreaming in Books: The Bibliographic Imagination in the Romantic Age*. Chicago: University of Chicago Press, 2009.
Plath, O. E. "Schiller's Influence on *Wilhelm Meisters Lehrjahre*." *Modern Language Notes* 31: 5 (1916), 257–67.
Plotz, John. *Portable Property: Victorian Culture on the Move*. Princeton: Princeton University Press, 2008.
Plotz, John. *Semi-Detached: The Aesthetics of Virtual Experience since Dickens*. Princeton: Princeton University Press, 2017.
Poovey, Mary. *A History of the Modern Fact*. Chicago: University of Chicago Press, 1998.
Postlethwaite, Diana. "George Eliot and Science." In *The Cambridge Companion to George Eliot*. Ed. George Levine, 98–118. Cambridge: Cambridge University Press, 2001.
Poulet, Georges. "The Role of Improvisation in *Corinne*." Trans. Kevin Clark and Richard Macksey. *ELH* 41: 4 (1974), 602–12.
Price, Leah. *The Anthology and the Rise of the Novel*. Cambridge: Cambridge University Press, 2000.
Prum, Richard O. *The Evolution of Beauty: How Darwin's Forgotten Theory of Mate Choice Shapes the Animal World—and Us*. New York: Anchor, 2017.
Psomiades, Kathy A. "The Marriage Plot in Theory." *Novel: A Forum on Fiction* 43: 1 (2010), 53–59.
Puckett, Kent. *Bad Form: Social Mistakes and the Nineteenth-Century Novel*. New York: Oxford University Press, 2008.
Pugh, David. "Schiller as Platonist." *Colloquia Germanica* 24: 4 (1991), 273–95.
Ragussis, Michael. *Figures of Conversion: The "Jewish Question" and English National Identity*. Durham, NC: Duke University Press, 1995.
Rajan, Supritha. "The Epistemology of Trust and Realist Effect in Charles Dickens's *Bleak House*." *Nineteenth-Century Literature* 72: 1 (2017), 64–106.
Rajan, Supritha. *A Tale of Two Capitalisms: Sacred Economics in Nineteenth-Century Britain*. Ann Arbor: University of Michigan Press, 2015.
Rancière, Jacques. "Aesthetics as Politics." In *Aesthetics and Its Discontents*, 19–44. Cambridge: Polity, 2009.
Rather, L. J. "Some Relations between Eighteenth-Century Fiber Theory and Nineteenth-Century Cell Theory." *Clio Medica* 4 (1969), 191–202.
Redfield, Marc. *Phantom Formations: Aesthetic Ideology and the Bildungsroman*. Ithaca: Cornell University Press, 1996.
Reese, Diana. "A Troubled Legacy: *Frankenstein* and the Inheritance of Human Rights." *Representations* 96: 1 (2006), 48–72.
Rehbock, Philip F. "Huxley, Haeckel and the Oceanographers: The Case of *Bathybius haeckelii*." *Isis* 66: 4 (1975), 504–35.
Reill, Peter Hanns. *Vitalizing Nature in the Enlightenment*. Berkeley: University of California Press, 2005.
Richards, Evelleen. *Darwin and the Making of Sexual Selection*. Chicago: University of Chicago Press, 2017.
Richards, Evelleen. "A Political Anatomy of Monsters, Hopeful and Otherwise: Teratogeny, Transcendentalism, and Evolutionary Theorizing." *Isis* 85: 3 (1994), 377–411.

Richards, Evelleen. "A Question of Property Rights: Richard Owen's Evolutionism Reassessed." *The British Journal for the History of Science* 20: 2 (1987), 129–71.
Richards, Robert J. *Darwin and the Emergence of Evolutionary Theories of Mind and Behavior*. Chicago: University of Chicago Press, 1987.
Richards, Robert J. "Kant and Blumenbach on the *Bildungstrieb*." *Studies in the History and Philosophy of Biology and the Biomedical Sciences* 31: 1 (2000), 11–32.
Richards, Robert J. *The Romantic Conception of Life: Science and Philosophy in the Age of Goethe*. Chicago: University of Chicago Press, 2002.
Richards, Thomas. *The Imperial Archive: Knowledge and the Fantasy of Empire*. London: Verso, 1993.
Richmond, Marsha L. "T. H. Huxley's Criticism of German Cell Theory: An Epigenetic and Physiological Interpretation of Cell Structure." *Journal of the History of Biology* 33: 2 (2000), 247–89.
Richmond, Marsha L. "Thomas Huxley's Developmental View of the Cell." *Nature Reviews* 3 (2002), 61–65.
Ricoeur, Paul. *Memory, History, Forgetting*. Trans. Kathleen Blamey and David Pellauer. Chicago: University of Chicago Press, 2009.
Robb, Graham. *Victor Hugo: A Biography*. New York: Norton, 1998.
Rosset, François. "Poétique des nations dans *Corinne ou l'Italie*." In *"Une mélodie intellectuelle": Corinne ou l'Italie de Germaine de Staël*. Ed. Christine Planté, Christine Pouzoulet, and Alain Vaillant, 139–58. Montpellier: Université Paul-Valéry, 2000.
Rossi, Paolo. *The Dark Abyss of Time: The History of the Earth and the History of Nations from Hooke to Vico*. Trans. Lydia G. Cochrane. Chicago: University of Chicago Press, 1984.
Rothfield, Lawrence. *Vital Signs: Medical Realism in Nineteenth-Century Literature*. Princeton: Princeton University Press, 1992.
Rudwick, Martin S. *Bursting the Limits of Time: The Reconstruction of Geohistory in the Age of Revolution*. Chicago: University of Chicago Press, 2005.
Rudwick, Martin S. *Georges Cuvier, Fossil Bones, and Geological Catastrophes: New Translations and Interpretations of the Primary Texts*. Chicago: University of Chicago Press, 2008.
Rudwick, Martin S. *Worlds before Adam: The Reconstruction of Geohistory in the Age of Reform*. Chicago: University of Chicago Press, 2008.
Ryan, Vanessa L. *Thinking without Thinking in the Victorian Novel*. Baltimore: John Hopkins University Press, 2012.
Saintsbury, George. *The English Novel*. London: J. M. Dent, 1913.
Salmon, Paul. "Herder's *Essay on the Origin of Language*, and the Place of Man in the Animal Kingdom." *German Life and Letters* 22: 1 (1968), 59–70.
Sammons, Jeffrey. "The Mystery of the Missing *Bildungsroman*; or, What Happened to Wilhelm Meister's Legacy?" *Genre* 14: 2 (1981), 229–46.
Sarafianos, Aris. "Pain, Labor, and the Sublime: Medical Gymnastics and Burke's Aesthetics." *Representations* 91: 1 (2005), 58–83.
Schaffer, Talia. *Romance's Rival: Familiar Marriage in Victorian Fiction*. Oxford: Oxford University Press, 2016.
Scheinberg, Cynthia. "'The Beloved Ideas Made Flesh': *Daniel Deronda* and Jewish Poetics." *ELH* 77: 3 (2010), 813–39.

Schiebinger, Londa. *The Mind Has No Sex? Women in the Origins of Modern Science.* Cambridge, MA: Harvard University Press, 1989.

Schiebinger, Londa. "Skeletons in the Closet: The First Illustrations of the Female Skeleton in Eighteenth-Century Anatomy." *Representations* 14 (Spring 1986), 42–82.

Schultz, Elizabeth. "Melville's Environmental Vision in *Moby-Dick.*" *Isle: Interdisciplinary Studies in Literature and Environment* 7: 1 (2000), 97–113.

Scott, Patrick. "Comparative Anatomies: Scott, Darwin, Eliot, Stevenson and the Legacy of 1820s Edinburgh." Unpublished paper, International Scott Conference, Eugene, OR, 1999.

Secord, James A. "Edinburgh Lamarckians: Robert Jameson and Robert E. Grant." *Journal of the History of Biology* 24: 1 (1991), 1–18.

Secord, James A. *Victorian Sensation: The Extraordinary Publication, Reception, and Secret Authorship of* Vestiges of the Natural History of Creation. Chicago: University of Chicago Press, 2000.

Secord, James A. *Visions of Science: Books and Readers at the Dawn of the Victorian Age.* Chicago: University of Chicago Press, 2014.

Sen, Sambudha. *London, Radical Culture, and the Making of the Dickensian Aesthetic.* Columbus: Ohio State University Press, 2012.

Sera-Shriar, Efram. *The Making of British Anthropology, 1813–1871.* London: Pickering and Chatto, 2013.

Seshadri, Kalpana Rahita. *HumAnimal: Race, Law, Language.* Minneapolis: University of Minnesota Press, 2012.

Shaffer, E. S. *"Kubla Khan" and the Fall of Jerusalem: The Mythological School in Biblical Criticism and Secular Literature, 1770–1880.* Cambridge: Cambridge University Press, 1980.

Sheehan, Jonathan, and Dror Wahrman. *Invisible Hands: Self-Organization and the Eighteenth Century.* Chicago: University of Chicago Press, 2015.

Shuttleworth, Sally. *George Eliot and Nineteenth-Century Science: The Make Believe of a Beginning.* Cambridge: Cambridge University Press, 1984.

Siegel, Jonah. *Haunted Museum: Longing, Travel and the Art-Romance Tradition.* Princeton: Princeton University Press, 2005.

Simmons, Clare. "A Man of Few Words: The Romantic Orang-Outang and Scott's *Count Robert of Paris.*" *Scottish Literary Journal* 17: 1 (1990), 21–34.

Simpson, David. *Fetishism and Imagination: Dickens, Melville, Conrad.* Baltimore: Johns Hopkins University Press, 1984.

Simpson, David. *Romanticism and the Question of the Stranger.* Chicago: University of Chicago Press, 2013.

Simpson, Erik. *Literary Minstrelsy, 1770–1830: Minstrels and Improvisers in British, Irish, and American Literature.* London: Palgrave Macmillan, 2008.

Siskin, Clifford. *System.* Cambridge: MIT Press, 2016.

Slater, Michael. *Charles Dickens.* New Haven: Yale University Press, 2009.

Slaughter, Joseph. *Human Rights, Inc.: The World Novel, Narrative Form, and International Law.* New York: Fordham University Press, 2007.

Sloan, Philip. "The Gaze of Natural History." In *Inventing Human Science: Eighteenth Century Domains.* Ed. Christopher Fox, Roy Porter, and Robert Wokler, 112–51. Berkeley: University of California Press, 1995.

Sluga, Glenda. "Gender and the Nation: Madame de Staël or Italy." *Women's Writing* 10: 2 (2003), 241–51.
Smith, Jonathan. "Dickens and Astronomy, Geology and Biology." In *The Oxford Handbook of Charles Dickens*. Ed. John Jordan, Robert L. Patten, and Catherine Waters, 404–19. Oxford: Oxford University Press, 2018.
Smith, Jonathan. *Fact and Feeling: Baconian Science and the Nineteenth-Century Literary Imagination*. Madison: University of Wisconsin Press, 1994.
Soare, Monica. "The Female Gothic Connoisseur: Reading, Subjectivity, and the Feminist Uses of Gothic Fiction." PhD diss., University of California–Berkeley, 2013.
Spary, E. C. *Utopia's Garden: French Natural History from Old Regime to Revolution*. Chicago: University of Chicago Press, 2000.
Stanzel, F. K. *A Theory of Narrative*. Trans. Charlotte Goedsche. Cambridge: Cambridge University Press, 1984.
Staum, Martin S. "Cabanis and the Science of Man." *Journal of the History of the Behavioral Sciences* 10: 2 (1974), 135–43.
Staum, Martin S. *Cabanis: Enlightenment and Medical Philosophy in the French Revolution*. Princeton: Princeton University Press, 1980.
Steinlight, Emily. "Dickens's 'Supernumeraries' and the Biopolitical Imagination of Victorian Fiction." *Novel: A Forum on Fiction* 43: 2 (2010), 227–50.
Steinlight, Emily. *Populating the Novel: Literary Form and the Politics of Surplus Life*. Ithaca: Cornell University Press, 2018.
Stepan, Nancy. *The Idea of Race in Science: Great Britain, 1800–1960*. London: Macmillan, 1982.
Stewart, Garrett. *Dickens and the Trials of the Imagination*. Cambridge, MA: Harvard University Press, 1974.
Stewart, Susan. *On Longing: Narratives of the Miniature, the Gigantic, the Souvenir, the Collection*. Durham: Duke University Press, 1993.
Stierstofer, Klaus. "Vestiges of English Literature: Robert Chambers." *Unmapped Countries: Biological Visions in Nineteenth-Century Literature and Culture*. Ed. Anne-Julia Zwierlein, 27–39. London: Anthem, 2005.
Stocking, George W., Jr. "French Anthropology in 1800." *Isis* 55: 2 (1964), 134–50.
Stocking, George W., Jr. *Victorian Anthropology*. New York: Free Press, 1987.
Storey, Graham. *Charles Dickens:* Bleak House. Cambridge: Cambridge University Press, 1987.
Sutherland, John. *The Life of Walter Scott: A Critical Biography*. Oxford: Blackwell, 1995.
Swales, Martin. *The German Bildungsroman from Wieland to Hesse*. Princeton: Princeton University Press, 1978.
Szmurlo, Karyna, ed. *The Novel's Seductions: Staël's* Corinne *in Critical Inquiry*. Lewisburg: Bucknell University Press, 1999.
Taylor, Jesse Oak. "Anthropocene." *Victorian Literature and Culture* 46: 3–4 (2018), 573–77.
Taylor, Jesse Oak. "The Novel as Climate Model: Realism and the Greenhouse Effect in *Bleak House*." *Novel: A Forum on Fiction* 46: 1 (2013), 1–25.
Tenenbaum, Susan. "*Corinne*: Political Polemics and the Theory of the Novel." In *The Novel's Seductions*. Ed. Karyna Szmurlo, 154–62. Lewisburg: Bucknell University Press, 1999.

Thomas, Keith. *Man and the Natural World: Changing Attitudes in England 1500–1800*. London: Penguin, 1983.
Tillotson, Kathleen. *Novels of the Eighteen-Forties*. Oxford: Clarendon Press, 1961.
Tracy, Robert. "Lighthousekeeping: *Bleak House* and the Crystal Palace." *Dickens Studies Annual: Essays on Victorian Fiction* 33 (2003), 25–53.
Tracy, Robert. "Maria Edgeworth and Lady Morgan: Legality versus Legitimacy." *Nineteenth-Century Fiction* 40: 1 (1985), 1–22.
Trotter, David. "Space, Movement and Sexual Feeling in *Middlemarch*." In *Middlemarch in the 21st Century*. Ed. Karen Chase, 34–63. Oxford: Oxford University Press, 2006.
Trower, Shelley. "Nerves, Vibrations and the Aeolian Harp." *Romanticism and Victorianism on the Net* 54 (May 2009), n.p. https://www.erudit.org/en/journals/ravon/2009-n54-ravon3401/038761ar/.
Trumpener, Katie. *Bardic Nationalism: The Romantic Novel and the British Empire*. Princeton: Princeton University Press, 1997.
Tucker, Irene. *The Moment of Racial Sight: A History*. Chicago: University of Chicago Press, 2012.
Uexküll, Jakob von. *A Foray into the Worlds of Animals and Humans; with a Theory of Meaning*. Trans. Joseph. D. O'Neill. Minneapolis: University of Minnesota Press, 2010.
Van den Berg, Hein. *Kant on Proper Science: Biology in the Critical Philosophy and the* Opus Postumum. Dordrecht: Springer, 2014.
Vanfasse, Nathalie. "'Grotesque, but Not Impossible': Dickens's Novels and Mid-Victorian Realism." *Revue électronique d'études sur le monde Anglophone* 2: 1 (2004), 1–12. https://erea.revues.org/500.
Van Ghent, Dorothy. *The English Novel: Form and Function*. New York: Harper & Row, 1953.
Vermeule, Blakey. *Why Do We Care about Literary Characters?* Baltimore: Johns Hopkins University Press, 2010.
Vincent, Patrick. *The Romantic Poetess: European Culture, Politics and Gender 1820–1840*. Hanover and London: University Press of New England, 2004.
Waldow, Anik. "Between History and Nature: Herder's Human Being and the Naturalization of Reason." In *Herder: Philosophy and Anthropology*. Ed. Anik Waldow and Nigel DeSouza, 148–65. Oxford: Oxford University Press, 2017.
Watt, James. "Scott, the Scottish Enlightenment, and Romantic Orientalism." In *Scotland and the Borders of Romanticism*. Ed. Leith Davis, Ian Duncan, and Janet Sorensen, 94–112. Cambridge: Cambridge University Press, 2004.
Weliver, Phyllis. "George Eliot and the Prima Donna's 'Script.'" *Yearbook of English Studies* 40 (1–2), 103–20.
Wellek, René. "German and English Romanticism: A Confrontation." *Studies in Romanticism* 4: 1 (1964), 35–56.
Welsh, Alexander. *Dickens Redressed: The Art of* Bleak House *and* Hard Times. New Haven: Yale University Press, 2000.
Wiesenfarth, Joseph. *George Eliot's Mythmaking*. Carl Winter: Heidelberg, 1977.
Wilkins, John S. *Species: A History of the Idea*. Berkeley: University of California Press, 2009.

Wilkinson, Ann Y. "*Bleak House*: From Faraday to Judgment Day." *ELH* 34: 2 (1967), 225–47.
Wills, David. *Dorsality: Thinking Back through Technology and Politics*. Minneapolis: University of Minnesota Press, 2008.
Woloch, Alex. "*Bleak House*: 19, 20, 21." *b20: boundary2online*. Uploaded October 24, 2016. https://www.boundary2.org/2016/10/alex-woloch-bleak-house-19-20-21/.
Woloch, Alex. *The One vs. the Many: Minor Characters and the Space of the Protagonist in the Novel*. Princeton: Princeton University Press, 2003.
Woolf, Virginia. *The Common Reader*. New York: Mariner Books, 2002.
Wormald, Mark. "Microscopy and Semiotic in *Middlemarch*." *Nineteenth-Century Literature* 50: 4 (1996), 501–24.
Wright, Daniel. "George Eliot's Vagueness." *Victorian Studies* 56: 4 (2014), 626–48.
Young, Robert M. Young. *Darwin's Metaphor*. Cambridge: Cambridge University Press, 1985.
Zalasiewicz, Jan, Sverker Sörlin, Libby Robin, and Jacques Grinevald. "Introduction: Buffon and the History of the Earth." In Georges-Louis Leclerc, Comte de Buffon, *The Epochs of Nature*, xiii–xxxiv. Trans. and ed. Jan Zalasiewicz, Anne-Sophie Milon, and Mateusz Zalasiewicz. Chicago: University of Chicago Press, 2018.
Zammito, John H. "Epigenesis: Concept and Metaphor in Herder's *Ideen*." In *Vom Selbstdenken: Aufklärung und Aufklärungskritik in Herders 'Ideen zur Philosophie der Geschichte der Menschheit.'* Ed. Regine Otto and John H. Zammito, 131–45. Heidelberg: Synchron Wissenschaftsverlag, 2001.
Zammito, John H. *The Genesis of Kant's* Critique of Judgment. Chicago: University of Chicago Press, 1992.
Zammito, John H. "Herder between Reimarus and Tetens." In *Herder: Philosophy and Anthropology*. Ed. Anik Waldow and Nigel DeSouza, 127–46. Oxford: Oxford University Press, 2017.
Zammito, John H. "Herder on Historicism and Naturalism." Conference on Herder and Anthropology. Center for Cultural Complexity in the New Norway, University of Oslo, May 2006.
Zammito, John H. *Kant, Herder, and the Birth of Anthropology*. Chicago: University of Chicago Press, 2002.
Zanone, Damien. "L'esthétique du 'tableau philosophique' dans *Corinne ou l'Italie*." In *"Une mélodie intellectuelle": Corinne ou l'Italie de Germaine de Staël*. Ed. Christine Planté, Christine Pouzoulet, and Alain Vaillant, 9–29. Montpellier: Université Paul-Valéry, 2000.
Zarifopol-Johnston, Ilinca M. "'Notre-Dame de Paris': The Cathedral in the Book." *Nineteenth-Century French Studies* 13: 2–3 (1985), 22–35.
Zimmerman, Virginia. *Excavating Victorians*. Albany: State University of New York Press, 2008.
Zunshine, Lisa. *Why We Read Fiction: Mind Reading and the Novel*. Columbus: Ohio State University Press, 2006.
Zwerdling, Alex. "Esther Summerson Rehabilitated." *PMLA* 88: 3 (1973), 429–39.

INDEX

Abrams, M. H., 148, 149
Addison, Joseph, 209n48
aesthetic, the, and aesthetics: as education, 56, 57, 63–66, 76–77, 79, 81–85, 216n47, 217n49, 217n50, 218n62; and human form, 20, 22–23, 24–25, 98, 109, 115, 118, 122, 135, 144; neoclassical, 21–22, 133–34; and nature, 127–28, 137, 174, 187; and the novel, 1, 19, 20–21, 24–25, 58, 72, 128, 134, 144, 218n74; and Romantic ideology, 147–49; as sensuous cognition, 20, 22, 46, 50, 58, 112, 113–15, 148, 194. *See also* form; grotesque, the; sublime, the
Agamben, Giorgio, 13–14, 15, 36–37, 38, 112
Alexander, J. H., 87
All the Year Round, 131, 233n44
allegory, 34–35, 40, 111, 126, 138, 152, 157, 179–81, 185; national, 22, 63, 65, 68, 71, 72, 82, 84, 222n126; and realism, 4, 26, 76, 78, 80, 126, 127, 152, 191, 193, 196, 228n90, 230n11, 233n48
Allen, Ralph, 23
analogy, 12, 40, 47, 50, 92, 133, 137, 161, 167, 170, 180–81, 190, 191, 203n39, 203n45, 208n17
Anderson, Amanda, 193, 237n105
Anderson, Benedict, 71
Anthropocene, 8, 30, 206n110, 230n10
anthropology: Enlightenment philosophical, 1–2, 3, 13–14, 22, 31–34, 37–38, 41–42, 45–46, 54, 58, 59, 61, 62, 63, 71, 72–73, 103, 119, 147, 171, 199, 203n45, 206n3, 208n33, 213n88; literary, 7, 9, 20, 55, 74, 88, 105, 106, 108, 116; and modernity, 1–2, 7–8, 201n3, 201n9; religious, 41, 173–74, 179–80; Victorian, 40, 107, 168, 179–80, 190–91, 242n72. *See also* history, natural; human nature; monogenesis; mythology; polygenesis; race
apes: and humans, 96, 106, 138–39, 176, 213n78, 225n35, 225n40, 227n74, 234n72, 234n74; orangutan, 15, 25, 49, 51, 87, 88, 94–98, 100, 104, 108–11, 114, 212n76, 223n8, 224n33, 225n39
Arendt, Hannah, 143
Aristotle, 23, 37, 133–34, 175, 208n32
Armstrong, Nancy, 56, 71, 82, 133, 227n71, 242n69
Arnold, Matthew, 193, 216n47
Augustine, St., 41, 99
Austen, Jane, 189, 149, 219n82, 221n124; *Emma*, 143–44, 236n103; free indirect discourse in, 143, 149, 236n101; and human nature, 18–19; *Northanger Abbey*, 18–19; realism and provincial life in, 18–19, 26, 69, 80, 189

Baer, Karl Ernst von, 136, 161, 162, 169, 225n35, 248n156
Bakhtin, Mikhail, 68, 69, 220n103
Balzac, Honoré de, 7, 59, 128, 171, 225n44, 231n21
Banfield, Ann, 142, 143, 185, 236n89
Barrell, John, 237n114
Bartfeld, Fernande, 225n43, 228n93
Barthes, Roland, 72
Bataille, Georges, 22
Baudelaire, Charles, 117
Beenstock, Zoe, 208n32
Beer, Gillian, 128, 160, 162, 166, 177, 243n85, 246n124, 247n142
Behler, Ernst, 219n76
Bender, John, 12, 207n17
Bentley's Miscellany, 135
Bentley's *Standard Novels*, 88, 214n6
Bergson, Henri, 28, 148, 200
Bernard, Claude, 28, 243n91, 244n94, 245n112
Bichat, Xavier, 179, 181
Bildung, 211n64, 216n35; individual, 21, 55–56, 66–67, 72, 78, 99–100, 139, 183, 186–87, 193; of humanity, 3, 4, 7, 10, 16, 17, 20, 21, 38, 46–48, 53–54, 57–58, 62–63, 71, 88, 105–6, 167–78, 173, 178–79, 183, 192, 211n64; and gender and marriage, 21–22, 53, 56–59,

[279]

Bildung (continued)
75–76, 82, 83–84, 105, 167–68, 177, 186; and nation, 21–22, 63, 70, 72, 76, 78, 82, 105–6. See also *Bildungstrieb*; development

Bildungsroman: criticism of, 21, 62–3, 66–68, 70–71, 72, 220n95; hero of, 3, 55, 192, 58–59, 61–62, 66, 72, 135, 192; and formation of humanity, 3, 7, 14, 21, 24, 53–54, 57–59, 61–63, 70–71, 171, 201n16, 219n88; women in, 5, 21, 57–58, 76, 77–79, 82, 83–84, 167, 240n39. *See also* realism

Bildungstrieb, 2, 40, 46–47, 76, 212n67

Blackwell, Thomas, 73, 220n104

Blackwood's Edinburgh Magazine, 88, 237n113

Blair, Kirstie, 243n92

Blumenbach, Johann Friedrich, 245n112; on the *Bildungstrieb*, 46, 47, 212n67; on the human species, 15, 16, 106, 209n36, 227n66

Bode, Christoph, 69, 218n68, 219n77

Boehm, Katharina, 131, 132, 136

Boes, Tobias, 217n52

book: of mankind, 119–21; of nature, 103, 226n56

Bougainville, Louis Antoine de, 16

Bowen, John, 126

Bowler, Peter, 161, 239n9, 239n13

Brassier, Ray, 29, 30, 206n104

Brewer, David, 82

Brombert, Victor, 121

Brown, Jane, 217n56

Brown, Laura, 223n8, 225n33

Brown, Marshall, 61

Browning, Elizabeth Barrett, 83

Bryant, Jacob, 179–80, 191, 242n76

Buckland, Adelene, 131, 132

Buffon, George-Louis Leclerc, Comte de: *Epochs of Nature*, 29, 35, 40, 60; on human nature, 10, 14, 36, 37, 41, 49, 94, 106, 114, 213n78; method and style of, 12, 32, 35, 39–40, 207n17; *Natural History*, 32, 36, 37, 203n45; and novelists, 59, 128, 131; on species, 10, 11, 27, 32, 37, 41, 43, 107, 122

Burke, Edmund, 16, 89; *Origin of the Sublime and Beautiful*, 23–24, 113, 115, 205n83

Burney, Frances, 80, 221n124

Butler, Samuel, 248n155

Buzard, James, 140, 220n107, 238n131

Cabanis, Pierre Jean Georges, 61, 75–76

Cadell, Robert, 87

Campbell, Timothy, 220n101

Campe, Rüdiger, 58

Canguilhem, Georges, 28–29, 116, 136, 204n53, 241n45

Carlyle, Thomas, 70, 214n4

Carroll, Siobhan, 202n27

Casanova, Pascale, 95

Cave, Terence, 82

cell biology, 11, 27, 166, 179, 181–82, 184–85, 187, 196–97, 198, 243n91, 244n103. *See also* protoplasm

Cervantes Saavedra, Miguel de, 69

Chambers, Robert, *Vestiges of the Natural History of Creation*, 15, 129–31; and Darwin, 137, 161; Dickens on, 132; evolutionary theory of, 129–31, 132, 136–37, 139, 161, 188, 225n35, 231n26, 246n134; on human nature, 9, 17, 130, 158, 163, 213n79; on the order of nature, 127, 156–57, 231n27; Owen on, 137, 225n35, 234n72

Chambers's Edinburgh Journal, 233n44

Chandler, James, 220n105, 226n55

Chase, Cynthia, 193, 247n142

Cheselden, William, 111

Chico, Tita, 12, 239n26

China, 100

city: as evolutionary environment, 4, 25–26, 87, 89, 95, 100–101, 106, 111, 116–17, 123, 126, 139, 155–56, 171, 228n90. *See also* Constantinople; cosmopolitanism; Paris; provincial life

cognition and consciousness, 24, 26–27, 28, 29, 40, 41, 45, 46, 50, 58, 96, 108, 111, 117, 142–44, 146–51, 167, 172–77, 178, 181, 187, 194–95, 200, 230n14, 236n89

Cohn, Dorrit, 144, 236n89

Cohn, Elisha, 177

Coleridge, Samuel Taylor, 20, 147, 151, 241n44; on allegory and symbol, 180, 185

Collins, Wilkie, 238n130

Comte, Auguste, 143, 161, 169, 194–95, 240n40, 240n42, 241n45
Condorcet, Marquis de, 42, 60, 61, 75, 219n76
Constantinople, 4, 87, 95, 100, 101, 102, 108
Cook, James, 16, 106
correlationism, 29, 150
Corsi, Pietro, 90, 91, 93, 224n18
cosmopolitanism, 4, 39, 76–77, 81, 82, 83–84, 95, 189
Crawford, Hugh, 229n4
Crawford, Robert, 220n107
Creuzer, Friedrich, 179–80, 243n76
Crystal Palace, 129, 132, 156, 230n9
Cullen, William, 113
Cuvier, Georges, 92, 94, 122, 131, 169, 202n37, 207n5, 233n52; debate with Geoffroy, 25, 86, 91–92, 93, 128; eulogy of Lamarck, 52, 91, 213n84

Dallas, E. S., 24
Dames, Nicholas, 24
Dante Alighieri, 176
Darwin, Charles: and Chambers's *Vestiges*, 137, 161, 232n43, 234n69; and development theory, 161–62, 163, 209n83, 239n9; and Dickens, 131, 234n73; and George Eliot, 128, 160–62, 177, 190–91, 242n67; on human nature, 16, 28, 38, 108, 192, 199; and Lamarck, 52, 93; Lewes on, 161–62, 196, 198–99; and Lyell, 52, 94; on natural selection, 15, 29, 127, 160, 162, 186, 189, 199–200; and Owen, 132, 137, 190; "Pangenesis hypothesis" in, 196–99; on scientific language, 163–65, 166
—works of: *The Descent of Man*, 16, 33, 160, 162, 163, 192, 242n67, 246n134; *The Expression of Emotions in Animals and Man*, 15, 108; *On the Origin of Species*, 27, 160, 162–63, 174, 190, 240n28; *The Variation of Animals and Plants under Domestication*, 163, 164
Dawson, Gowan, 131, 133, 233n49, 233n52
De Man, Paul, 36, 38, 68, 70
deep time. *See* geology, temporal scale of
Derrida, Jacques, 8, 201n3

Desmond, Adrian, 93, 94
development: organic, 2, 11, 13, 15, 21, 22, 37, 45–47, 49, 50–52, 61, 62, 63, 64, 69, 89, 90–91, 94, 97–99, 102, 106–7, 108, 112, 116, 125–26, 127, 128–30, 132, 136–37, 161–63, 167–68, 169, 171, 172, 177–78, 186, 187–88, 190–91, 193, 196–200, 211n65, 212n76, 216n32, 225n35, 225n47, 246n134, 248n156; progressive, 3–4, 14, 16, 42, 45–51, 52–53, 57, 58, 62–65, 69, 72, 75, 94, 104–6, 161, 163, 169–70, 187, 189, 192, 195–96, 198, 199–200, 211n63; retrogressive, diverted or arrested, 4, 74, 89, 97, 100–102, 103–4, 108–9, 111, 117, 125–26, 136–40, 168, 171, 172, 177–78, 186, 188, 190–92, 193, 196, 198, 227n73, 230n14, 246n134; sociohistorical or cultural, 7, 14, 18–19, 21, 38, 42–44, 48, 49–51, 52–53, 57, 58–61, 62–65, 70, 72, 100–102, 104–5, 139–40, 169–71, 172, 179–81, 191–92, 215n27; theories of, 11, 13, 14, 37, 38, 42–49, 51–53, 58–59, 62–63, 64, 69, 72, 90–92, 94, 97–99, 102, 106–7, 108, 111–12, 116, 127, 128–30, 132, 136–37, 139, 160, 161–63, 179–81, 190–91, 194, 196–99, 211n63, 215n27, 219n76, 224n18, 225n35, 225n47, 231n26, 234n72, 239n9, 239n13, 240n42, 246n134, 248n156. *See also Bildung*; *Bildungstrieb*; Chambers, Robert; Darwin, Charles; embryology; epigenesis; Lamarck, Jean-Baptiste; monogenesis; Owen, Richard; perfectibility; polygenesis; preformation; Saint-Hilaire, Geoffroy; teratology
Dickens, Charles: and Chambers's *Vestiges*, 131, 132; character in, 15, 126, 135, 137–40, 234n63, 236n104; and Darwin, 131, 234n73; Lewes on, 24, 138; and novelistic form, 24, 25, 127, 133–34, 141–45, 151–52, 156, 231n16; and realism, 4, 5, 25, 30, 126, 127–29, 142–43, 156–57, 230n14; and science, 127, 128, 131–32, 136, 138, 230n7; and Wordsworth, 145–47, 155–56
—*Bleak House*, 125–29, 132–47, 151–57; character system of, 126, 127, 135, 137–40; double narrative in, 141–42, 144–45,

Dickens, Charles (continued)
  152–54, 156, 238n130; as historical novel, 237n113; nature in, 30, 126, 127, 128–29, 139–40, 156–57, 231n23; serial form of, 133–34, 153; visionary prospect in, 127–28, 145–47, 151–56
  —*Our Mutual Friend*, 131, 135, 136
  —"The Poetry of Science," 131, 132. See also *All the Year Round*
Diderot, Denis, 16, 54, 111–12, 113, 205n101, 227n78
Dilthey, Wilhelm, 62, 70, 216n36
Diogenes, 42, 103
domestic fiction, 22, 54, 57, 66, 69, 76, 77, 79–80, 82, 84, 167–68, 177, 190. See also provincial life; realism
Duff, David, 218n74
Durkheim, Émile, 208n30

Eliot, George: and the Bildungsroman, 21, 83–85, 167–68, 171–72, 177, 178–79, 191–92; and Darwin, 160–61, 163, 177, 190–91, 196–99; and development theory, 161–63, 167, 169–70, 190–91, 196–99, 240n37, 240n40; figural and literal language in, 165–66, 182–83, 184–86, 190–93, 199, 240n36; mythology and religion in, 179–81, 187–88, 190–92, 194, 242n76, 245n116; and novelistic form, 26, 167–68, 169–70, 179, 183–86; and provincial life, 167–68, 171, 174, 189–90; and realism, 26, 127, 128, 143, 149–50, 167–68, 189–90, 199, 236n104; and scientific knowledge, 160–62, 166, 179–83, 185, 190–91, 196–98, 199, 242n73, 243n91, 244n94, 244n95, 244n97, 244n103, 245n112; translations of Strauss and Feuerbach by, 172–74; and women's destiny, 83–85, 158–59, 167–68, 186–88
  —*Adam Bede*, 127, 160, 171
  —*Daniel Deronda*, 26, 83–85, 167, 168, 189–99; art in, 83–84; avatars of Corinne in, 83–85; Jewish theme in, 84, 168, 189, 192–8; organic form and development in, 150, 168, 190–93, 196–99
  —*Middlemarch*, 158–60, 177–89; epiphany in, 149–50, 178–9; 184–86; heroine of, 158–59, 177–79, 184–87; and the history of man, 158–60, 162–63, 178–79; organic mutation in, 27, 167, 184–88; scientific vocation in, 179–83, 243n86, 243n92, 243n93
  —*The Mill on the Floss*, 161, 171–77; Bildung in, 83, 167, 172, 177, 240n37; and species consciousness, 172, 174–77; tragedy in, 175, 176–77
  —"The Natural History of German Life," 170–71, 236n104
  —"Notes on Form in Art," 170, 183
  —"The Progress of the Intellect," 179–80
  —"Shadows of the Coming Race," 28, 168, 199–200
embryology, 15, 45, 46, 90, 98–99, 129–30, 136–37, 161, 162, 169, 181, 196, 197, 225n35, 226n47, 234n70, 243n86, 246n132, 246n134, 248n156. See also epigenesis; preformation; teratology
epic, 21, 67, 69, 71, 72, 73, 114, 119, 124, 126n50, 184, 229n102, 231n26
epigenesis, 2, 21, 38, 40, 45, 46, 64, 98–99, 161–62, 169, 182, 209n44, 212n70, 216n32, 216n32, 211n63, 226n47. See also development
Ermarth, Elizabeth Deeds, 134, 142, 144
Esty, Jed, 63
evolution. See development
Examiner, The, 131, 232n28
extinction, 10, 22, 29, 75, 122, 125, 163, 199–200, 206n110

Ferguson, Adam, *History of Civil Society*, 13, 32, 36, 42, 43–44, 103, 104, 119
Ferguson, Frances, 236n94, 236n101
Festa, Lynn, 17
Feuerbach, Ludwig, *Essence of Christianity*, 27, 172–74, 175–76, 178, 179, 192, 201n11, 210n55, 213n88, 241n60, 242n62, 242n65, 242n66
Fichte, Johann Gottlieb, 68, 71
fiction and figurative language (versus scientific knowledge), 12–13, 32, 34–36, 38–40, 48, 50, 52, 53, 92, 112–13, 132, 164–66, 179, 183–84, 189, 194–95, 198–99, 207n14, 207n17, 239n26

Fielding, Henry: *Joseph Andrews*, 10–11, 202n33; *Tom Jones*, 10–11, 12, 17, 18, 22–23, 144, 159–60, 201n1, 204n63, 226n56
Flaubert, Gustave, *Madame Bovary*, 144, 236n94, 236n103, 237n106
Fleishman, Avrom, 243n93, 244n100
Fludernik, Monika, 236n94
form: archetypal, classical, providential, total, 11, 17, 23, 26, 27, 37, 53, 63–64, 67, 71, 72, 76, 103, 113, 120, 130, 133, 142, 147, 150, 155, 156–57, 170–71, 174, 179, 183–84, 185, 225n47, 233n49, 244n100, 245n110; indefinite, irregular, variable, 1–3, 6, 9, 14–15, 17, 20–21, 22–24, 26–28, 40, 44–46, 47–49, 50–53, 57–59, 61–62, 64, 67–69, 72–74, 97, 108, 113–14, 116–17, 119, 133–34, 162–63, 187–88, 193, 196–98, 219n77, 246n117; mixed, 49, 57, 61–62, 68, 76–77, 88, 109, 113–15, 123, 126–27, 130, 133, 136–37, 168, 188, 196–98, 200, 218n74; organic, 47, 134, 156, 167, 168, 184–87, 244n100; *See also* aesthetic, the; *Bildunsgtrieb*; development; grotesque, the; monsters; preformation; realism; sublime, the; teratology
Forster, E. M., 234n63
Forster, Georg, 16, 106
Foucault, Michel, 1, 7–8, 37, 38, 201n3, 202n37, 207n5, 208n33
Frank, Lawrence, 232n38
Franklin, Caroline, 221n126
free indirect discourse, 19, 26, 29, 128, 141, 142–45, 149–51, 153–54, 156, 179, 185–86, 189, 236n89, 236n94, 236n101, 236n103, 237n104, 238n128. *See also* narration, omniscient
Freedgood, Elaine, 245n110
Fuseli, Henry (Johann Heinrich Füssli), 207n11

Gallagher, Catherine, 27, 187, 203n40, 245n107
Gaukroger, Stephen, 50, 203n45, 206n3, 213n88
Génette, Gerard, 142, 143, 247n142
geology, 93, 132, 134, 230n7, 230n9; temporal scale of, 2, 22, 29–30, 121, 123–24, 129, 160, 162–63, 197, 239n6

Germany: Goethe and, 55, 65, 217n51; life sciences in, 27, 38, 45–47, 91, 92; literature in, 53–54, 55, 57, 68, 70; philosophy and philology in, 32, 165, 170–71, 173, 179; Staël and, 55, 76, 214n3
Ghosh, Amitav, 8
Gigante, Denise, 49, 212n67, 224n15, 233n53
Godwin, William, 42
Goethe, Johann Wolfgang: on the Geoffroy-Cuvier debate, 86, 91–92; and Herder, 49, 62, 215n32; on Hugo, 224n20
— *Italian Journey*, 214n17, 221n115
— *Metamorphosis of Plants*, 246n117
— *Wilhelm Meister's Apprenticeship*, form of, 21, 58, 61–62, 64–66, 68, 78–79, 216n46; Hegel on, 66–67; and human development, 62–63, 66; Lukács on, 67; Mignon in, 56, 78–79, 82, 214n4; and nationality, 64–65, 71, 217n51; Novalis on, 66, 217n58; protagonist of, 3, 55, 58, 62, 66, 83; Schlegel on, 62, 67–68; and Schiller, 63–66, 216n43; Mme. de Staël and, 55–56, 57, 214n3; women and sexuality in, 78–79
— *Wilhelm Meister's Journeyman Years*, 62, 66, 78
Goldstein, Amanda Jo, 40, 50, 206n101, 209n44, 216n32
Goodlad, Lauren, 26, 106
Goodman, Kevis, 211n58
Grant, Robert Edmond, 52, 93, 94, 95, 246n117
Graver, Suzanne, 240n40, 240n42
Greif, Mark, 7, 8
Greiner, Rae, 134–35, 143, 144
Griffiths, Devin, 135, 167, 203n45, 208n17
Grosz, Elizabeth, 27, 28, 163, 200
grotesque, the, 20, 24–25, 88, 109, 113–16, 118, 122, 123, 125, 126–27, 133, 135, 136–37, 140, 228n88, 228n93, 230n11, 231n18. *See also* form, mixed; monsters
Guerlac, Suzanne, 221n126

Haeckel, Ernst, 161, 185, 186, 239n13, 245n106
Hall, Basil, 226n51
Haller, Albrecht von, 46
Hamann, Johann Georg, 46
Hamilton, Emma, 221n115
Hamilton, Paul, 217n48, 217n50
Hardy, Barbara, 244n97
Hartley, David, 205n83
Hawkins, Waterhouse, 132
Heffernan, Julián Jiménez, 29, 246n124
Hegel, G.W.F.: *Aesthetics*, 56, 66–67, 218n62, 218n64; on epic and novel, 21, 66–67, 69, 71, 84, 119, 149; *Phenomenology of Spirit*, 53, 58, 63, 173
Heidegger, Martin, 210n55, 211n59
Henchman, Anna, 238n132
Hensley, Nathan K., 129
Herder, Johann Gottfried: and anthropology, 1, 33, 213n88; and the Bildungsroman, 58, 62, 63; in debate with Kant, 13, 32–35, 46–49, 92, 199, 206n3; evolutionary speculation in, 46–51, 206n101, 209n38, 211n63, 211n65, 212n72, 248n156; and Goethe, 215n32, 217n51; on human nature, 17, 19, 34, 38, 44–46, 48, 49–53, 69, 105, 129, 130, 147, 173; on human-ape division, 49, 96, 100, 176, 234n74; *Ideas for a History of the Philosophy of Man*, 33, 40, 46–48, 52–53, 211n65, 231n26; *On the Origin of Language*, 44–46, 210n53; and Rousseau, 44, 210n53; on sensation, instinct and cognition, 40, 44–45, 46, 50, 173–74, 210n54, 210n55, 241n58; on society and culture, 14, 16, 38; and use of figurative language, 34, 39–40, 50, 58
Hibberd, Sarah, 223n13, 228n88
Hilles, Frederick W., 23, 205n81
Hillman, Susanne, 222n128
historical novel, 3–4, 14, 22, 16, 57, 71–75, 87–88, 100–107, 135, 220n103, 237n113. *See also* history
history: conjectural, 13, 18, 21, 32, 33–36, 39, 51, 72, 100–101, 103, 114, 119, 159, 203n45, 206n3, 207n15, 231n26; natural, 1–3, 4, 10, 11, 12–13, 14–16, 18, 20, 22, 28, 31–33, 35–37, 40, 41–43, 47, 49–50, 52, 59, 60, 72, 73, 90–94, 95, 100, 102, 114–15, 119, 123–24, 126–28, 129–32, 136, 156–57, 161–63, 164, 170–72, 179, 187, 190–91, 202n37, 203n45, 206n3, 207n5, 213n79, 231n26, 241n49. *See also* geology, temporal scale of; historical novel; human nature; science of man
Hoffmann, E.T.A., 68, 70
Hölderlin, Friedrich, 148, 237n120
Homo duplex (Buffon), 14, 15, 37, 41, 50, 114, 173, 208n30. *See also* human nature
Hugo, Victor: *Cromwell* (preface), 25, 114–15, 126; *Hernani*, 96; and revolution, 89, 96
—*Notre-Dame de Paris*: aesthetic categories in, 115, 118, 119–20, 121–22; architecture and media theory in, 117, 118–22; English reception and influence of, 96, 133, 223n10; evolutionary speculation in, 97–99, 116, 127; monsters and monstrosity in, 88–89, 97, 115–18; Goethe on, 224n20; Quasimodo in, 88–89, 97–100, 115, 122; sensation and sentiment in, 99–100, 122; urban life in, 25, 116–17
human nature: as constant, uniform, universal, 3, 10–11, 17–19, 21, 22–23, 72–73, 102–3, 104–5, 118–20, 121–22, 134–35, 159, 170–71, 189, 210n54; as double, divided, 2, 13–14, 36–37, 38–39, 41–44, 50, 112; erect form of, 50–51, 52–53, 138, 147, 213n81; and language, 44–45, 96, 110–11, 227n75; as mixed, variable, plastic, 1–2, 14–17, 19–21, 22, 24–28, 32, 45–53, 57–59, 60–61, 62–63, 69, 74–75, 82, 87–88, 94–100, 103–4, 106–9, 114, 119–21, 126, 127–28, 135–36, 137–40, 160, 162–63, 167–68, 171–72, 177–78, 187–89, 190–93, 195–98, 199, 210n57, 212n76, 215n27, 246n134; and reason or consciousness, 14, 37, 44–45, 96, 151, 172–77, 178, 180–81, 183–84, 194–95; as social or antisocial, 38, 42–44, 45, 48, 53, 143–44, 175–77, 208n32. *See also* anthropology; *Bildung*; development; history, natural; middle; science of man
Humboldt, Wilhelm von, 63, 217n47

Hume, David: *Enquiry Concerning Human Understanding*, 17–18, 226n55; "Natural History of Religion," 35, 203n45; "On National Character," 19; *Treatise of Human Nature*, 11, 12, 31, 33, 70, 108, 159, 165, 203n40, 207n15
Hunt, Leigh, 237n113
Hunt, Robert, 131
Hutton, James, 22, 205n76
Huxley, T. H., 12, 160, 206n101; "The Cell-Theory," 181–82, 242n73; "On the Physical Basis of Life," 185, 244n105, 245n106

instinct, 2, 14, 38, 39, 43, 44–45, 52, 74, 119, 174, 192, 198, 210n54, 242n67
irony, 19, 65, 67–68, 70–71, 219n77
Italy and Italians, 18, 56, 80–81, 84, 221n126

Jacyna, L. S., 181
Jaffe, Audrey, 141, 144, 235n86
James, Henry: on Dickens, 123, 135, 144; on novelistic form, 24, 28, 133, 134, 150
James, John Patrick, 129
Jameson, Fredric, 26, 61, 66, 148, 215n14, 216n46, 236n103
Jameson, Robert, 93
Jeffrey, Francis, 221n121
Jews and Judaism, 84, 104–6, 168, 189, 192–93, 193–96, 247n141, 247n147
Johnson, Joseph, 33, 207n11
Jones, Catherine, 220n109, 222n127
Jordan, John, 145
Joshi, Priya, 204n67

Kames, Lord (Henry Home), 16, 32, 106
Kant, Immanuel: and anthropology, 1, 14, 16, 33–34, 213n88; *Anthropology from a Pragmatic Point of View*, 34, 38–39, 213n81; on constitutive versus regulative causality, 13, 39, 47, 207n8; critical philosophy of, 21, 33, 39, 52, 150; in debate with Herder, 13, 32–35, 40, 46–49, 50, 52, 92, 199, 206n3; on human nature and history, 14, 16, 21, 33–35, 38–39, 47, 48–49, 54, 106, 213n81; "Idea for a Universal History with a Cosmopolitan Aim," 7, 33, 39; on science and fiction, 34, 36, 39, 46–47, 199, 207n14; on transformism, 48–49, 50, 212n76
Kidd, Colin, 243n76, 247n135
Klancher, Jon, 95, 207n7
Klettke, Cornelia, 221n113
Knox, Robert, 95, 107, 227n70
Kucich, John, 175–76

Lacoue-Labarthe, Philippe, and Jean-Luc Nancy, 68, 69, 218n68
Lamarck, Jean-Baptiste: and Chambers's *Vestiges*, 129, 130, 139; and Cuvier, 52, 91–92, 169; evolutionary process in, 15, 25, 51, 98, 163, 169, 112, 187, 224n18; and Geoffroy, 90–92, 128, 224n18; on man, 17, 51, 94, 138, 147, 202n27; natural scale, 51, 88, 97, 107, 162; reception of, British, 52, 92–95, 127; reception of, French, 90–91, 94
Landon, Letitia Elizabeth, 214n6
Landor, Walter Savage, 237n113
Laplace, Pierre-Simon, 129
Lawrence, William, 106
Leavis, F. R., 26, 126, 236n103
Lecourt, Sebastian, 247n141
Levine, Caroline, 134, 235n79, 235n84
Levine, George, 131, 132, 134, 156, 230n7, 246n124, 247n146
Lewes, George Henry: on Comte, 169, 194, 240n42; and Darwin, 161–62, 196, 198; on development, 11, 161–62, 169, 182, 185, 186, 225n47, 240n42, 243n91, 244n94, 245n112; on Dickens, 24, 138; on German literature, 70, 217n57, 219n85
Lightman, Bernard, 132, 231n26, 232n37, 233n45
Linnaeus, Carolus (Carl von Linné), 37
Locke, John, 111
Lockhart, John Gibson, 87
Lokke, Kari, 57, 221n120
Long, Edward, 17, 106
Lootens, Tricia, 221n126
Lovecraft, H.P., 188
Lucretius and Lucretianism, 112, 206n101, 209n44, 228n79, 231n26, 245n112
Lukács, Georg (György): *The Historical Novel*, 72; *Studies in European Realism*, 7, 28, 126; *Theory of the Novel*, 21, 67, 69, 71, 72, 218n65, 219n77

Lyell, Charles, 29, 52, 93–94, 125, 129, 131, 160, 177, 213n87, 224n30
Lynch, Deidre, 221n119
Lyotard, Jean-François, 30
lyric, 56, 83, 127–28, 147–51, 154–55, 238n122. *See also* poetry
Lytton, Edward Bulwer, 199

MacDuffie, Allen, 230n10, 231n23, 239n9
Mackay, Robert William, 179–80, 191
MacLeay, William Sharp, 213n79, 231n27
Malraux, André, 8
Malthus, Thomas, 27, 101
marriage plot, 22, 56–57, 66, 76, 78–79, 82–84, 106, 167–68, 177, 190. *See also* domestic fiction
Martin, John, 146
Martineau, Harriet, 169
Marx, Karl, 9, 14, 30, 175, 211n59, 241n60, 242n61, 242n62
Mason, Michael York, 182, 240n34, 243n86
Maugham, Somerset, 8
Maxwell, James Clerk, 245n116
Maxwell, Richard, 24, 223n12, 228n90, 231n15, 233n48
McGann, Jerome, 101
McLane, Maureen, 19–20, 202n31
McMaster, Graham, 225n39
megalosaurus, 30, 125–26, 129, 131, 132, 133, 230n9, 230n10, 233n49, 233n52. *See also* monsters
Meillassoux, Quentin, 29, 150
Melville, Herman, *Moby-Dick*, 124, 126, 229n3, 229n4
Mendel, Gregor, 161, 198
middle: as condition or scale of human life, 19, 37, 49, 64, 73, 121, 134–35, 150. *See also* form, mixed; realism
Miller, D. A., 235n80
Miller, J. Hillis, 245n110, 246n116
Millgate, Jane, 220n95
Mitchell, James, 111, 227n77
Mitchell, Robert, 224n18, 241n45
Modernism, 7, 148–49, 150, 188
Mohl, Hugo von, 244n105
Moi, Toril, 58, 82
Monboddo, Lord (James Burnett), 15, 51, 74, 94, 104, 109, 220n106, 223n8

monogenesis, 16, 17, 102, 106, 108, 109. *See also* anthropology; human nature; polygenesis; race
monsters and monstrosity: human, 4, 10, 13, 20, 22, 25, 37, 88–89, 97–99, 110, 115–18, 123, 126, 135–36, 137–39, 140, 188–89, 191, 198, 223n13, 226n53; and literary form, 24–25, 28, 32, 127, 131, 133–34, 233n49; nonhuman, 117, 120, 123–26, 186, 233n53; in organic development, 15, 49, 89–90, 97–99, 116, 136–37, 196, 198, 204n53, 224n15, 225n47, 228n87, 233n53. *See also* teratology
Moretti, Franco, 21, 60, 62, 116, 134, 143, 144, 215n14, 216n46, 218n60, 220n95
Morgenstern, Karl, 62–63, 67, 172
Muhammad XII, Abu Abdullah (Boabdil), 105
Müller, F. Max, 179, 180, 187
Müller-Sievers, Helmut, 212n70, 213n88
Murphy, Arthur, 22–23, 205n79
Musil, Robert, 70, 219n88
Muthu, Sankar, 16.
mythology, comparative, 11, 109, 166, 179–81, 183, 185, 187, 191–92, 246n117. *See also* anthropology; religion

narration, omniscient, 19, 128, 141, 143, 144–45, 149, 151, 152, 154–56, 168, 179, 235n82, 235n86, 238n131. *See also* free indirect discourse; realism
nation and nationalism, 3–4, 5, 19, 22, 54, 57, 64–65, 69–73, 76, 77, 79–82, 84, 87, 95, 101–3, 105, 147, 159, 168, 170–71, 172, 189–90, 192–93, 196, 204n67, 217n51, 221n126, 222n128
national tale, 21–22, 57, 82
Nazar, Hina, 242n62
Novalis (Georg Philipp Friedrich von Hardenberg), 66, 68, 158, 217n58
novel. *See* Bildungsroman; domestic fiction; free indirect discourse; historical novel; marriage plot; narration, omniscient; national tale; realism; science fiction; serial form
Noyes, John K., 50, 209n41

Odyssey, the, 110, 246n116
orangutan. *See* apes

Ortiz-Robles, Mario, 134
Otis, Laura, 182, 243n92
Otter, Samuel, 229n3
Ovid, 187
Owen, Richard: and Chambers's *Vestiges*, 15, 132, 137, 225n35, 232n39, 234n72; and Darwin, 137, 190, 232n31, 234n73; and Dickens, 131; and evolutionary theory, 94, 131, 132, 136, 137, 161, 188, 199, 225n35, 232n39, 234n70, 234n72, 248n156; and the megalosaurus, 129, 131, 132, 233n52
Owenson, Sydney (Lady Morgan), 82, 222n129

Pacini, Giulia, 221n126
Palladio, Andrea, 23
Palmeri, Frank, 36, 203n45, 237n106, 241n50
Paltock, Robert, 87
Paris, 4, 89–90, 93–95, 100, 116–17, 126
Pascal, Blaise, 41
Pascal, Roy, 142, 143, 144, 236n89, 236n94
pastoral, 140, 150,155, 184, 186
Paul, St., 41
Paxton, Nancy, 240n39, 240n40
perfectibility and perfection, 14, 39, 42–44, 45, 48–50, 51–53, 58, 60–61, 62–64, 79, 75–76, 82–83, 89, 96, 111, 119, 162–63, 173, 192, 199–200, 202n27, 209n36, 209n48, 219n76, 224n18. See also *Bildung*; development
Pfau, Thomas, 216n35
Pico della Mirandola, 45, 210n57
Plotz, John, 192, 196, 246n125
poetry: as human discourse, 20, 57, 66, 219n76; and the novel, 20, 56, 68–69, 127, 147–49. See also epic; lyric; prospect
polygenesis, 15, 54, 104, 106–8, 227n70. See also anthropology; monogenesis; race
posthuman, 2, 6, 8, 10, 163, 199–200
preformation, 2, 14, 20, 45, 64, 97–99, 197, 212n76, 225n47, 226n49. See also epigenesis
Priestley, Joseph, 42
prospect, lyric or visionary, 51, 111–12, 127–28, 145–51, 153–56, 178, 184–86

protoplasm, 184–87, 244n105, 245n106
provincial life, as preserve of English realism, 4, 19, 26, 54, 79–80, 82, 84, 135, 144, 150, 163, 167, 168, 171–72, 178, 186, 189–90, 192, 221n121
Puckett, Kent, 184

Quetelet, Adolphe, 135, 233n60

race and racial difference, 8, 15, 16–17, 26, 54, 84, 87, 100, 102–8, 116, 163, 168, 170–71, 192–93, 196–98, 227n70
Radcliffe, Ann, 18, 147–8, 149
Rancière, Jacques, 217n50
Reade, William Winwood, 163, 239n20
realism, novelistic: alternative modes and disruptions of, 4, 8, 15, 19, 25–26, 122, 126–29, 135–36, 160, 168, 189–90, 193–96, 199, 230n11, 247n142; as anthropomorphic project, 3, 6, 7, 25–26, 29, 58, 126, 133, 134–35, 150–51, 156, 167, 186, 189; character and, 126, 135–36, 140, 231n18; and development, 3, 7, 29–30; as European preserve, 4, 8, 30, 204n67, 214n91; genres of, 57, 66–67, 69–70, 80, 82; narrative techniques of, 141–44, 149–51, 167–68, 179, 236n103; provincial, 4, 19, 21, 30, 168, 189. See also Bildungsroman; domestic fiction; free indirect discourse; historical novel; narration, omniscient; provincial life; science fiction
realism, scientific, 28–30, 156–57, 206n104
Redfield, Marc, 217n54
Reid, Thomas, 207n15
Reimarus, Hermann Samuel, 210n54
religion, 13, 32, 41, 100, 104–5, 106, 109, 114, 158, 172–73, 176, 179–180, 192–96, 241n60, 247n141
reversion or regression. See development
revolutions: French, 19, 25, 42, 60–61, 65, 67, 76, 86, 89, 90, 92, 93, 96–97, 113, 119, 215n27, 217n50; literary, 21, 57, 60, 149, 233n45; scientific, 2, 3, 11–12, 20, 22, 32, 37, 41, 126, 160, 164, 179
Reynolds, G.W.M., 25
Richards, Evelleen, 225n35, 232n39, 234n70
Richards, Robert J., 210n54, 216n32, 245n112

Richardson, Samuel, 13, 80, 207n14
Richter, Jean Paul, 68, 70
Ricoeur, Paul, 73
Riehl, Wilhelm Heinrich, 170–71, 241n52
Robertson, William, 16, 54
Romanticism, 37, 57, 68–70, 76, 88, 114–15, 127–28, 147–49, 180, 198, 224n10
Rothfield, Lawrence, 246n132
Rousseau, Jean-Jacques: and conjecture, 12–13, 36, 38, 208n22; on human nature and history, 14, 16, 21, 32, 38, 41–43, 44, 58, 104, 171, 216n36; on orangutan as natural man, 15, 51, 94, 223n8; on perfectibility, 14, 42, 60–61, 210n48
 —works of: *Discourse on the Origins of Inequality*, 12–13, 32, 36, 41–43, 44, 103, 210n53; *Julie*, 77, 80; "On the Origin of Languages," 33, 210n53; *The Social Contract*, 38, 60–61
Rudwick, Martin S., 203n45, 205n76, 231n24, 239n6
Ruskin, John, 30, 185

Saint-Beuve, Charles Augustin, 96
Saint-Hilaire, Etienne Geoffroy: debate with Cuvier, 25, 52, 91–93, 224n24; and Lamarck, 90–93, 224n18; on organic form and development, 21, 15, 88, 90, 91–2, 97, 98, 99, 116, 127, 128, 133, 136, 188, 196, 224n14, 225n44, 225n46, 226n47, 234n70, 234n74
Saint-Hilaire, Isidore Geoffroy, 15, 88, 90, 116, 127, 136, 188, 196, 223n14, 226n49, 228n87
Saint-Vincent, Bory de, 94
Saintsbury, George, 219n82
Scheinberg, Cynthia, 196, 247n147
Schiller, Friedrich, 180, 213n88, 214n3; *Aesthetic Education of Man*, 57, 63–64, 65–66, 218n62, 228n83; the aesthetic state in, 64, 81, 217n49, 217n50; and Goethe's *Wilhelm Meister*, 63, 65–66, 216n43, 217n54; the political state in, 63–64, 216n47
Schlegel, August Wilhelm, 76
Schlegel, Dorothea, 76
Schlegel, Friedrich: on Goethe's *Wilhelm Meister*, 62, 67–8; on the grotesque, 113–14; on irony, 68, 70, 71, 218n65, 219n76, 219n77, 247n142; on the novel, 21, 66, 68–70, 71, 76, 83, 218n74; and "progressive universal poetry," 57, 69, 76, 119
Schleiden, Matthias, 27, 179, 181–82, 242n73
Schultz, Elizabeth, 229n4
Schwann, Theodor, 27, 179, 181–82, 242n73
science and sciences: human (modern), 1–2, 11, 32–33, 95, 201n3; of man (Enlightenment), 1, 11–12, 31, 32, 61, 111, 112, 159, 203n45, 215n18. *See also* anthropology; history, natural; human nature
science fiction, 53, 88, 168, 187–88, 198–200, 231n23
Scotland: Enlightenment in, 4, 14, 35, 36, 38, 43, 72–75, 93, 94, 96, 106, 113, 192, 206n3, 220n109; historical novel in, 21–22, 57, 71–75, 102–6
Scott, Walter: excluded heroines of, 83, 105–6, 176; on *Frankenstein*, 88; on historical change, 18, 19, 22, 72–5, 89, 95; and the historical novel, 3–4, 22, 57, 69, 71–75, 87, 100, 219n82; on human nature, 18, 57, 72–73, 74–75, 102–5, 112, 126, 220n105
 —*Count Robert of Paris*, 87–88, 100–102, 106–13; Anna Comnena in, 102, 226n53; evolution in, 93, 95–96, 102, 108, 111, 112; as literary experiment, 87–88, 106, 113; orangutan in, 25, 87, 88, 95–96, 102, 108–10, 111, 114, 116, 223n8, 225n39; race and species in, 87, 106–8, 110–12; world history in, 95, 100–102, 226n51, 226n52
 —*Ivanhoe*, 75, 104–6, 201n6
 —*Rob Roy*, 103–4
 —*Waverley*, 18, 72–5, 82, 84, 102–3, 202n26; and the Bildungsroman, 71–72, 76, 220n95; historical change in, 72–73, 74, 102–3, 220n101; protagonist of, 72, 76, 83, 148
Secord, James A., 93, 231n26, 232n39
Selbin, Jesse Cordes, 245n112

INDEX [289]

serial form, 23–25, 26, 125, 131, 133–34
Shakespeare, William, 69, 114, 139, 147, 229n105
Shelley, Mary: *Frankenstein*, 10, 14, 20, 21, 22, 25, 70, 88, 99, 149, 223n8; *The Last Man*, 9–10, 14
Shelley, Percy Bysshe, 20, 147, 148, 237n120
Shuttleworth, Sally, 171, 203n39, 241n49, 244n94
Simmons, Clare, 110, 223n3
Slater, Michael, 232n28
Slaughter, Joseph, 201n16, 217n52
slavery and the slave trade, 15, 16–17, 20, 54, 106, 107, 110
Sluga, Glenda, 222n128
Smith, Adam, 32, 35, 96
Spencer, Herbert, 11, 27, 52, 143, 161, 162, 163, 167, 169, 170, 239n9, 240n39, 240n40, 241n44, 241n50
Spenser, Edmund, *The Faerie Queene*, 110
Sprat, Thomas, 164
Staël, Germaine de: and England, 79–80, 221n121, 221n124; and Goethe, 55–56, 57, 76, 214n3; and humanity, 33, 57, 75–6, 82, 220n109
—*Corinne*, 56–57, 76–77, 79–83; afterlife and translations of, 76, 83–85, 105, 176, 214n6; art and the aesthetic in, 56, 76–77, 81, 95, 221n126; *Bildung* in, 5, 21–22, 56–57, 76, 80–82, 167; and domesticity, 22, 79–80; heroine of, 22, 56–57, 76–77, 79–82, 83; Italy in, 77, 80–81, 82–83, 95, 214n7, 221n126; and national history, 22, 57, 76, 79–82, 221n126, 222n128, 222n129; and novelistic form, 57, 67, 77, 79–80; as Schlegelian "romantic book," 57, 76, 77, 221n113
—*Germany*, 55–56, 81
—*Literature*, 76, 80
Stanzel, F. K., 142, 235n88
Steinlight, Emily, 230n10
Sterne, Laurence, *Tristram Shandy*, 215n18
Stewart, Dugald, 32, 35, 227n77
Stewart, Garrett, 237n108

Stewart, Susan, 235n81
Stocking, George, 242n72
Strauss, David Friedrich, 172, 173, 179, 192.
sublime, the, 23–24, 106, 113–15, 118, 120–21, 125–26, 129, 133–34, 146, 147, 162–63, 178, 186, 190, 228n93
Sue, Eugène, 25
Sutherland, John, 227n73
sympathy, 9–10, 17, 20, 70, 73, 74, 191–92, 195

Taylor, Jesse Oak, 129, 206n110, 230n11
Tenenbaum, Susan, 81, 221n126
Tennenhouse, Leonard, 71
teratology, 15, 25, 88–90, 116, 136–37, 139, 188, 196, 234n72. *See also* monsters
Thackeray, W. M., 24, 144
Thomson, William (Lord Kelvin), 160
Tiedemann, Friedrich, 246n134
Tillotson, Kathleen, 231n16
tragedy and the tragic, 167, 175–77, 186
transformism or transmutation. *See* Chambers, Robert; Darwin, Charles; development; form, variable; Herder, Johann Gottfried; Lamarck, Jean-Baptiste; Lewes, George Henry; Owen, Richard; Saint-Hilaire, Geoffroy
Tylor, Edward Burnett, 179, 180, 190–91
Tyndall, John, 12, 182, 199, 206n101, 244n94, 244n95, 248n156
Tyson, Edward, 227n74

Uexküll, Jakob, 174, 210n55, 241n58

Vaihinger, Hans, 209n38
Vanfasse, Nathalie, 231n18
Vermeule, Blakey, 151
Vincent, Patrick, 221n126, 222n134
Vinci, Leonardo da, 23
Virchow, Rudolf, 179, 243n91, 243n92
Vitruvius and "Vitruvian man," 23, 53, 113
Volney, Constantin, 33, 61

Wallace, Alfred Russel, 163
Weismann, August, 198
Westminster Review, 161, 169, 170, 179
Whewell, William, 12, 165, 182, 183, 185, 194, 198, 203n41, 240n29

Wills, William Henry, 233n44
Wolff, Caspar Friedrich, 46, 161–62, 182, 209n41, 212n67
Woloch, Alex, 135, 234n63, 235n82
Wood, John, Prior Park, 23
Woolf, Virginia, 148–49
Wordsworth, William: on poetry and humanity, 20; Preface to *The Excursion*, 150–51; *The Prelude*, 145–46, 147; visionary prospect and reverie in, 145–46, 147–48, 156, 177; "Upon Westminster Bridge," 155
Wormald, Mark, 244n103
Wright, Daniel, 240n36

Zammito, John F. 48, 211n63
Zola, Emile, 7, 28
Zunshine, Lisa, 151

A NOTE ON THE TYPE

THIS BOOK has been composed in Miller, a Scotch Roman typeface designed by Matthew Carter and first released by Font Bureau in 1997. It resembles Monticello, the typeface developed for The Papers of Thomas Jefferson in the 1940s by C. H. Griffith and P. J. Conkwright and reinterpreted in digital form by Carter in 2003.

Pleasant Jefferson ("P. J.") Conkwright (1905–1986) was Typographer at Princeton University Press from 1939 to 1970. He was an acclaimed book designer and AIGA Medalist.

The ornament used throughout this book was designed by Pierre Simon Fournier (1712–1768) and was a favorite of Conkwright's, used in his design of the *Princeton University Library Chronicle*.